Bifurcation and Buckling in Structures

Bifurcation and Buckling in Structures

Kiyohiro Ikeda
Kazuo Murota

CRC Press
Taylor & Francis Group
Boca Raton London New York

CRC Press is an imprint of the
Taylor & Francis Group, an **informa** business

First edition published 2022
by CRC Press
6000 Broken Sound Parkway NW, Suite 300, Boca Raton, FL 33487-2742

and by CRC Press
2 Park Square, Milton Park, Abingdon, Oxon, OX14 4RN

Library of Congress Cataloging-in-Publication Data

Names: Ikeda, Kiyohiro, author. | Murota, Kazuo, 1955- author.
Title: Bifurcation and buckling in structures / Kiyohiro Ikeda, Kazuo Murota.
Description: First edition. | Boca Raton, FL : CRC Press, 2022. | Includes bibliographical references and index.
Identifiers: LCCN 2021031070 (print) | LCCN 2021031071 (ebook) | ISBN 9780367631611 (hbk) | ISBN 9780367631604 (pbk) | ISBN 9781003112365 (ebk)
Subjects: LCSH: Buckling (Mechanics) | Deformations (Mechanics) | Bifurcation theory.
Classification: LCC TA656.2 .I43 2022 (print) | LCC TA656.2 (ebook) | DDC 624.1/76--dc23
LC record available at https://lccn.loc.gov/2021031070
LC ebook record available at https://lccn.loc.gov/2021031071

ISBN: 978-0-367-63161-1 (hbk)
ISBN: 978-0-367-63160-4 (pbk)
ISBN: 978-1-003-11236-5 (ebk)

DOI: 10.1201/9781003112365

Publisher's note: This book has been prepared from camera-ready copy provided by the authors.

Contents

Part II Buckling of Structures

Preface

This book describes the theory and analysis of bifurcation and buckling in structures, targeting a wide range of readers from undergraduate students to professionals. An emphasis is not placed on individual topics for particular structures but on a general procedure for solving nonlinear governing and equilibrium equations and an analysis procedure related to the finite-element method.

Bifurcation leads to buckling of structures. Buckling is related to physical phenomena in the real world, whereas bifurcation is a mathematical concept of the loss of uniqueness of the solution of a governing equation when the value of a parameter is changed. The multiplicity of solutions of the governing equation of a structure often emerges when the value of a load parameter is changed. This multiplicity is related to the buckling behavior of this structure.

Structures in the real world are inevitably imperfect, possessing deviations from the designed values. Nonetheless, the bifurcation mechanism of the perfect system without imperfections would be an idealized guideline for the buckling behavior of the structures. This shows the importance of theory on bifurcation and buckling advanced in this book.

This book is divided into two parts: Part I deals with bifurcation in structures in a general setting. To begin with, the general mathematical framework for bifurcation and buckling are presented. Then procedures for the buckling load and buckling mode analyses of discrete systems are introduced. We present numerical analysis procedures to trace the solution curves and to switch to bifurcation solutions. We introduce, as advanced topics, asymptotic theory of bifurcation in the neighborhood of a bifurcation point and bifurcation theory of symmetric systems. Part II deals with the buckling of the perfect and imperfect structures. We present the member buckling of columns and beams and buckling of truss and frame structures. The buckling of a structural system with initial imperfections is studied; in particular, the worst and the random imperfections are studied as advanced topics. A review of the history of buckling is presented.

Bifurcation is studied in a broad field of engineering and science, and its mechanism has been elucidated by bifurcation theory in nonlinear mathematics. This book places emphasis on consistency with this theory. In lectures in structural mechanics, mathematical description tends to be avoided from time to time. Nonetheless, there are several mathematical backgrounds that are essential in the understanding of the mechanism of bifurcation and buckling in structures. Mathematical background is placed at the end of some chapters. To assist the understanding of the readers, problems are provided at the end of each chapter, and their answers are given at the end of the book.

The academic advice of Professor Isaac Elishakoff, Professor Makoto Ohsaki, and Professor Fumio Fujii was indispensable for publishing this book. The joint lecture on "Stability of Structures" at Tohoku University with Dr. Yamakawa was useful in developing materials of this book and is greatly appreciated.

April 2021

Kiyohiro Ikeda
Kazuo Murota

The Authors

Kiyohiro Ikeda is Professor Emeritus at Tohoku University, Japan. His research interests include the application of bifurcation theory to bifurcation and buckling in structures and materials, as well as the theoretical and numerical analysis of structures with initial imperfections.

Kazuo Murota is a Project Professor at the Institute of Statistical Mathematics, Japan, as well as Professor Emeritus at the University of Tokyo, Kyoto University, and Tokyo Metropolitan University, Japan. His research interests include bifurcation theory, discrete optimization, and numerical analysis.

Part I

Bifurcation in Structures

Bifurcation is studied in broad fields of engineering and science, and the mechanism of bifurcation in symmetric systems has been elucidated by bifurcation theory in nonlinear mathematics. Placing emphasis on the consistency with this theory, Part I of this book introduces elastic bifurcation theory in structural mechanics. Chapter 1 presents the general mathematical framework for bifurcation and buckling. Chapter 2 introduces the buckling load and the buckling mode analyses of discrete systems. Chapters 3 and 4 present numerical analysis procedures to trace the solution curves and to switch to bifurcating solutions. Chapter 5 advances asymptotic theory of bifurcation in the neighborhood of a bifurcation point. Chapter 6 introduces an advanced theory for symmetric systems.

1 Introduction to Buckling and Bifurcation

1.1 SUMMARY

This chapter introduces buckling and bifurcation behaviors of structures. Physically, these behaviors arise from the loss of shapes of the structures. Mathematically, buckling and bifurcation behaviors can be ascribed to the nonlinearity of the governing equation. While the general framework of the mathematical treatment of buckling and bifurcation problems is presented in Chapter 2, the concepts of bifurcation and buckling and the basic mathematical issues are introduced in this chapter.

Section 1.2 explains what buckling and bifurcation are by several illustrative structural examples. The mathematical framework is introduced in a simple setting in Section 1.3 without mathematical sophistication. Snap buckling and bifurcation buckling of structures are described in Sections 1.4 and 1.5, respectively.

Keywords: • Bifurcation • Buckling • Critical point • Governing equation • Snap through • Stability • Total potential energy

1.2 WHAT ARE BUCKLING AND BIFURCATION?

Typical ways how things break are depicted in Fig. 1.1. If a structure tumbles, it loses its function even if it is not broken itself. If a log bridge slides and falls into water, it loses its function as a bridge. If a glass breaks, it is of no use. If a steel member is pulled off, it cannot transmit a tensile force anymore. If a steel cylindrical shell buckles, it loses its shape and strength.

Shape is closely related to strength. For example, sand in a sandbag is strong, but its strength is lost when its shape is lost. The strength of sand thus arises from its shape. A pyramid and an arch have shapes that are superior in strength. Many structures have symmetric shapes as symmetry is often a result of optimization of their strength. Symmetric structures subjected to excessive loads often undergo buckling and bifurcation en route to collapses.

When a structure or its member is subjected to a load that increases gradually from a small value, the deformation progresses slowly when the load is small. As the load approaches a critical level, the deformation of a structure or its member often increases progressively, leading to a sudden change in shape. This kind of behavior is called *buckling* and the structure and its member are said to be buckled.

It is an important task in structural engineering to design or control the buckling of structures. Buckling is mostly harmful as it leads to a decrease or loss of the member's load-carrying capacity, often en route to a failure of a member or the whole

DOI: 10.1201/9781003112365-1

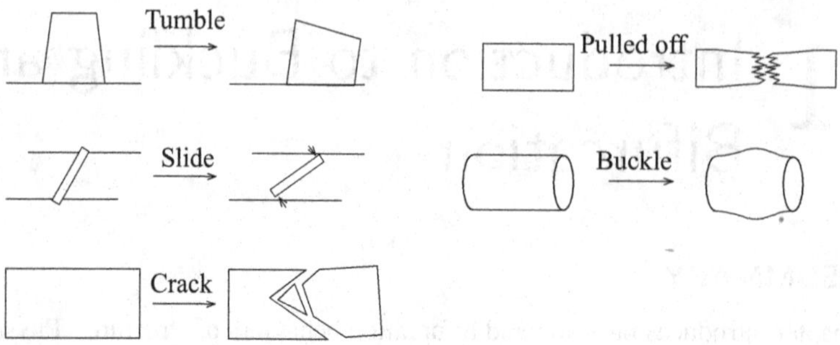

Figure 1.1 Typical ways things break

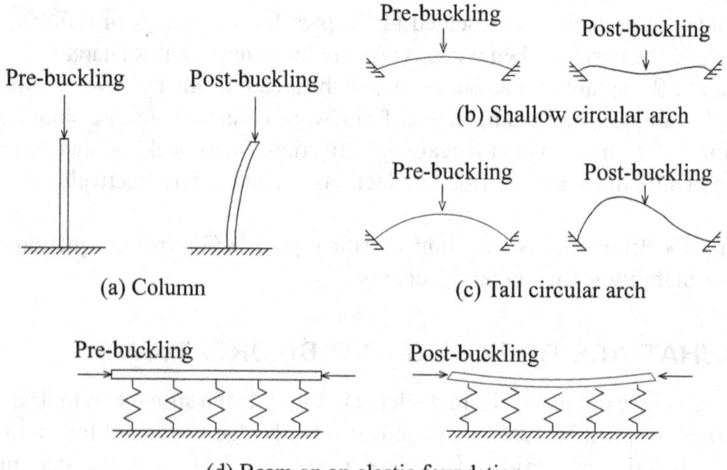

Figure 1.2 Typical examples of buckling of structures

structure. As modern structures become slender, the risk of undergoing buckling increases. Nonetheless, buckling is not always harmful. For example, braces in framed structures are sometimes designed to buckle so as to dissipate energy during a large earthquake.

Typical examples of buckling of structures are illustrated in Fig. 1.2.

- A vertical column subjected to a compression load retains its upright state when the load is small but undergoes a failure due to horizontal deformation for an excessively large load (Fig. 1.2(a)).
- A shallow circular arch buckles vertically (Fig. 1.2(b)), and a tall circular arch buckles horizontally (Fig. 1.2(c)).
- A beam on an elastic foundation subjected to an axial compression undergoes buckling and deforms vertically (Fig. 1.2(d)).

Behaviors of a structure, in general, can be described mathematically by a system of equations (or differential equations). This is also the case with the buckling behavior of a structure. In the following, we consider some simple equations to reveal the mathematical nature of buckling behaviors.

As a simplest example, let us consider the following system of linear equations

$$\begin{cases} F_1 \equiv 2x + y - 3f = 0, \\ F_2 \equiv x + 2y - 3f = 0 \end{cases} \tag{1.1}$$

in (x, y) with a parameter f. The variable (x, y) is supposed to represent the state of some structure, and f is interpreted as a loading parameter. This system of equations has a solution curve of $x = y = f$ in the 3-dimensional space of (x, y, f). The curve is depicted in Fig. 1.3(a), in which the left is a 3-dimensional plot for (x, y, f) and the others are 2-dimensional ones for (x, f) and (y, f). In particular, a solution (x, y) exists uniquely for any value of f, and there is nothing peculiar that seems to correspond to the buckling for excessively large f.

As an example of a more complex system of equations, let us consider

$$\begin{cases} F_1 \equiv x - y^2 - f = 0, \\ F_2 \equiv x - y = 0 \end{cases} \tag{1.2}$$

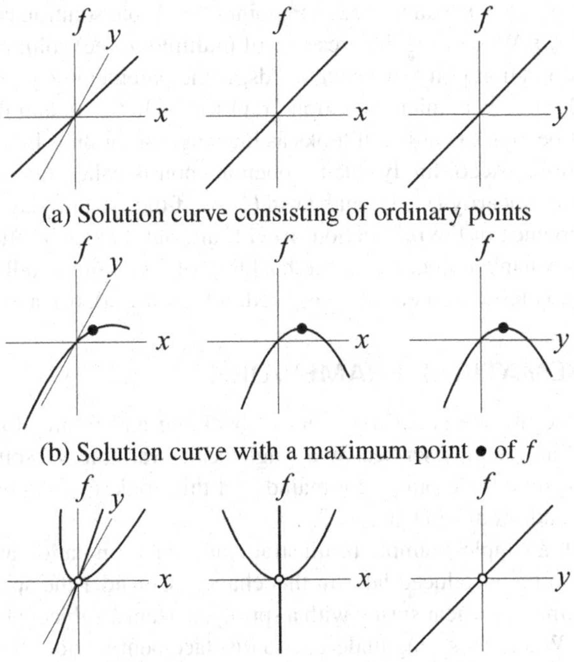

(a) Solution curve consisting of ordinary points

(b) Solution curve with a maximum point • of f

(c) Solution curve with a bifurcation point ○

Figure 1.3 Typical solution curves

with a quadratic nonlinear term y^2. Substituting $y = x$, which is obtained from $F_2 = 0$, into $F_1 = 0$, we obtain $f = x - x^2$. This solution curve is plotted in Fig. 1.3(b). There is a *maximum point of bifurcation parameter* f shown as (•). If f is larger than this maximal value, there is no solution (x, y) corresponding to a given f, whereas two solutions exist for a given f below this maximum value. As we have seen, the nonlinearity (y^2 in (1.2)) of the system plays an important role in the emergence of the maximum point of the load parameter f. It will turn out that this is the mathematical mechanism underlying the buckling of a shallow circular arch illustrated in Fig. 1.2(b).

As a third example of a system of equations, we consider the equation

$$\begin{cases} F_1 \equiv x^3 + xy - 2fx = 0, \\ F_2 \equiv y - f = 0 \end{cases} \tag{1.3}$$

with a cubic nonlinear term x^3 and cross terms xy and fx. This system of equations has a more complex or intriguing solution curve. Substituting $y = f$, which is obtained from $F_2 = 0$, into $F_1 = 0$, we obtain $x(x^2 - f) = 0$. This equation thus has two solutions curves:

$$\begin{cases} x = 0,\ y = f, & \text{solution curve 1,} \\ f = x^2,\ y = f, & \text{solution curve 2.} \end{cases}$$

These curves, which are plotted in Fig. 1.3(c), intersect at the origin $(x, y, f) = (0, 0, 0)$ shown as (○). Such an intersection point of multiple solution curves is called a *bifurcation point*. We can see the presence of multiple (three) solutions for a given f above this bifurcation point. In other words, if the parameter f is increased from $f < 0$ to $f > 0$, there is a unique solution (x, y) for each $f \leq 0$, and three solutions emerge when f becomes positive. It looks as if a single solution is bifurcated or split into three solutions. Accordingly, such a phenomenon is called *bifurcation*. As we have seen, the nonlinearity (x^3, xy, and fx in (1.3)) of the system plays an important role in the emergence of the bifurcation. It will turn out that such bifurcation is the mathematical mechanism underlying the buckling of a column, a tall circular arch, and a beam on an elastic foundation illustrated in Figs. 1.2(a), (c), and (d).

1.3 MATHEMATICAL FRAMEWORK

An outline of the mathematical treatment of buckling and bifurcation phenomena is presented for a system described by a single scalar variable. Despite its simplicity, this scalar system underpins the remainder of this book, especially "Analysis of Buckling Load and Mode" in Chapter 2.

We start with a simple example to illustrate our general notation and mathematical formalism to be introduced later in this chapter. Consider the spring system in Fig. 1.4, consisting of a linear spring with a spring constant k subjected to an external tensile force f. When the spring undergoes a displacement u under this force f, the

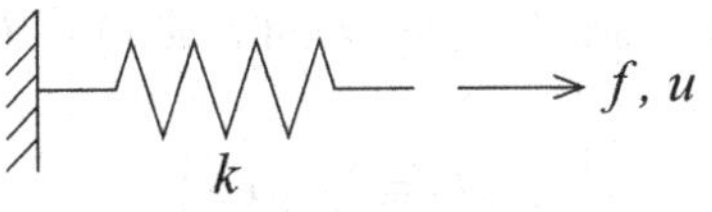

Figure 1.4 A spring system

total potential energy is given by

$$\begin{aligned} U(u,f) &= \text{(Internal potential energy)} + \text{(External potential energy)} \\ &= \int \text{(Spring force) d(Spring displacement)} \\ &\quad - \text{(External force)} \times \text{(Distance external force traveled)} \\ &= \int_0^u ku\mathrm{d}u - fu = \frac{1}{2}ku^2 - fu. \end{aligned} \tag{1.4}$$

The potential U should be minimal at the equilibrium state u, and therefore,

$$\frac{\partial U}{\partial u} = ku - f = 0 \tag{1.5}$$

gives the equilibrium equation for this spring system. Note that the derived equilibrium equation (1.5) is nothing but *Hooke's law*.

This simple example demonstrates the general principle that the equilibrium equation, say, $F(u,f) = 0$ can be obtained from the potential function U by defining $F(u,f) = \partial U/\partial u$. In this example, however, the resulting equation (1.5) is linear in u with $F(u,f) = ku - f$, and therefore, no bucking or bifurcation occurs. For buckling and bifurcation phenomena, nonlinearity of $F(u,f)$ is an essential ingredient.

We now consider a general nonlinear *governing equation*:

$$F(u,f) = 0 \tag{1.6}$$

for a *scalar system*, which is described by a single (scalar) variable u and a bifurcation parameter f. We assume that the function $F(u,f)$ is sufficiently smooth and is derived from a potential function $U(u,f)$ as

$$F(u,f) = \frac{\partial U}{\partial u}(u,f). \tag{1.7}$$

A system equipped with a potential function is called a *potential system*.

In search of the solution curves, it is customary to start from the initial state (in structural mechanics, the initial state is usually the state of $(u,f) = (0,0)$ with no load and displacement) and, in turn, to obtain the solution curve containing the initial state, which is called the *fundamental path*.

Since the governing equation (1.6) is nonlinear, there may be several solutions u for a given value of parameter f. The possibility of non-unique solutions u can be detected with the aid of the partial derivative of $F(u,f)$ with respect to u. We denote this by $J(u,f)$, that is,

$$J(u,f) \equiv \frac{\partial F}{\partial u}(u,f), \tag{1.8}$$

which is called the *Jacobian*[1] hereinafter. Since $F = \partial U/\partial u$, we have

$$J(u,f) = \frac{\partial^2 U}{\partial u^2}(u,f). \tag{1.9}$$

Consider a solution (u_*, f_*) to the equilibrium equation (1.6). If $J(u_*, f_*) \neq 0$, the variable u is uniquely determined from[2] f. In this case, (u_*, f_*) is called an *ordinary point*. Otherwise, that is, if $J(u_*, f_*) = 0$, the solution (u_*, f_*) is called a *critical point*. In the latter case, u is (usually) not determined uniquely from f. Accordingly, we have the following classification for a solution point (u_*, f_*) of $F(u,f) = 0$:

$$\begin{cases} J(u_*, f_*) \neq 0 & \Rightarrow \quad (u_*, f_*) \text{ is an } \textit{ordinary point}, \\ J(u_*, f_*) = 0 & \Rightarrow \quad (u_*, f_*) \text{ is a } \textit{critical point}. \end{cases} \tag{1.10}$$

In this book, we usually denote a critical point by $(u_\mathrm{c}, f_\mathrm{c})$, where the subscript $(\cdot)_\mathrm{c}$ indicates the value of the variable at a critical point. A critical point $(u_\mathrm{c}, f_\mathrm{c})$ can be detected as a solution that simultaneously satisfies the governing equation (1.6) and the *criticality condition*

$$J(u_\mathrm{c}, f_\mathrm{c}) = 0. \tag{1.11}$$

Since an equilibrium solution (u_*, f_*) satisfies $F(u_*, f_*) = (\partial U/\partial u)(u_*, f_*) = 0$ by (1.6) with (1.7), $u = u_*$ is a *stationary point* of the total potential energy $U(u, f_*)$ when regarded as a function in u with f fixed to f_*. An equilibrium solution (u_*, f_*) is said to be *stable* if the function $U(u, f_*)$ in u is strictly minimal (local minimum) at $u = u_*$; otherwise, it is *unstable*.[3]

If (u_*, f_*) is an ordinary point, we have $J(u_*, f_*) \neq 0$, while $J = \partial^2 U/\partial u^2$ by (1.9). Hence the stability of an ordinary point (u_*, f_*) is determined by the sign of the Jacobian as follows:

$$\begin{cases} J(u_*, f_*) > 0 & \Rightarrow \quad (u_*, f_*) \text{ is a stable ordinary point}, \\ J(u_*, f_*) < 0 & \Rightarrow \quad (u_*, f_*) \text{ is an unstable ordinary point}. \end{cases} \tag{1.12}$$

Figure 1.5 shows the increment of the potential $\delta U = U(u, f_*) - U(u_*, f_*)$ against the incremental displacement $\delta u = u - u_*$.

---

[1] The Jacobian in (1.8) corresponds to the *tangent stiffness* in structural mechanics.

[2] To be precise, the unique existence of u as a function of f is guaranteed in a neighborhood of (u_*, f_*) and we have $u \approx u_* - J(u_*, f_*)^{-1}(\partial F/\partial f)(u_*, f_*)\,(f - f_*)$.

[3] In this book, we set up a dichotomy between "stable" and "unstable" so as to simplify the statements without entering into subtle mathematical issues concerning the neutral case. According to our definition, a point is unstable if it is not stable. For example, if $U(u) = 0$ for all u, every point u is categorized as an unstable point according to our definition.

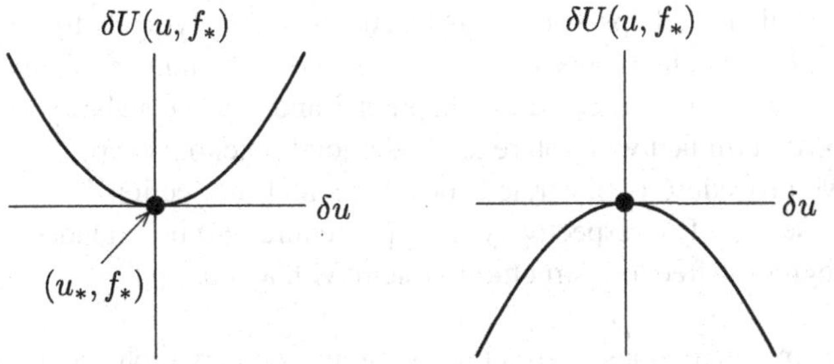

(a) Minimal point: stable (b) Maximal point: unstable

Figure 1.5 Distribution of potential $U(u, f_*)$ and stability of an ordinary point (u_*, f_*) .

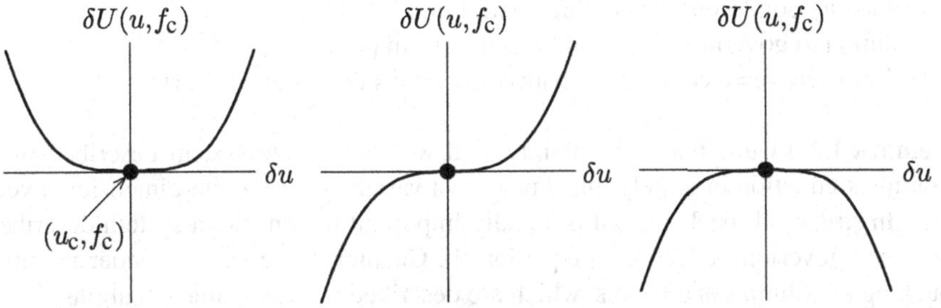

(a) Minimal point: stable (b) Inflection point: unstable (c) Maximal point: unstable

Figure 1.6 Distribution of potential $U(u, f_c)$ and stability of a critical point (u_c, f_c)

For a critical point (u_c, f_c), we have $J(u_c, f_c) = 0$. In this case, the stationary point $u = u_c$ is either a minimal point, an inflection point, or a maximal point of $U(u, f_c)$, as depicted in Fig. 1.6 showing $\delta U = U(u, f_c) - U(u_c, f_c)$ against $\delta u = u - u_c$. The stability of a critical point (u_c, f_c) can be classified as

$$\begin{cases} \text{minimal point of } U(u, f_c) & \Rightarrow \quad \text{stable,} \\ \text{inflection point or maximal point of } U(u, f_c) & \Rightarrow \quad \text{unstable.} \end{cases} \tag{1.13}$$

To sum up, the buckling analysis of a scalar system may be conducted by the following procedure:

Step 1: Derive the total potential energy of the system.
Step 2: Derive the governing equation and the Jacobian from the total potential energy.
Step 3: Obtain the solution curves, comprising the fundamental and bifurcating paths, from the governing equation.
Step 4: Using the Jacobian, find critical points, classify these critical points, and determine the stability of ordinary points.

Step 1 is based on the mechanical property of the system. Step 2 is a straightforward mathematical calculus. Steps 3 and 4 can be conducted analytically for a simple equilibrium equation or numerically for a complex one. (The numerical analysis procedures in these steps are presented in Chapters 3 and 4.) The stability of the critical points can be determined with reference to the total potential energy.

The above procedure is illustrated for a truss arch in Section 1.4 and a propped cantilever in Section 1.5, respectively. This procedure will be extended in Chapter 2 to a multi-degree-of-freedom structural system with a potential.

Remark 1.1 In applications to structural systems, we may rephrase the mathematical terms as follows:
governing equation ⇒ equilibrium equation,
Jacobian (matrix) ⇒ tangent stiffness (matrix),
criticality condition ⇒ buckling condition,
solution to governing equation ⇒ equilibrium point,
solution curve ⇒ equilibrium path, load versus displacement curve. □

Remark 1.2 In this book, we mainly deal with a discrete system described by a nonlinear equation in finitely many unknown variables (or a finite-dimensional vector). In structural mechanics, it is equally important to consider a system described by some governing differential equation. In Chapter 7, we shall consider member bucking of columns and beams, which are described by differential equations. □

1.4 SNAP BUCKLING OF A STRUCTURE

In this section, we introduce a buckling phenomenon called *snap buckling*, which typically occurs at a maximal point of load.

Consider the truss arch subjected to a vertical load depicted in Fig. 1.7. This truss arch comprises two identical truss members (called members 1 and 2) that are fixed at nodes 1 and 2 to the ground by hinges that allow rotation but restrict translation in any direction. This arch is a scalar system only with a vertical (y-directional) displacement u of the crown (node 3), while its horizontal (x-directional) displacement is constrained. By bilateral symmetry, the truss members have the same length even after deformation.

In the following, we carry out the analysis procedure presented at the end of Section 1.3, which consists of four steps.

Step 1: We obtain the total potential energy $U(u, f)$ from the mechanical properties of the system. For this purpose, we introduce several properties of the truss members: the modulus E of elasticity (Young's modulus), the cross-sectional area A, the initial member length $L = \sqrt{2}$, and the member length $\hat{L} = \sqrt{1+(u-1)^2}$ after deformation; in addition, we introduce a vertical load f normalized by EA. Then we

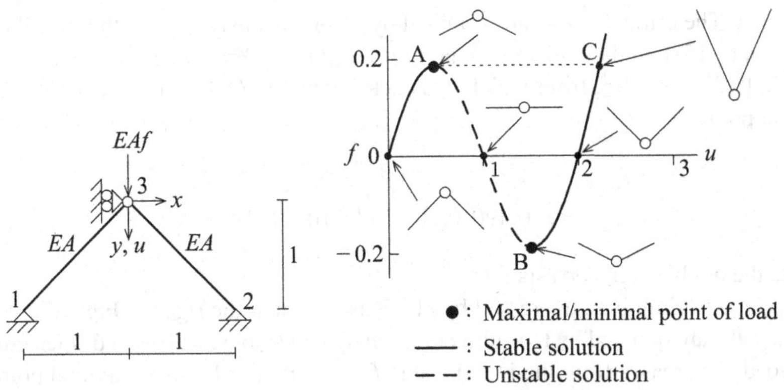

Figure 1.7 A truss arch with two truss members, its equilibrium path (load versus displacement curve), and typical deformation patterns

have

$$\begin{aligned} U(u,f) &= \sum_{m=1}^{2} \int (\text{Axial force})\ \mathrm{d}(\text{Axial displacement}) \\ &\qquad - (\text{External force}) \times (\text{Distance external force traveled}) \\ &= \frac{EA}{L}(\hat{L}-L)^2 - EAfu \\ &= EA\left[\frac{1}{\sqrt{2}}\left(\sqrt{1+(u-1)^2}-\sqrt{2}\right)^2 - fu\right]. \end{aligned} \tag{1.14}$$

Step 2: The differentiation of $U(u,f)$ in (1.14) with respect to u leads to the equilibrium equation

$$F(u,f) \equiv \frac{\partial U}{\partial u} = EA\left[\left(\sqrt{2}-\frac{2}{\sqrt{1+(u-1)^2}}\right)(u-1) - f\right] = 0. \tag{1.15}$$

The differentiation of this equation with respect to u leads to the Jacobian (tangent stiffness)

$$J(u,f) \equiv \frac{\partial F}{\partial u} = EA\left(\sqrt{2}-\frac{2}{[1+(u-1)^2]^{3/2}}\right). \tag{1.16}$$

Step 3: The equilibrium path (load versus displacement curve) satisfying the equation (1.15) is obtained as

$$f = \left(\sqrt{2}-\frac{2}{\sqrt{1+(u-1)^2}}\right)(u-1). \tag{1.17}$$

This is the fundamental path and a bifurcating path does not exist for this truss arch.

Step 4: The critical points are obtained by simultaneously solving the equilibrium equation (1.15) and the criticality condition $J(u,f)=0$. From $J(u,f)=0$, we obtain $u_\mathrm{c}=1\pm(2^{1/3}-1)^{1/2}$, from which f_c is determined by (1.17). Thus we obtain two critical points

$$\begin{aligned}(u_\mathrm{c},f_\mathrm{c}) &= (1\mp(2^{1/3}-1)^{1/2}, \pm\sqrt{2}(2^{1/3}-1)^{3/2}) \\ &\approx (0.490, 0.187),\ (1.510, -0.187), \end{aligned} \tag{1.18}$$

where the double sign corresponds.

The equilibrium path expressed by (1.17) is shown at the right of Fig. 1.7, where stable paths are denoted by the solid curves and an unstable one by the dashed curve. The load f increases from the initial state ($f=0$) and reaches the maximal point A of load and then continuously decreases and arrives at the minimal point B. These maximal and minimal points are the ones obtained in (1.18). After this minimal point B, the load increases continuously, exceeding the load of the maximal point A. This system is stable until reaching the maximal point A and also after the minimal point B since the criterion $J>0$ in (1.12) is satisfied therein. The buckling load of this system is $f_\mathrm{c}=\sqrt{2}(2^{1/3}-1)^{3/2}\approx 0.187$ at the maximal point A. These points A and B are both unstable.[4]

In an experiment of increasing the load f from zero to the load limit that the truss arch can sustain, the load increases stably until reaching the maximal point A. Upon exceeding this maximal point, the crown of the arch jumps downward to another stable equilibrium point C and the load can further increase stably. This is called a *snap through*. Yet, it is often the case that structures after a snap through lose their functions even they are stable mathematically. That is, a snap through usually induces collapse of structures, which is called *snap-through buckling* or simply *snap buckling*.

1.5 BIFURCATION BUCKLING OF A STRUCTURE

In this section, we introduce another kind of buckling phenomenon called *bifurcation buckling*, which occurs at a bifurcation point.

We investigate the behavior of the propped cantilever,[5] comprising a rigid bar[6] supported by a spring, depicted in Fig. 1.8. The bottom of this bar is fixed to the ground by a hinge that allows only rotation, and its top is supported horizontally by the spring and is subjected to the vertical load kLf, where f represents the load normalized by kL using the length L of the member and the spring constant k.

When the load f is small, the bar retains its upright state, supported by the spring. When the load f is increased to some extent, the bar cannot resist the load any further and starts to rotate, undergoing the bifurcation buckling, as shown at the right

[4] In general, maximal and minimal points of load are unstable; see Section 5.4.1.

[5] This propped cantilever is taken from Thompson and Hunt [1].

[6] A rigid bar neither undergoes any deformation nor stores internal energy.

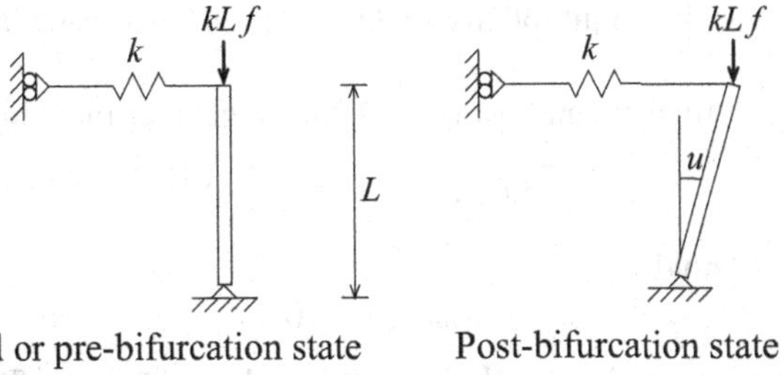

Figure 1.8 A propped cantilever

of Fig. 1.8. Note that the displacement of the bar is horizontal, but the load is vertical. It is a characteristic of the bifurcation phenomenon that the direction of the displacement is perpendicular to the direction of the load.

The following analysis reveals that this system with a linear spring undergoes buckling due to bifurcation, a different type of buckling phenomenon than a snap buckling. We hereafter denote by u the rotation of the bar and assume $-\pi/2 \leq u \leq \pi/2$.

In the following, we carry out the analysis procedure presented at the end of Section 1.3, which consists of four steps.

Step 1: The total potential energy U is

$$\begin{aligned} U = & \int (\text{Spring force})\, \mathrm{d}(\text{Spring displacement}) \\ & - (\text{External force}) \times (\text{Distance external force traveled}) \\ = & \frac{1}{2}k(L\sin u)^2 - kLf \cdot L(1-\cos u). \end{aligned} \tag{1.19}$$

Step 2: The equilibrium equation and the Jacobian (tangent stiffness) are obtained, respectively, as

$$F(u,f) \equiv \frac{\partial U}{\partial u} = kL^2 \sin u(\cos u - f) = 0, \tag{1.20}$$

$$J(u,f) \equiv \frac{\partial F}{\partial u} = kL^2(\cos 2u - f\cos u). \tag{1.21}$$

Step 3: From the equilibrium equation (1.20), we obtain two kinds of equilibrium paths

$$\begin{cases} u = 0, & \text{fundamental path (trivial solution)}, \\ f = \cos u, & \text{bifurcated path}. \end{cases} \tag{1.22}$$

(If the fundamental path is obtained easily, e.g., $u = 0$ as in this example, it is often referred to as the *trivial solution.*) As shown in Fig. 1.9(a), these two equilibrium paths intersect at point A with $(u,f) = (0,1)$, and therefore this point is a bifurcation point. A solution u is unique for each $f \geq 1$, whereas there are three solutions

u for each $f < 1$. This multiplicity of the solution u for each f is engendered by bifurcation.

Step 4: On the fundamental path $u = 0$, the tangent stiffness in (1.21) becomes

$$J(0, f) = kL^2(1 - f).$$

Since $J(0, 1) = 0$ holds,

$$(u_c, f_c) = (0, 1) \tag{1.23}$$

is a critical point, at which two solution paths in (1.22) intersect. Then an equilibrium point on the fundamental path $u = 0$ is classified by (1.10) and (1.12) as

$$\begin{cases} f < 1 & \Rightarrow \quad \text{stable ordinary point,} \\ f = 1 & \Rightarrow \quad \text{critical point (bifurcation point),} \\ f > 1 & \Rightarrow \quad \text{unstable ordinary point.} \end{cases}$$

Accordingly, the fundamental path is stable when the load is small ($f < 1$).

On the bifurcated path $f = \cos u$, the tangent stiffness in (1.21) becomes

$$J(u, \cos u) = kL^2(\cos 2u - \cos^2 u) = -kL^2 \sin^2 u < 0 \qquad (u \neq 0).$$

Accordingly, the solutions on this bifurcated path are unstable.

At the bifurcation point $(u_c, f_c) = (0, 1)$, the total potential energy U with f fixed to $f_c = 1$ is obtained from (1.19) as

$$U(u, 1) = 2kL^2 \left(\sin^2 \frac{u}{2} \cos^2 \frac{u}{2} - \sin^2 \frac{u}{2} \right) = -2kL^2 \sin^4 \frac{u}{2}.$$

This function takes a local maximum at u_c of the bifurcation point; accordingly, this bifurcation point is unstable (see the potential distribution in Fig. 1.9(b)).

This type of bifurcation is called *unstable–symmetric bifurcation* as the bifurcated paths are unstable and bilaterally symmetric. (See Fig. 1.9(a), in which a stable path is shown by the solid line and unstable ones by the dashed curves.)

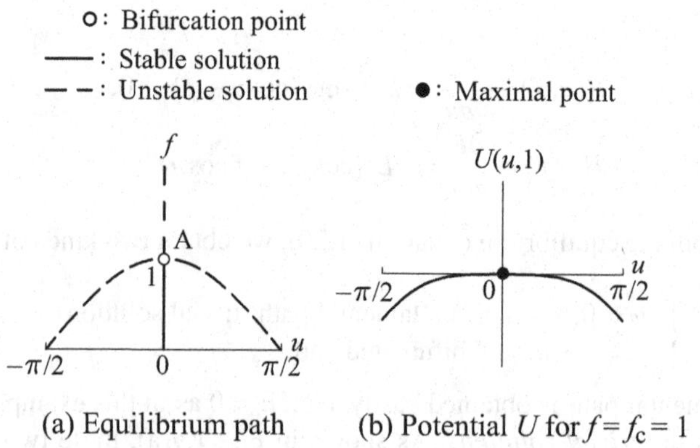

Figure 1.9 Bifurcation properties of the propped cantilever with the linear spring

When the load f increases from the initial state $(u, f) = (0, 0)$, the system remains stable until reaching the bifurcation load $f_c = 1$ and the bar retains the upright state. At $f_c = 1$, bifurcation occurs and the bar rotates leftward or rightward. The states of rotating leftward $(u < 0)$ and that of rotating rightward $(u > 0)$ are mathematically identical. The bilateral symmetry of the structural system is thus reflected in the bilateral symmetry of the bifurcation behavior. The bifurcation point, the trivial solution beyond the bifurcation point, and bifurcated paths are all unstable; accordingly, there is no stable state beyond the bifurcation load $(f > 1)$.

1.6 PROBLEMS

Problem 1.1 Consider the linear spring system with spring constant k in Fig. 1.4. (1) Investigate the stability using the Jacobian. (2) Investigate the stability using the total potential energy.

Problem 1.2 Consider a nonlinear spring that develops the force $F_s(u) = k(u - u^2)$ against displacement u. (1) Obtain the total potential energy $U(u, f)$. (2) Determine the stability of this system using the Jacobian.

Problem 1.3 Obtain the total potential energy, the equilibrium equation, and the Jacobian for the rigid-bar-spring system depicted in Fig. 1.10. In the rotational spring, a bending moment of $k_\theta \theta$ develops for rotational displacement θ. We set $k_\theta = kL^2$, $k_1 = 3k$, and $k_2 = 2k$. Employ the rotational displacement u of the rigid bar as the independent variable.

Problem 1.4 Obtain the governing equation of a system with potential

$$U(x, f) = x^3 + x^2 f + f^2,$$

where f is a parameter.

Problem 1.5 Obtain the governing equation of a system with potential

$$U(x, f) = x^3 - x^2 f - x f^2,$$

where f is a parameter.

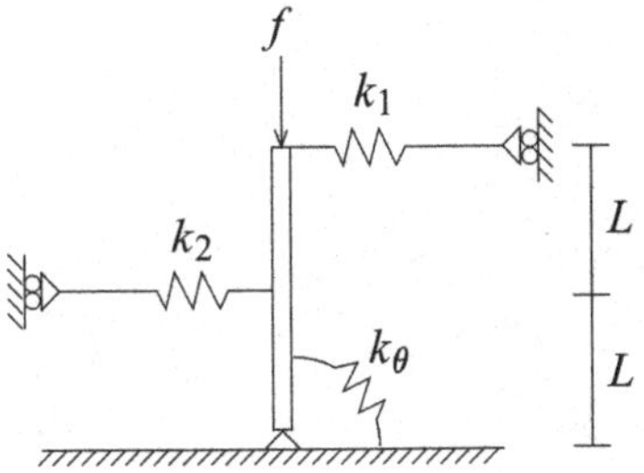

Figure 1.10 A rigid-bar-spring system

REFERENCE

1. Thompson JMT, and Hunt GW (1973), *A General Theory of Elastic Stability*, Wiley, New York.

2 Analysis of Buckling Load and Mode

2.1 SUMMARY

This chapter advances the general framework for the mathematical treatment of buckling and bifurcation problems as a foundation of the whole book. The principle of stationary total potential energy is introduced as a systematic means to derive the governing equation. The concepts of critical point and stability are presented using the Jacobian matrix of the governing equation. A procedure to obtain the buckling load and the buckling mode of a structure is introduced.

When the load is increased from the initial state of the system, the buckling takes place at a critical point, at which the Jacobian matrix of the governing equation becomes singular. The eigenvector of this matrix for the zero eigenvalue gives bifurcation mode. In particular, the buckling load at a critical point is of great engineering interest.

Buckling and bifurcation behaviors arise from the nonlinearity of the governing equation, and the theoretical methods for these behaviors have been established in nonlinear mathematics. Accordingly, we employ mathematical terminology in this chapter, while the corresponding terminology in structural mechanics is also introduced.

This chapter deals with fundamental issues of buckling. In this chapter, we do not deal either with the tracing of an equilibrium path (Chapter 3), the branch switching to a bifurcating path (Chapter 4), or the analysis of the post-buckling behavior (Chapter 5).

The mathematical framework is introduced in Section 2.2, followed by the procedure of buckling analysis in Section 2.3. Critical points, total potential energy, and stability are discussed in Sections 2.4, 2.5, and 2.6, respectively. Structural examples of buckling load analysis are given in Section 2.7, and those of buckling mode analysis are presented in Section 2.8.

Keywords: • Bifurcation • Buckling analysis • Buckling load • Buckling mode • Critical point • Multi-degree-of-freedom system • Stability • Total potential energy

2.2 MATHEMATICAL FRAMEWORK

In Section 1.3, we have focused on a scalar system (*single-degree-of-freedom system*) to present fundamental tools and concepts for bucking analysis. In this section, we extend them to a *multi-degree-of-freedom system*, which is common in realistic structures. Mathematical details of the ingredients of the procedure are explained in Sections 2.4, 2.5, and 2.6.

DOI: 10.1201/9781003112365-2

A buckling and bifurcation problem for a *discrete* or *discretized system* is given in the form of some nonlinear *governing equation*

$$\boldsymbol{F}(\boldsymbol{u}, f) = \boldsymbol{0}, \tag{2.1}$$

where $\boldsymbol{u} = (u_1, \ldots, u_n)^\top$ is an *unknown variable vector*, f is a *bifurcation parameter*, and

$$\boldsymbol{F}(\boldsymbol{u}, f) = (F_1(u_1, \ldots, u_n, f), \ldots, F_n(u_1, \ldots, u_n, f))^\top \tag{2.2}$$

is a vector of nonlinear functions assumed to be sufficiently smooth; $(\cdot)^\top$ denotes the transpose of a vector (or a matrix). In structural mechanics, the governing equation (2.1) typically expresses an *equilibrium equation*, the unknown variable $\boldsymbol{u}$ denotes displacements, and the bifurcation parameter f stands for a load.

In this book, we mostly consider a *potential system*, in which the functions $F_1, \ldots,$ F_n of the governing equation (2.1) are derived from a potential function U as $F_i = \partial U / \partial u_i$; see Section 2.5.1 for details. Note that the existence of potential does not directly govern the occurrence/non-occurrence of bifurcation.

The trajectory of the solutions $(\boldsymbol{u}, f)$ obtained by continuously changing the parameter f is called a *solution curve*. In search of the solution curves, it is customary to start from an initial state[1] and, in turn, to obtain the solution curve containing the initial state, which is called the *fundamental path*.

The $n \times n$ matrix comprising the partial derivatives $\partial F_i / \partial u_j$ $(i, j = 1, \ldots, n)$ is called the *Jacobian matrix*, denoted by $J(\boldsymbol{u}, f)$ throughout this book (cf., (2.6) in Section 2.4). This corresponds to the *tangent stiffness matrix* in structural mechanics. In a potential system, the Jacobian matrix is a symmetric matrix (Section 2.5.2) and has real eigenvalues.

A *critical point* is defined as a point $(\boldsymbol{u}_\mathrm{c}, f_\mathrm{c})$ that simultaneously satisfies the governing equation (2.1) and the criticality condition

$$\det J(\boldsymbol{u}_\mathrm{c}, f_\mathrm{c}) = 0. \tag{2.3}$$

In other words, the tangent stiffness matrix $J(\boldsymbol{u}, f)$ becomes singular at a critical point $(\boldsymbol{u}_\mathrm{c}, f_\mathrm{c})$. Physically, buckling is expected to occur at a critical point. Accordingly, condition (2.3) serves as the condition for buckling of a structural system, and the *buckling load* is given by the load f_c at the critical point $(\boldsymbol{u}_\mathrm{c}, f_\mathrm{c})$. When the critical point is a bifurcation point, f_c is also called the *bifurcation load*.

The buckling condition in (2.3) is equivalent to the existence of a zero eigenvalue, that is, $J(\boldsymbol{u}_\mathrm{c}, f_\mathrm{c})\boldsymbol{\eta}_\mathrm{c} = \boldsymbol{0}$ for some nonzero vector $\boldsymbol{\eta}_\mathrm{c}$. We call such $\boldsymbol{\eta}_\mathrm{c}$ a *critical eigenvector*. The *buckling mode* is given by a critical eigenvector. When the critical point is a bifurcation point, the critical eigenvector is called the *bifurcation mode*.

[1] In structural mechanics, an initial state is usually the state of $(\boldsymbol{u}, f) = (\boldsymbol{0}, 0)$ with no load and displacement.

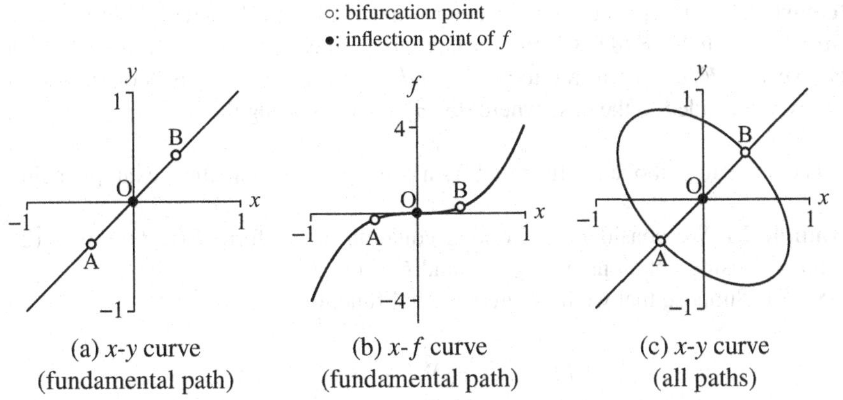

(a) x-y curve (fundamental path)
(b) x-f curve (fundamental path)
(c) x-y curve (all paths)

Figure 2.1 The solution curves of the system in Example 2.1

2.3 PROCEDURE OF BUCKLING ANALYSIS

With the concepts and terminology introduced above, we can present the following buckling analysis procedure to obtain the buckling load and the buckling mode. Steps 1–4 below are basically the same as those for a single-degree-of-freedom system in Section 1.3.

Step 1: Derive the total potential energy of the system.

Step 2: Derive the governing equation and the tangent stiffness matrix (Jacobian matrix) from the total potential energy.

Step 3: Obtain the solution curves, comprising the fundamental and bifurcating paths, from the governing equation.

Step 4: Using the tangent stiffness matrix $J(\boldsymbol{u}, f)$, find the locations $(\boldsymbol{u}_\mathrm{c}, f_\mathrm{c})$ of critical points, find the buckling modes $\boldsymbol{\eta}_\mathrm{c}$ by conducting the eigenanalysis of tangent stiffness matrix $J(\boldsymbol{u}_\mathrm{c}, f_\mathrm{c})$, classify the critical points, and determine the stability of ordinary points.

Step 1 is based on the mechanical property of the system. Step 2 is a straightforward mathematical calculus. Steps 3 and 4 can be conducted analytically for a simple equilibrium equation or numerically for a complex one. (The numerical analysis procedures in these steps are presented in Chapters 3 and 4.) In Step 4, the critical points can be classified based on the orthogonality of the buckling mode $\boldsymbol{\eta}_\mathrm{c}$ and the *load pattern vector* $\partial \boldsymbol{F}/\partial f$. That is, a critical point is a bifurcation point if the vectors $\boldsymbol{\eta}_\mathrm{c}$ and $\partial \boldsymbol{F}/\partial f$ are orthogonal; otherwise, it is a stationary point of f (cf., (2.13) in Section 2.4). The stability of the critical points can be determined with reference to the total potential energy (Section 2.6). Step 1 can be skipped if it is possible to obtain the equilibrium equation without reference to the potential. The analysis of post-buckling behaviors is not covered by the above procedure and is to be treated in Chapter 5.

Remark 2.1 It is customary in structural mechanics to deal with a *proportional loading*, in which $\partial \boldsymbol{F}/\partial f$ is kept constant along an equilibrium path; then, the constant vector $\partial \boldsymbol{F}/\partial f$ is referred to as the *load pattern vector*. In this book, this terminology is extended to the case where $\partial \boldsymbol{F}/\partial f$ changes along the path. □

The procedure above is illustrated for a pedagogic mathematical example below.

Example 2.1 We consider a governing equation of the form $\boldsymbol{F}(\boldsymbol{u}, f) = \boldsymbol{0}$ in (2.1) with $n = 2$ using notations $\boldsymbol{u} = (x, y)^\top$ and $\boldsymbol{F} = (F_x, F_y)^\top$.

Step 1: Suppose that we have the potential function

$$U(x,y,f) = \frac{1}{2}(x-y)^2 - x^4 - y^4 + (x+y)f,$$

where f is a parameter.

Step 2: The equilibrium equations are

$$F_x = \frac{\partial U}{\partial x} = x - y - 4x^3 + f, \tag{2.4}$$

$$F_y = \frac{\partial U}{\partial y} = y - x - 4y^3 + f, \tag{2.5}$$

and the Jacobian matrix is

$$J(x,y,f) = \begin{pmatrix} \partial F_x/\partial x & \partial F_x/\partial y \\ \partial F_y/\partial x & \partial F_y/\partial y \end{pmatrix} = \begin{pmatrix} 1-12x^2 & -1 \\ -1 & 1-12y^2 \end{pmatrix}.$$

Step 3: On noting that $(x,y,f) = (0,0,0)$ is a solution to $F_x = F_y = 0$, we see that the fundamental path is given by $x = y$ and $f = 4x^3$, that is, $(x,y,f) = (x,x,4x^3)$. This fundamental path is shown in Figs. 2.1(a) and (b), where (a) shows the x-y relation and (b) the x-f relation.

Step 4: To obtain critical points on the fundamental path, we set $(x,y,f) = (x,x,4x^3)$ in $J(x,y,f)$ to obtain

$$\det J(x,x,4x^3) = (1-12x^2)^2 - 1 = 24x^2(6x^2-1).$$

The buckling condition $\det J(x,x,4x^3) = 0$ results in $x = 0$, $x = \pm 1/\sqrt{6}$. Thus we have three critical points

$$(x,y,f) = (0,0,0), \qquad \left(-\frac{1}{\sqrt{6}}, -\frac{1}{\sqrt{6}}, -\frac{2}{3\sqrt{6}}\right), \qquad \left(\frac{1}{\sqrt{6}}, \frac{1}{\sqrt{6}}, \frac{2}{3\sqrt{6}}\right),$$

which are the points O, A, and B, respectively, indicated in Fig. 2.1. At these points, the Jacobian matrix $J = J(x,y,f)$ obtained in Step 2 is evaluated to

$$J_{\mathrm{O}} = \begin{pmatrix} 1 & -1 \\ -1 & 1 \end{pmatrix}, \qquad J_{\mathrm{A}} = J_{\mathrm{B}} = \begin{pmatrix} -1 & -1 \\ -1 & -1 \end{pmatrix}.$$

Their critical eigenvectors, that is, the buckling modes associated with the points O, A, and B, respectively, are

$$\boldsymbol{\eta}_{\mathrm{O}} = \frac{1}{\sqrt{2}}(1,1)^{\top}, \qquad \boldsymbol{\eta}_{\mathrm{A}} = \boldsymbol{\eta}_{\mathrm{B}} = \frac{1}{\sqrt{2}}(1,-1)^{\top}.$$

These critical points are classified into a stationary (inflection) point O and bifurcation points A and B, since $\boldsymbol{\eta}_{\mathrm{O}}$ is not orthogonal to the load pattern vector $\partial \boldsymbol{F}/\partial f = (1,1)^{\top}$, and $\boldsymbol{\eta}_{\mathrm{A}}$ and $\boldsymbol{\eta}_{\mathrm{B}}$ are orthogonal to it.

With some additional calculation, we can capture all solution paths, shown in Fig. 2.1(c). The x-y relation of the bifurcated curve is described by $x^2+xy+y^2=1/2$. This shows, in particular, that A and B are bifurcation points, indeed. □

Remark 2.2 The governing equation (2.1) involves many variables, and it is often difficult to depict a solution curve $(\boldsymbol{u}, f)$ involving all variables $(u_1,\ldots,u_n,f)$. Therefore, it is customary to express a solution curve as a relation between the parameter f and a particular displacement as in Fig. 2.1(b). This curve is called the *load versus displacement curve* in structural mechanics. Such a representation is convenient to express the solution curves in a succinct way, but the appearance of the curves is affected substantially by the displacement to be employed. For example, in Fig. 2.2(c) in Section 2.4, the x-f curve and the y-f curve are qualitatively different. □

2.4 JACOBIAN MATRIX AND CRITICAL POINT

For a family of functions $\boldsymbol{F}(\boldsymbol{u}, f) = (F_1,\ldots,F_n)$ in variables $\boldsymbol{u} = (u_1,\ldots,u_n)$ and parameter f, the $n\times n$ matrix

$$J = J(\boldsymbol{u}, f) = \frac{\partial \boldsymbol{F}}{\partial \boldsymbol{u}} = \begin{pmatrix} \dfrac{\partial F_1}{\partial u_1} & \cdots & \dfrac{\partial F_1}{\partial u_n} \\ \vdots & \ddots & \vdots \\ \dfrac{\partial F_n}{\partial u_1} & \cdots & \dfrac{\partial F_n}{\partial u_n} \end{pmatrix}, \tag{2.6}$$

comprising the partial derivatives $\partial F_i/\partial u_j$ $(i,j=1,\ldots,n)$, is called the *Jacobian matrix*[2] of $\boldsymbol{F}$ with respect to $\boldsymbol{u}$. This is an extension of the derivative to multivariate functions and plays a vital role in the description and analysis of bifurcation and stability.

Remark 2.3 When the functions $(F_1,\ldots,F_n)$ are derived from a potential function, the Jacobian matrix $J(\boldsymbol{u}, f)$ is a symmetric matrix for all $(\boldsymbol{u}, f)$. In particular, the eigenvalues of the Jacobian matrix of a potential system are all real. □

---

[2]The Jacobian matrix in (2.6) corresponds to the *tangent stiffness matrix* in structural mechanics. In this book, we often use the latter when we have a structure in mind.

The Jacobian matrix is fundamental in investigating the local properties of a solution $(\boldsymbol{u}, f)$ of the governing equation $\boldsymbol{F}(\boldsymbol{u}, f) = \boldsymbol{0}$. Consider another solution $(\boldsymbol{u}+\tilde{\boldsymbol{u}}, f+\tilde{f})$ in a neighborhood of $(\boldsymbol{u}, f)$, where $\tilde{\boldsymbol{u}}$ and $\tilde{f}$ are incremental values. By the Taylor expansion we have

$$\begin{aligned}\tilde{\boldsymbol{F}}(\tilde{\boldsymbol{u}}, \tilde{f}) &\equiv \boldsymbol{F}(\boldsymbol{u}+\tilde{\boldsymbol{u}}, f+\tilde{f}) - \boldsymbol{F}(\boldsymbol{u}, f)\\ &= \frac{\partial \boldsymbol{F}}{\partial \boldsymbol{u}}\tilde{\boldsymbol{u}} + \frac{\partial \boldsymbol{F}}{\partial f}\tilde{f} + (\text{h.o.t.})\\ &= J(\boldsymbol{u}, f)\tilde{\boldsymbol{u}} + \frac{\partial \boldsymbol{F}}{\partial f}\tilde{f} + (\text{h.o.t.}),\end{aligned}$$

where (h.o.t.) denotes higher-order terms in $\tilde{\boldsymbol{u}}$ and $\tilde{f}$. Noting that both $(\boldsymbol{u}, f)$ and $(\boldsymbol{u}+\tilde{\boldsymbol{u}}, f+\tilde{f})$ satisfy the governing equation $\boldsymbol{F} = \boldsymbol{0}$, we can derive a governing equation in incremental variables $(\tilde{\boldsymbol{u}}, \tilde{f})$ as

$$\tilde{\boldsymbol{F}}(\tilde{\boldsymbol{u}}, \tilde{f}) = J(\boldsymbol{u}, f)\tilde{\boldsymbol{u}} + \frac{\partial \boldsymbol{F}}{\partial f}\tilde{f} + (\text{h.o.t.}) = \boldsymbol{0}, \tag{2.7}$$

which is sometimes called the *incremental governing equation.*

When the Jacobian matrix $J = J(\boldsymbol{u}, f)$ is nonsingular, it has the inverse J^{-1}, and hence the equation (2.7) can be solved uniquely for $\tilde{\boldsymbol{u}}$ as

$$\tilde{\boldsymbol{u}} = -J^{-1}\frac{\partial \boldsymbol{F}}{\partial f}\tilde{f} + (\text{h.o.t.}). \tag{2.8}$$

In this case, the solution $(\boldsymbol{u}, f)$ is called an *ordinary point*. Otherwise, that is, when $J = J(\boldsymbol{u}, f)$ is singular, the solution $(\boldsymbol{u}, f)$ is called a *critical point*. In the latter case, $\tilde{\boldsymbol{u}}$ is (usually) not determined uniquely from (2.7). Accordingly, we have the following classification for a solution point $(\boldsymbol{u}, f)$ of $\boldsymbol{F} = \boldsymbol{0}$:

$$\begin{cases} J(\boldsymbol{u}, f) \text{ is nonsingular} & \Rightarrow \quad (\boldsymbol{u}, f) \text{ is an } \textit{ordinary point},\\ J(\boldsymbol{u}, f) \text{ is singular} & \Rightarrow \quad (\boldsymbol{u}, f) \text{ is a } \textit{critical point}.\end{cases} \tag{2.9}$$

In this book, we usually denote a critical point by $(\boldsymbol{u}_\mathrm{c}, f_\mathrm{c})$, where the subscript $(\cdot)_\mathrm{c}$ indicates the value of the variable at a critical point.

A critical point can be detected as a point $(\boldsymbol{u}_\mathrm{c}, f_\mathrm{c})$ that simultaneously satisfies the governing equation $\boldsymbol{F}(\boldsymbol{u}_\mathrm{c}, f_\mathrm{c}) = \boldsymbol{0}$ and the *criticality condition*

$$\det J(\boldsymbol{u}_\mathrm{c}, f_\mathrm{c}) = 0. \tag{2.10}$$

Note that there are, in total, $n+1$ equations in $n+1$ variables $(\boldsymbol{u}_\mathrm{c}, f_\mathrm{c})$.

The *multiplicity* of a critical point $(\boldsymbol{u}_\mathrm{c}, f_\mathrm{c})$ is defined by

$$M = n - \operatorname{rank}(J(\boldsymbol{u}_\mathrm{c}, f_\mathrm{c})), \tag{2.11}$$

where $\operatorname{rank}(\cdot)$ means the rank of the matrix in the parentheses. On the basis of this multiplicity, a critical point is classified as

$$\begin{cases} M = 1 & \Rightarrow \quad \textit{simple} \text{ critical point},\\ M \geq 2 & \Rightarrow \quad \textit{multiple} \text{ critical point}.\end{cases} \tag{2.12}$$

The multiplicity M is equal, by (2.11), to the number of zero eigenvalues[3] of $J(\boldsymbol{u}_{\rm c}, f_{\rm c})$.

A simple critical point can be classified into a bifurcation point and a stationary point of f on the basis of the orthogonality between the critical eigenvector $\boldsymbol{\eta}_{\rm c}$ and the load pattern vector $\partial \boldsymbol{F}/\partial f$. A critical point is a bifurcation point if these vectors are orthogonal; otherwise, it is a stationary point of f. That is,[4]

$$\begin{cases} \boldsymbol{\eta}_{\rm c}^\top \left(\dfrac{\partial \boldsymbol{F}}{\partial f} \right)_{\rm c} = 0 \quad \Rightarrow \quad \text{bifurcation point}, \\ \boldsymbol{\eta}_{\rm c}^\top \left(\dfrac{\partial \boldsymbol{F}}{\partial f} \right)_{\rm c} \neq 0 \quad \Rightarrow \quad \text{stationary point of } f. \end{cases} \tag{2.13}$$

A complex bifurcation phenomenon occurs at multiple critical points, which usually arises as a result of the (geometrical) symmetry of the system in question. This is the case, for example, with axisymmetric domes (see Chapter 6).

Various kinds of critical points are illustrated in the following examples. We use notations $\boldsymbol{u} = (x, y)^\top$ and $\boldsymbol{F} = (F_x, F_y)^\top$ in place of $\boldsymbol{u} = (u_1, u_2)^\top$ and $\boldsymbol{F} = (F_1, F_2)^\top$.

Example 2.2 We investigate the critical points on the solution curves that we considered[5] in Section 1.2.

(a) The system of equations

$$\begin{cases} F_x \equiv 2x + y - 3f = 0, \\ F_y \equiv x + 2y - 3f = 0 \end{cases} \tag{2.14}$$

in (x, y) has the Jacobian matrix

$$J = \begin{pmatrix} \partial F_x/\partial x & \partial F_x/\partial y \\ \partial F_y/\partial x & \partial F_y/\partial y \end{pmatrix} = \begin{pmatrix} 2 & 1 \\ 1 & 2 \end{pmatrix}.$$

The eigenvalues of this matrix are 1 and 3 and are nonzero for any value of f; accordingly, all solutions on the curve $x = y = f$ depicted in Fig. 2.2(a) are ordinary points.

(b) The system of equations

$$\begin{cases} F_x \equiv x - y^2 - f = 0, \\ F_y \equiv x - y = 0 \end{cases} \tag{2.15}$$

[3]Precisely speaking, when $J(\boldsymbol{u}_{\rm c}, f_{\rm c})$ is not a symmetric matrix, $M = n - \text{rank}(J(\boldsymbol{u}_{\rm c}, f_{\rm c}))$ is equal to the geometric multiplicity of the eigenvalue 0 of $J(\boldsymbol{u}_{\rm c}, f_{\rm c})$.

[4]This classification, to be derived in (5.35) in Remark 5.2 in Section 5.4, is quite convenient from the engineering point of view. Mathematically, however, there is also a critical point that satisfies the condition $\boldsymbol{\eta}_{\rm c}^\top \left(\frac{\partial \boldsymbol{F}}{\partial f} \right)_{\rm c} = 0$ but has no bifurcating solutions (Footnote 2 in Section 5.4.2).

[5]Equations (2.14), (2.15), and (2.17) correspond to (1.1), (1.2), and (1.3) in Section 1.2, respectively.

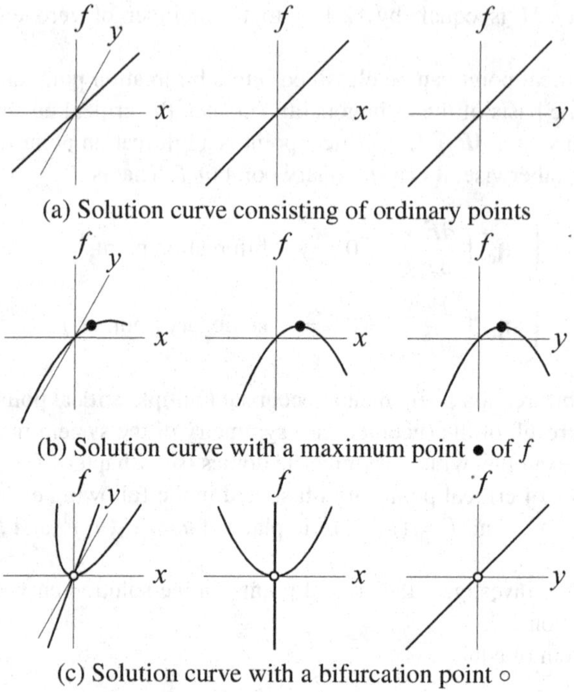
(a) Solution curve consisting of ordinary points
(b) Solution curve with a maximum point • of f
(c) Solution curve with a bifurcation point ○

Figure 2.2 Typical solution curves

in (x,y) has the Jacobian matrix

$$J = \begin{pmatrix} 1 & -2y \\ 1 & -1 \end{pmatrix}.$$

The criticality condition is

$$\det J = -1 + 2y = 0. \tag{2.16}$$

By solving simultaneously the governing equation (2.15) and the criticality condition (2.16), we can obtain a critical point $(x_c, y_c, f_c) = (1/2, 1/2, 1/4)$. As can be seen from the solution curve in Fig. 2.2(b), this critical point is a maximum point of the bifurcation parameter f.

(c) The system of equations

$$\left\{ \begin{array}{l} F_x \equiv x^3 + xy - 2fx = 0, \\ F_y \equiv y - f = 0 \end{array} \right. \tag{2.17}$$

in (x,y) has the Jacobian matrix

$$J = \begin{pmatrix} 3x^2 + y - 2f & x \\ 0 & 1 \end{pmatrix}.$$

The criticality condition is

$$\det J = 3x^2 + y - 2f = 0. \tag{2.18}$$

The simultaneous solution of (2.17) and (2.18) gives one and the only critical point $(x_c, y_c, f_c) = (0,0,0)$, while other points are all ordinary points. As can be seen from the solution curve in Fig. 2.2(c), this critical point is a bifurcation point. □

Example 2.3 As an example of a double bifurcation point, we consider the system of equations

$$\begin{cases} F_x \equiv xf + xy^2 = 0, \\ F_y \equiv yf + x^2 y = 0 \end{cases}$$

in (x,y). This system has five solution curves:

$$\begin{cases} x = y = 0, & \text{solution curve 1 } (f\text{-axis}), \\ x = f = 0, & \text{solution curve 2 } (y\text{-axis}), \\ y = f = 0, & \text{solution curve 3 } (x\text{-axis}), \\ x = y,\ f = -y^2, & \text{solution curve 4}, \\ x = -y,\ f = -y^2, & \text{solution curve 5}, \end{cases}$$

which intersect at the origin $(x,y,f) = (0,0,0)$. These solution curves are depicted in Fig. 2.3. The Jacobian matrix is given by

$$J = \begin{pmatrix} f + y^2 & 2xy \\ 2xy & f + x^2 \end{pmatrix}.$$

At the origin $(x,y,f) = (0,0,0)$, the Jacobian matrix J is equal to the zero matrix, having two zero eigenvalues; accordingly, the origin is a double critical point with $M = 2$. Bifurcation occurs at this point, as shown in Fig. 2.3. □

2.5 TOTAL POTENTIAL ENERGY

2.5.1 DERIVATION OF GOVERNING EQUATION

As is already mentioned in Section 1.3 for a scalar system with one degree of freedom, the total potential energy plays an important role in the description of static

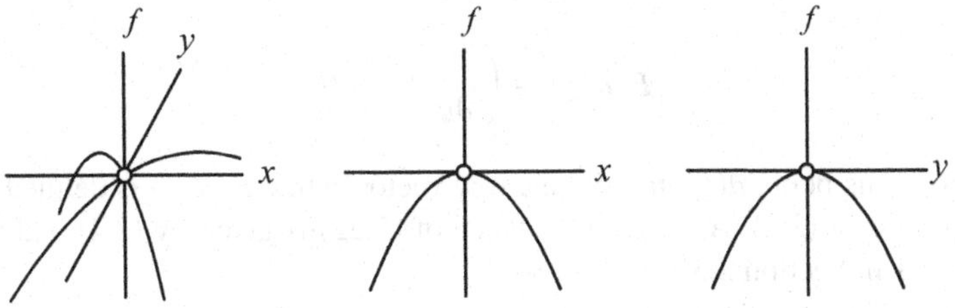

Figure 2.3 Solution curves at a double critical point ○

problems. In general, the *total potential energy*, to be denoted by U, is defined as

$$U = (\text{Internal potential energy}) + (\text{External potential energy}). \tag{2.19}$$

Internal potential energy is the energy stored in the system. *External potential energy* is the work (energy) done by the external force. A concrete instance of (2.19) is given in (1.4) for the simple spring system in Section 1.3. We often abbreviate "total potential energy" to "potential energy" or "potential" in this book. A system equipped with potential will be called a *potential system.*

For a potential system, the governing equation (equilibrium equation) can be derived from the potential by the following principle (e.g., Oden and Ripperger [1]).

Principle 2.1 (Principle of stationary total potential energy) The total potential energy takes a stationary value at static equilibrium. □

Usually, it is much easier to obtain the governing equation via potential than to obtain the governing equation directly from static equilibrium.

For a discrete system described by a vector $\boldsymbol{u} = (u_1, \ldots, u_n)$ with a bifurcation parameter f, the total potential energy U is expressed as

$$U = U(\boldsymbol{u}, f) = U(u_1, \ldots, u_n, f)$$

as a function of $\boldsymbol{u}$ and f. We define functions $F_1(\boldsymbol{u}, f), \ldots, F_n(\boldsymbol{u}, f)$ as the partial derivatives of U with respect to $u_1, \ldots, u_n$, that is,

$$F_i(u_1, \ldots, u_n, f) \equiv \frac{\partial U(u_1, \ldots, u_n, f)}{\partial u_i}, \qquad i = 1, \ldots, n. \tag{2.20}$$

Then the governing equation is obtained by Principle 2.1 as

$$F_i(u_1, \ldots, u_n, f) = 0, \qquad i = 1, \ldots, n.$$

Using a vector representation

$$\frac{\partial U}{\partial \boldsymbol{u}} = \left(\frac{\partial U}{\partial u_1}, \ldots, \frac{\partial U}{\partial u_n} \right), \tag{2.21}$$

we can summarize the above in the following expression:

$$\boldsymbol{F}(\boldsymbol{u}, f) \equiv \left(\frac{\partial U}{\partial \boldsymbol{u}} \right)^{\top} = \boldsymbol{0}. \tag{2.22}$$

Throughout this book, $\partial U / \partial \boldsymbol{u}$ will be a row vector, while $\boldsymbol{F}(\boldsymbol{u}, f)$ is defined to be a column vector in (2.2). A concrete instance of (2.22) is given by (2.4) and (2.5) in Example 2.1 in Section 2.3.

2.5.2 EXISTENCE OF A POTENTIAL

While a potential function U induces a family of functions $(F_1,\ldots,F_n)$ by $F_i = \partial U/\partial u_i$ in (2.20), not every family $(F_1,\ldots,F_n)$ can be induced from a potential function in this way. The condition for the existence of a potential for $(F_1,\ldots,F_n)$ is expressed in terms of *reciprocity*, as stated in the following theorem.

Theorem 2.1: Poincaré lemma

For a system of functions $F_1,\ldots,F_n$, a total potential energy $U(u_1,\ldots,u_n)$ satisfying $F_i = \partial U/\partial u_i$ $(i = 1,\ldots,n)$ exists if and only if the *reciprocity* condition

$$\frac{\partial F_j}{\partial u_i} = \frac{\partial F_i}{\partial u_j}, \qquad i,j = 1,\ldots,n \tag{2.23}$$

is satisfied.

Proof: The necessity of the condition (2.23) is easy to see. Indeed, if a total potential energy $U(u_1,\ldots,u_n)$ exists, we have

$$\frac{\partial}{\partial u_i}\left(\frac{\partial U}{\partial u_j}\right) = \frac{\partial}{\partial u_j}\left(\frac{\partial U}{\partial u_i}\right), \qquad i,j = 1,\ldots,n.$$

Since $\partial U/\partial u_j = F_j$ and $\partial U/\partial u_i = F_i$, this relation implies $\partial F_j/\partial u_i = \partial F_i/\partial u_j$, which is the reciprocity in (2.23).

The proof of the converse is based on the construction of a potential function U via integration. Instead of giving a formal proof in the general setting, the basic idea is shown in Example 2.4 for a concrete example, followed by a supplementary discussion in Remark 2.4. ■

Example 2.4 Consider a pair of functions

$$\begin{cases} F_x \equiv 4x^3 + 2xy^2, \\ F_y \equiv 2x^2 y, \end{cases} \tag{2.24}$$

which satisfies the reciprocity condition (2.23) since

$$\frac{\partial F_x}{\partial y} = \frac{\partial F_y}{\partial x} = 4xy.$$

Therefore, by Theorem 2.1, a potential U satisfying $\partial U/\partial x = F_x$ and $\partial U/\partial y = F_y$ should exist for this system. We can construct the potential U by integrating F_x in (2.24) as

$$U(x,y) \equiv \int F_x \mathrm{d}x = \int (4x^3 + 2xy^2)\mathrm{d}x = x^4 + x^2y^2 + C(y). \tag{2.25}$$

The partial derivative of this function with respect to y is

$$\frac{\partial U}{\partial y} = 2x^2y + C'(y).$$

A combination of this equation with F_y in (2.24) gives $C'(y) = 0$, i.e., $C(y) = c$, where c is a constant (an integration constant). Then from (2.25), we can obtain the potential

$$U(x,y) = x^4 + x^2y^2 + c,$$

where we may choose $c = 0$. For this U we have

$$\frac{\partial U}{\partial x} = 4x^3 + 2xy^2 = F_x, \qquad \frac{\partial U}{\partial y} = 2x^2y = F_y,$$

as desired. □

Remark 2.4 The general idea underlying the construction of U in Example 2.4 is explained briefly by formal calculus. In (2.25), we have defined U by $U = \int F_x \, \mathrm{d}x$. This immediately ensures $\partial U/\partial x = F_x$. The reciprocity plays a key role to ensure $\partial U/\partial y = F_y$. Indeed, we have

$$\frac{\partial U}{\partial y} = \frac{\partial}{\partial y}\int F_x \, \mathrm{d}x = \int \frac{\partial F_x}{\partial y} \, \mathrm{d}x = \int \frac{\partial F_y}{\partial x} \, \mathrm{d}x = F_y,$$

where the integration constant is omitted for simplicity. □

2.6 STABILITY

An equilibrium state of a potential system is characterized by the stationarity of the total potential energy (Principle 2.1 in Section 2.5). An equilibrium solution $(\boldsymbol{u}, f)$ is *stable* if the total potential energy U is minimal at $(\boldsymbol{u}, f)$ in the sense that the potential energy U increases for all perturbations of $\boldsymbol{u}$ when f is kept constant.

The stability of an equilibrium point $(\boldsymbol{u}, f)$ can be investigated as follows. Consider an arbitrary incremental displacement $\delta\boldsymbol{u}$. By the Taylor expansion, the increment of the total potential energy $U(\boldsymbol{u}, f)$ is given, in a neighborhood of $(\boldsymbol{u}, f)$, as

$$\begin{aligned}\delta U(\boldsymbol{u}, f) &\equiv U(\boldsymbol{u} + \delta\boldsymbol{u}, f) - U(\boldsymbol{u}, f) \\ &= \frac{\partial U}{\partial \boldsymbol{u}}\delta\boldsymbol{u} + \frac{1}{2}\delta\boldsymbol{u}^\top \frac{\partial \boldsymbol{F}}{\partial \boldsymbol{u}}\delta\boldsymbol{u} + (\text{h.o.t.}).\end{aligned}$$

Since $(\boldsymbol{u}, f)$ is an equilibrium point, we have $(\partial U/\partial \boldsymbol{u})^\top = \boldsymbol{F}(\boldsymbol{u}, f) = \boldsymbol{0}$ by (2.22), whereas $\partial \boldsymbol{F}/\partial \boldsymbol{u}$ is equal, by definition (2.6), to the Jacobian matrix $J(\boldsymbol{u}, f)$. Therefore, we obtain

$$\delta U(\boldsymbol{u}, f) = \frac{1}{2}\delta\boldsymbol{u}^\top J(\boldsymbol{u}, f)\,\delta\boldsymbol{u} + (\text{h.o.t.}). \tag{2.26}$$

Here the Jacobian matrix $J = J(\boldsymbol{u}, f)$ is symmetric as we are dealing with a potential system.

It follows from the expression (2.26) that we have $\delta U(\boldsymbol{u}, f) > 0$ for all nonzero perturbations $\delta \boldsymbol{u}$ if the Jacobian matrix J is positive definite. Note that the positive definiteness of J is not a necessary condition for $\delta U(\boldsymbol{u}, f) > 0$ due to the higher order terms (h.o.t.). For example, if $n = 1$ and $U(u, f) = u^4$, we have $\delta U(u, f) = U(u, f) - U(0, f) > 0$ for all $u \neq 0$, but $J = 0$ at $u = 0$.

As is well known in linear algebra, a symmetric matrix is *positive definite* if and only if all of its eigenvalues are positive. In contrast, if the Jacobian matrix J has at least one negative eigenvalue, then $\delta U(\boldsymbol{u}, f) < 0$ for some $\delta \boldsymbol{u}$. With reference to the eigenvalues, the stability of an equilibrium state is classified as follows:[6]

$$\begin{cases} \text{all eigenvalues of } J \text{ are positive} & \Rightarrow \quad \text{stable}, \\ \text{at least one eigenvalue of } J \text{ is negative} & \Rightarrow \quad \text{unstable}. \end{cases} \tag{2.27}$$

If J has a zero eigenvalue and others are positive, the stability cannot be determined based on J alone.

Example 2.5 The stability analysis is illustrated for simple examples. The bifurcation parameter f is suppressed as it is kept constant.

(a) A system with the total potential energy

$$U(x, y) = x^2 + 2y^2$$

has the governing equation

$$\frac{\partial U}{\partial x} = 2x = 0, \qquad \frac{\partial U}{\partial y} = 4y = 0.$$

The origin $(x, y) = (0, 0)$ is the (unique) solution to this equation, and the Jacobian matrix at this point reads

$$J(0,0) = \left. \begin{pmatrix} \partial^2 U / \partial x^2 & \partial^2 U / \partial y \partial x \\ \partial^2 U / \partial x \partial y & \partial^2 U / \partial y^2 \end{pmatrix} \right|_{(x,y)=(0,0)} = \begin{pmatrix} 2 & 0 \\ 0 & 4 \end{pmatrix}.$$

Since all the eigenvalues of this matrix are positive, this system is stable at the origin $(0, 0)$ by (2.27). This is seen from Fig. 2.4(a), which displays a local minimum of the potential at the origin.

(b) A system with the total potential energy

$$U(x, y) = x^2 - y^2$$

has the governing equation

$$\frac{\partial U}{\partial x} = 2x = 0, \qquad \frac{\partial U}{\partial y} = -2y = 0.$$

[6] As already mentioned in Section 1.3, we set up a dichotomy between "stable" and "unstable" so as to simplify the statements without entering into subtle mathematical issues concerning the neutral case. According to our definition, a point is unstable if it is not stable.

The origin $(x,y) = (0,0)$ is the (unique) solution to this equation, and the Jacobian matrix at this point reads

$$J(0,0) = \begin{pmatrix} 2 & 0 \\ 0 & -2 \end{pmatrix}.$$

Since this matrix has a negative eigenvalue, this system is unstable at the origin $(0,0)$ by (2.27). This is apparent from Fig. 2.4(b), which displays a saddle point of the potential at the origin.

(c) A system with the total potential energy

$$U(x,y) = \frac{1}{2}(x-y)^2 - x^4 - y^4 \tag{2.28}$$

has the governing equation

$$\frac{\partial U}{\partial x} = x - y - 4x^3 = 0, \qquad \frac{\partial U}{\partial y} = y - x - 4y^3 = 0.$$

The origin $(x,y) = (0,0)$ is a solution to this equation (not a unique solution). The Jacobian matrix at this point reads

$$J(0,0) = \begin{pmatrix} 1 & -1 \\ -1 & 1 \end{pmatrix}$$

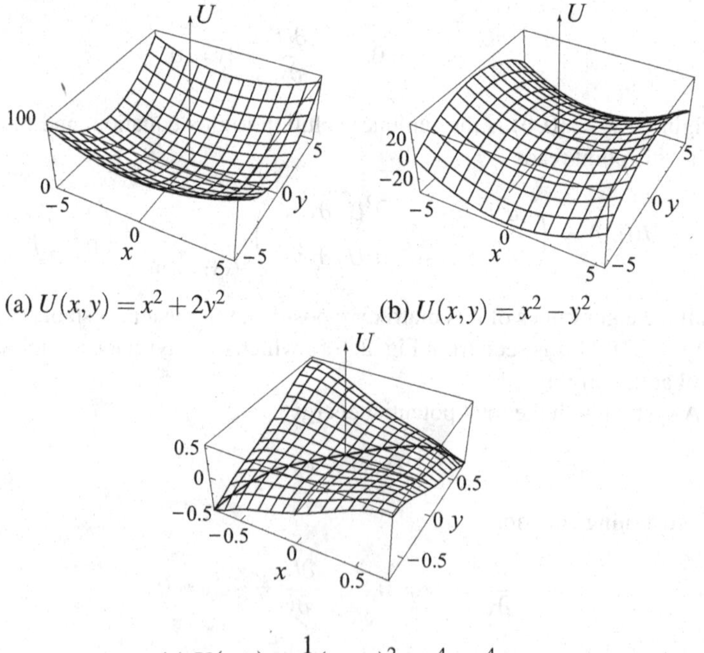

(a) $U(x,y) = x^2 + 2y^2$

(b) $U(x,y) = x^2 - y^2$

(c) $U(x,y) = \frac{1}{2}(x-y)^2 - x^4 - y^4$

Figure 2.4 Distributions of the total potential energy around solution points

and its eigenvalues are determined from

$$\det\begin{pmatrix} 1-\lambda & -1 \\ -1 & 1-\lambda \end{pmatrix} = (1-\lambda)^2 - 1 = \lambda(\lambda-2) = 0$$

as $\lambda = 0$ and $\lambda = 2$. Since one of them is zero and the other is positive, the stability of this system cannot be determined from the Jacobian matrix. We, therefore, investigate the potential distribution. From (2.28), in the direction of $x = y$, we have

$$U(x,x) = -2x^4,$$

which has a local maximum at the origin $(0,0)$ as depicted in Fig. 2.4(c); therefore, this system is unstable at the origin. □

2.7 EXAMPLE OF BUCKLING LOAD ANALYSIS

The buckling load analysis is illustrated for a structural example. Both snap buckling and bifurcation buckling occur, depending on the shape of the structure.

We consider the two-bar truss arch depicted in Fig. 2.5(a). Nodes 1 and 2 of this arch are fixed, and node 3 is free to move. The displaced location (x,y) of node 3 is the unknown vector and initial location of the nodes are $(x_1,y_1) = (-1,h)$, $(x_2,y_2) = (1,h)$, and $(x_3,y_3) = (0,0)$. The arch comprises two identical elastic truss members 1 and 2 both with the modulus of elasticity (Young's modulus) E, the cross-sectional area A, and the initial member length $L = \sqrt{1+h^2}$. The arch is subjected to the vertical load EAf at the crown node 3, where f is a non-dimensional normalized load.

Step 1: The total potential energy is given by

$$U(x,y,f) = \sum_{m=1}^{2} \frac{EA}{2L}(\hat{L}^{(m)} - L)^2 - EAf(y - y_3).$$

Here $\hat{L}^{(m)}$ is the length of the mth member after deformation, being given as

$$\hat{L}^{(m)} = \sqrt{(x-x_m)^2 + (y-y_m)^2}, \quad m = 1,2.$$

Step 2: The equilibrium equation $\boldsymbol{F} = (F_x, F_y)^\top = (\partial U/\partial x, \partial U/\partial y)^\top$ is obtained as

$$\begin{pmatrix} F_x \\ F_y \end{pmatrix} = \begin{pmatrix} \sum_{m=1}^{2} EA\left(\frac{1}{L} - \frac{1}{\hat{L}^{(m)}}\right)(x - x_m) \\ \sum_{m=1}^{2} EA\left(\frac{1}{L} - \frac{1}{\hat{L}^{(m)}}\right)(y - y_m) - EAf \end{pmatrix}. \tag{2.29}$$

The tangent stiffness matrix is

$$J(x,y,f) = EA\sum_{m=1}^{2} \begin{pmatrix} \frac{1}{L} - \frac{(y-y_m)^2}{(\hat{L}^{(m)})^3} & \frac{(x-x_m)(y-y_m)}{(\hat{L}^{(m)})^3} \\ \frac{(x-x_m)(y-y_m)}{(\hat{L}^{(m)})^3} & \frac{1}{L} - \frac{(x-x_m)^2}{(\hat{L}^{(m)})^3} \end{pmatrix}, \tag{2.30}$$

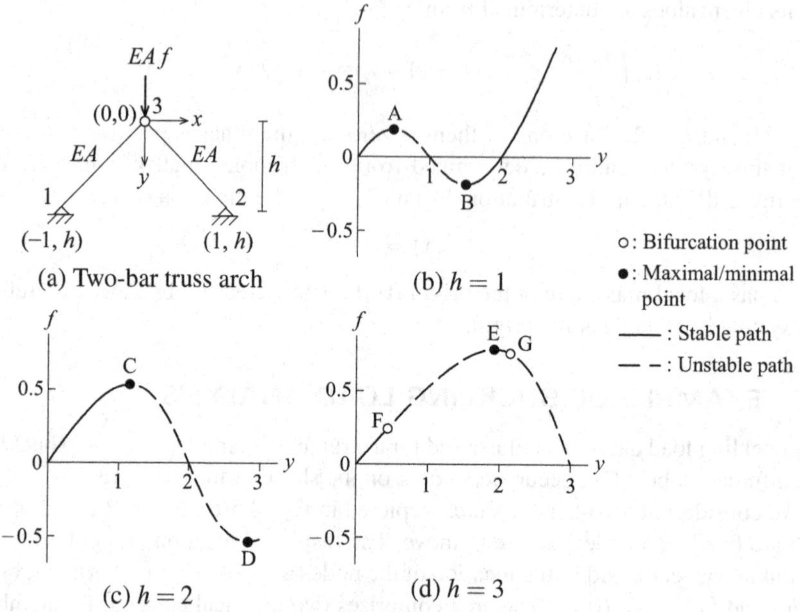

Figure 2.5 A two-bar truss arch and its fundamental paths for various heights $h = 1, 2, 3$

whereas the load pattern vector is

$$\frac{\partial \boldsymbol{F}}{\partial f} = \begin{pmatrix} \partial F_x/\partial f \\ \partial F_y/\partial f \end{pmatrix} = EA \begin{pmatrix} 0 \\ -1 \end{pmatrix}, \tag{2.31}$$

which is a constant vector.

Step 3: The fundamental path can be obtained analytically from the equilibrium equations $F_x = 0$ and $F_y = 0$ with (2.29) as

$$x = 0, \quad f = 2\left(\frac{1}{\sqrt{1+h^2}} - \frac{1}{\sqrt{1+(y-h)^2}}\right)(y-h). \tag{2.32}$$

The curves of this solution for $h = 1, 2, 3$ are plotted in Figs. 2.5(b)–(d).

Step 4: We investigate critical points on the fundamental path. On the fundamental path, where $x = 0$ by (2.32), the tangent stiffness matrix in (2.30) is a diagonal matrix

$$J(0, y, f) = \begin{pmatrix} \lambda_x & 0 \\ 0 & \lambda_y \end{pmatrix}$$

with

$$\lambda_x = 2EA\left(\frac{1}{\sqrt{1+h^2}} - \frac{(y-h)^2}{[1+(y-h)^2]^{3/2}}\right), \tag{2.33}$$

$$\lambda_y = 2EA\left(\frac{1}{\sqrt{1+h^2}} - \frac{1}{[1+(y-h)^2]^{3/2}}\right). \tag{2.34}$$

The eigenvalues of $J(0,y,f)$ are given by λ_x and λ_y.

The location (x_c, y_c, f_c) of a critical point can be determined from $\lambda_x = 0$ with (2.33) or $\lambda_y = 0$ with (2.34). Specifically, y_c is determined from these equations. Then f_c is determined from y_c by (2.32), while $x_c = 0$ is already known.

In the case of $\lambda_x = 0$ we have $J(x_c, y_c, f_c) = \begin{pmatrix} 0 & 0 \\ 0 & \lambda_y \end{pmatrix}$, and the critical eigenvector is given by $\boldsymbol{\eta}_{xc} = (1,0)^\top$. This represents a horizontal sway of the arch, as depicted in Fig. 2.6(a). Since the critical eigenvector $\boldsymbol{\eta}_{xc} = (1,0)^\top$ and the load pattern vector $\partial \boldsymbol{F}/\partial f = EA(0,-1)^\top$ in (2.31) are orthogonal, the point with $\lambda_x = 0$ is a bifurcation point by (2.13). Thus we will have a bifurcation buckling for such critical point.

In the case of $\lambda_y = 0$, in contrast, we have $J(x_c, y_c, f_c) = \begin{pmatrix} \lambda_x & 0 \\ 0 & 0 \end{pmatrix}$, and the critical eigenvector is $\boldsymbol{\eta}_{yc} = (0,1)^\top$. This represents a vertical displacement of the arch, as depicted in Fig. 2.6(b). Since the critical eigenvector and the load pattern vector are not orthogonal, the point with $\lambda_y = 0$ is a maximal/minimal point of f by (2.13). Thus we will have a snap buckling for such critical point.

It remains to identify the location (x_c, y_c, f_c) of a critical point. We first deal with the equation $\lambda_y = 0$, which is easier than $\lambda_x = 0$. By the expression (2.34) of λ_y, we see that

$$\lambda_y = 0 \iff \sqrt{1+h^2} = [1+(y-h)^2]^{3/2} \iff y = h \pm \sqrt{(1+h^2)^{1/3} - 1}.$$

We denote these two values of y by y_{c1}^{snap} and y_{c2}^{snap}, that is,

$$y_{c1}^{\text{snap}} = h - \sqrt{(1+h^2)^{1/3} - 1}, \qquad y_{c2}^{\text{snap}} = h + \sqrt{(1+h^2)^{1/3} - 1}. \tag{2.35}$$

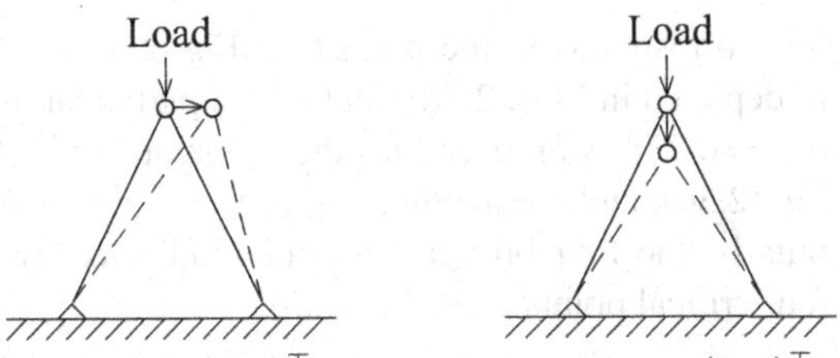

(a) $\boldsymbol{\eta}_{xc} = (1,0)^\top$
(bifurcation buckling)

(b) $\boldsymbol{\eta}_{yc} = (0,1)^\top$
(snap buckling)

Figure 2.6 Deformation modes of the truss arch for the critical eigenvectors

Next, we solve the equation $\lambda_x = 0$ with the expression (2.33) of λ_x. On defining $\xi = (y-h)^2$, we see that

$$\lambda_x = 0 \iff (1+\xi)^{3/2} = \xi\sqrt{1+h^2}.$$

It can be shown[7] that this equation admits a solution if and only if $h \geq h_0$, where $h_0 = \sqrt{23}/2 \approx 2.40$. When $h > h_0$, there are two solutions ξ_1 and ξ_2 with $0 < \xi_1 < 2 < \xi_2$, while these solutions coincide at $\xi_1 = \xi_1 = 2$ when $h = h_0$. We define

$$y_{c1}^{bif} = h - \sqrt{\xi_2}, \quad y_{c2}^{bif} = h - \sqrt{\xi_1}, \quad y_{c3}^{bif} = h + \sqrt{\xi_1}, \quad y_{c4}^{bif} = h + \sqrt{\xi_2}. \qquad (2.36)$$

From the engineering point of view, we are most interested in the smallest among the six critical values of y in (2.35) and (2.36), which is equal to the smaller of y_{c1}^{snap} and y_{c1}^{bif}. To determine which of the two is smaller for a given h, we consider the critical case where $y_{c1}^{snap} = y_{c1}^{bif}$, that is, $\lambda_x = \lambda_y = 0$ for the same y. This is equivalent to

$$\frac{(y-h)^2}{(1+(y-h)^2)^{3/2}} = \frac{1}{(1+(y-h)^2)^{3/2}} = \frac{1}{(1+h^2)^{1/2}},$$

from which we obtain $h = \sqrt{7} \approx 2.65$ and $y = \sqrt{7}-1 \approx 1.65$.

If $h < \sqrt{7}$, we have $y_{c1}^{snap} < y_{c1}^{bif}$, which means that, as the load f is gradually increased from zero, a snap buckling occurs before a bifurcation buckling. For $h = 1$, for example, the first critical point is given by

$$(x_c, y_c, f_c) = (0, 1-(2^{1/3}-1)^{1/2}, \sqrt{2}(2^{1/3}-1)^{3/2}) \approx (0,\ 0.490,\ 0.187),$$

where $y_c = y_{c1}^{snap}$. This corresponds to point A in Fig. 2.5(b). In particular, the buckling load for $h = 1$ is equal to $f_c = \sqrt{2}(2^{1/3}-1)^{3/2} \approx 0.187$. Moreover, the minimal point B in Fig. 2.5(b) corresponds to the other critical point y_{c2}^{snap} in (2.35).

If $h > \sqrt{7}$, we have $y_{c1}^{bif} < y_{c1}^{snap}$, which means that, as the load f is gradually increased from zero, a bifurcation buckling occurs before a snap buckling. Note that $\sqrt{7} > \sqrt{23}/2 = h_0$. For $h = 3$, for example, the first critical point y_{c1}^{bif} is found numerically as

$$(x_c, y_c, f_c) = (0,\ 0.447,\ 0.248),$$

where $y_c = y_{c1}^{bif}$. This corresponds to the point F in Fig. 2.5(d), which is indeed a bifurcation point, as depicted in Figs. 2.7(b) and (c). In particular, the buckling load for $h = 3$ is equal to $f_c = 0.248$. Moreover, the other bifurcation point G in Fig. 2.5(d) corresponds to y_{c2}^{bif} in (2.36), and the maximal point E in Fig. 2.5(d) corresponds to y_{c1}^{snap} in (2.35). Similarly, the four bifurcation points indicated by $\circ$ in Fig. 2.7(b) correspond to the four critical points given in (2.36).

---

[7] Define $g(\xi) = (1+\xi)^{3/2}$. In the (ξ, ζ)-plane, consider the graph of $\zeta = g(\xi)$ and the line $\zeta = \xi\sqrt{1+h^2}$ through the origin with slope $\sqrt{1+h^2}$. Then observe that $g'(\xi) = g(\xi)/\xi = \sqrt{1+h^2}$ for $\xi = 2$ and $h = \sqrt{23}/2$.

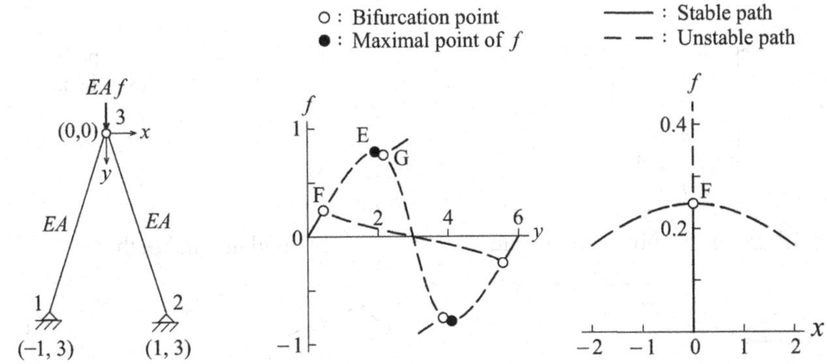

(a) Two-bar truss arch (b) f versus y curve (c) f versus x curve

Figure 2.7 A two-bar truss arch with the height of $h = 3$ and its equilibrium paths

2.8 EXAMPLES OF BUCKLING MODE ANALYSIS

Buckling mode analysis is illustrated for structural examples. In this section, we focus on critical points on the fundamental path.

2.8.1 BAR-SPRING SYSTEM WITH TWO DEGREES OF FREEDOM

As a simple example of the buckling mode analysis, we consider the bar-spring system[8] depicted in Fig. 2.8(a). It comprises two rigid bars of an identical length L, supported vertically by linear springs with a spring constant k, and subjected to a horizontal force kLf. A post-bifurcation state in Fig. 2.8(b) is characterized by the rotational displacement u_i of the ith bar $(i = 1, 2)$.

Step 1: The total potential energy is given by

$$
\begin{aligned}
U(\boldsymbol{u}, f) &= \int (\text{Spring force})\, \mathrm{d}(\text{Spring displacement}) \\
&\qquad - (\text{External force}) \times (\text{Distance external force traveled}) \\
&= kL^2 \left[\frac{1}{2} \sin^2 u_1 + \frac{1}{2} (\sin u_1 + \sin u_2)^2 - (2 - \cos u_1 - \cos u_2) f \right],
\end{aligned}
$$

where $\boldsymbol{u} = (u_1, u_2)^\top$ is the displacement vector comprising rotational displacements u_1 and u_2 of the bars.

Step 2: The partial differentiation of this potential energy with respect to u_1 and u_2 leads to the equilibrium equation

$$
\boldsymbol{F}(\boldsymbol{u}, f) = kL^2 \begin{pmatrix} \sin 2u_1 + \cos u_1 \sin u_2 - f \sin u_1 \\ \sin u_1 \cos u_2 + \frac{1}{2} \sin 2u_2 - f \sin u_2 \end{pmatrix}. \tag{2.37}
$$

[8] This system with rigid bars supported by vertical springs is similar to that with rotational springs studied by Thompson and Hunt [2].

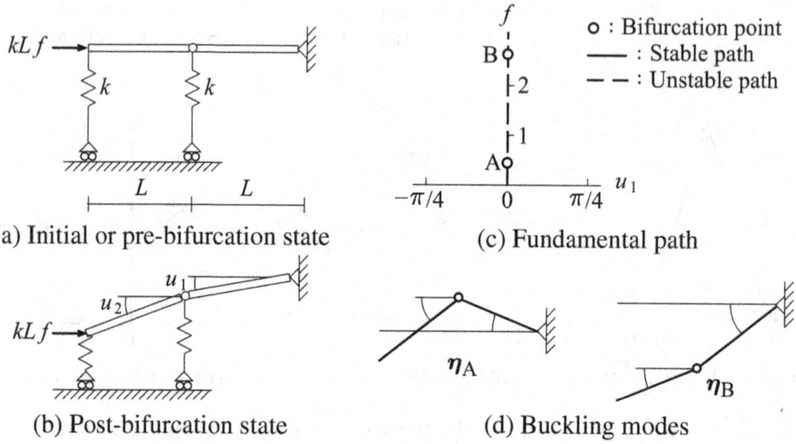

Figure 2.8 A bar-spring system with two degrees of freedom and its buckling modes

A further differentiation gives the tangent stiffness matrix

$$J(\boldsymbol{u}, f) = \frac{\partial \boldsymbol{F}}{\partial \boldsymbol{u}} = \begin{pmatrix} J_{11} & J_{12} \\ J_{21} & J_{22} \end{pmatrix}$$

with

$$\begin{aligned} J_{11} &= kL^2 \left(2\cos 2u_1 - \sin u_1 \sin u_2 - f \cos u_1\right), \\ J_{12} = J_{21} &= kL^2 \cos u_1 \cos u_2, \\ J_{22} &= kL^2 \left(-\sin u_1 \sin u_2 + \cos 2u_2 - f \cos u_2\right). \end{aligned}$$

Step 3: From the equilibrium equation $\boldsymbol{F}(\boldsymbol{u}, f) = \boldsymbol{0}$ with (2.37), the fundamental path is obtained as the trivial solution $\boldsymbol{u} = (u_1, u_2)^\top = \boldsymbol{0}$, which corresponds to the undeformed state.

Step 4: On the fundamental path $\boldsymbol{u} = \boldsymbol{0}$, the tangent stiffness matrix becomes

$$J(\boldsymbol{0}, f) = kL^2 \begin{pmatrix} 2-f & 1 \\ 1 & 1-f \end{pmatrix} \tag{2.38}$$

and

$$\det J(\boldsymbol{0}, f) = (kL^2)^2 \det \begin{pmatrix} 2-f & 1 \\ 1 & 1-f \end{pmatrix} = (kL^2)^2 (f^2 - 3f + 1).$$

From the buckling condition $\det J(\boldsymbol{0}, f) = 0$, the buckling loads are obtained as

$$f_{\mathrm{A}} = \frac{3-\sqrt{5}}{2} \approx 0.382, \qquad f_{\mathrm{B}} = \frac{3+\sqrt{5}}{2} \approx 2.618.$$

The fundamental path and the critical points on this path are shown in Fig. 2.8(c). The bifurcation points A and B correspond to buckling loads f_{A} and f_{B}, respectively.

On the fundamental path, the system is stable until reaching the bifurcation point A ($f < f_A$), but becomes unstable beyond this point ($f > f_A$). The (normalized) eigenvectors for these buckling loads are

$$\boldsymbol{\eta}_A = \frac{1}{\sqrt{10-2\sqrt{5}}}\begin{pmatrix}-\sqrt{5}+1\\ 2\end{pmatrix} \approx \begin{pmatrix}-0.5257\\ 0.8507\end{pmatrix},$$
$$\boldsymbol{\eta}_B = \frac{1}{\sqrt{10-2\sqrt{5}}}\begin{pmatrix}2\\ \sqrt{5}-1\end{pmatrix} \approx \begin{pmatrix}0.8507\\ 0.5257\end{pmatrix}.$$

Buckling modes expressed by these eigenvectors are depicted in Fig. 2.8(d). We see that the buckling mode $\boldsymbol{\eta}_A$ expresses the kink of the system and $\boldsymbol{\eta}_B$ denotes an overall bend downwards of the system. To see whether the buckling at the minimum buckling load f_A leads to the collapse of the system or not, we need to investigate the post-bifurcation behavior with respect to stability.

2.8.2 BAR-SPRING SYSTEM WITH MANY DEGREES OF FREEDOM

We consider the n-degree-of-freedom bar-spring system depicted in Fig. 2.9(a) as a realistic structural example for the buckling mode analysis. It comprises $n+1$ rigid bars of the same length L supported vertically by linear springs and subjected to a horizontal force kLf. The nodal vertical displacements $\boldsymbol{u} = (u_1, \ldots, u_n)^\top$ in Fig. 2.9(b) are employed as the unknown variables.

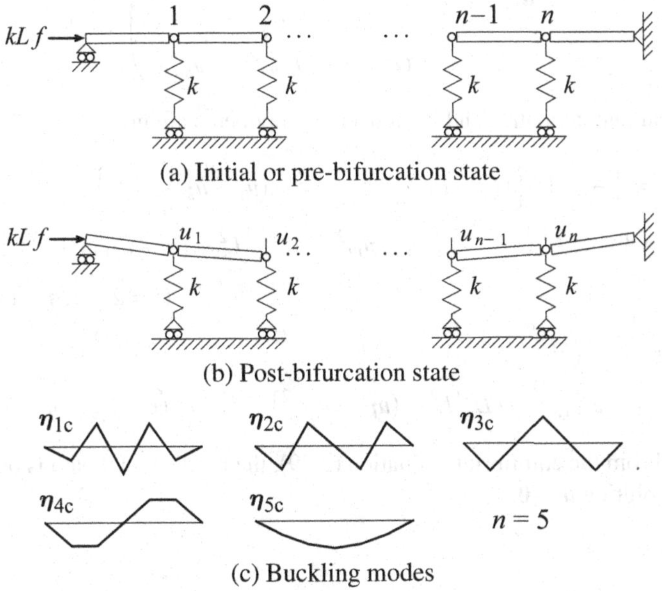

Figure 2.9 A bar-spring system with n degrees of freedom and its buckling modes

Step 1: The total potential energy of this system is given by

$$U(\boldsymbol{u},f) = \frac{k}{2}\sum_{i=1}^{n} u_i{}^2 - fkL\left[(n+1)L - \sqrt{L^2-u_1{}^2} - \sum_{i=1}^{n-1}\sqrt{L^2-(u_i-u_{i+1})^2} - \sqrt{L^2-u_n{}^2}\right].$$

Step 2: The partial differentiation of this potential with respect to $u_1,\ldots,u_n$ leads to the equilibrium equation

$$\begin{aligned}
&ku_1 + fkL\left(-\frac{u_1}{\sqrt{L^2-u_1{}^2}} + \frac{u_2-u_1}{\sqrt{L^2-(u_1-u_2)^2}}\right) = 0,\\
&ku_i + fkL\left(\frac{u_{i-1}-u_i}{\sqrt{L^2-(u_{i-1}-u_i)^2}} + \frac{u_{i+1}-u_i}{\sqrt{L^2-(u_i-u_{i+1})^2}}\right) = 0\\
&\qquad\qquad (i=2,\ldots,n-1),\\
&ku_n + fkL\left(\frac{u_{n-1}-u_n}{\sqrt{L^2-(u_{n-1}-u_n)^2}} - \frac{u_n}{\sqrt{L^2-u_n{}^2}}\right) = 0.
\end{aligned} \tag{2.39}$$

A further differentiation gives the tangent stiffness matrix

$$J(\boldsymbol{u},f) = \begin{pmatrix} J_{11} & J_{12} & & O \\ J_{21} & \ddots & \ddots & \\ & \ddots & \ddots & J_{n-1,n} \\ O & & J_{n,n-1} & J_{nn} \end{pmatrix}.$$

This is a tridiagonal matrix and its nonzero components are given by

$$\begin{aligned}
J_{11} &= k - fkL^3\left\{(L^2-u_1{}^2)^{-3/2} + \left[L^2-(u_1-u_2)^2\right]^{-3/2}\right\},\\
J_{ii} &= k - fkL^3\left\{\left[L^2-(u_{i-1}-u_i)^2\right]^{-3/2} + \left[L^2-(u_i-u_{i+1})^2\right]^{-3/2}\right\}\\
&\qquad\qquad (i=2,\ldots,n-1),\\
J_{nn} &= k - fkL^3\left\{\left[L^2-(u_{n-1}-u_n)^2\right]^{-3/2} + (L^2-u_n{}^2)^{-3/2}\right\},\\
J_{i,i+1} &= J_{i+1,i} = fkL^3\left[L^2-(u_i-u_{i+1})^2\right]^{-3/2} \qquad (i=1,\ldots,n-1).
\end{aligned}$$

Step 3: From the equilibrium equation (2.39), the fundamental path is obtained as the trivial solution $\boldsymbol{u}=\boldsymbol{0}$.

Step 4: On the fundamental path $\boldsymbol{u} = \boldsymbol{0}$, the tangent stiffness matrix becomes

$$J(\boldsymbol{0}, f) = k \begin{pmatrix} 1-2f & f & & O \\ f & \ddots & \ddots & \\ & \ddots & \ddots & f \\ O & & f & 1-2f \end{pmatrix}$$

$$= k \begin{pmatrix} 1 & & O \\ & \ddots & \\ O & & 1 \end{pmatrix} - fk \begin{pmatrix} 2 & -1 & & O \\ -1 & \ddots & \ddots & \\ & \ddots & \ddots & -1 \\ O & & -1 & 2 \end{pmatrix}. \tag{2.40}$$

By the generalized eigenvalue analysis corresponding to $J(\boldsymbol{0}, f)$ (cf., Remark 2.5 below), all buckling loads and buckling modes can be obtained numerically. For example, the analysis for $n = 5$ gives the buckling loads

$$f_{1c} \approx 0.2679, \quad f_{2c} \approx 0.3333, \quad f_{3c} \approx 0.5000, \quad f_{4c} \approx 1.000, \quad f_{5c} \approx 3.732,$$

and the associated eigenvectors $\boldsymbol{\eta}_{jc}$ $(j = 1, \ldots, 5)$ depicted in Fig. 2.9(c). Thus we have succeeded in the buckling load/mode analysis.

Remark 2.5 The tangent stiffness matrix $J(\boldsymbol{u}, f)$ often takes the form of

$$J(\boldsymbol{u}, f) = J_0(\boldsymbol{u}) - fJ_1(\boldsymbol{u}), \tag{2.41}$$

where $J_0(\boldsymbol{u})$ and $J_1(\boldsymbol{u})$ are symmetric matrices and $J_1(\boldsymbol{u})$ is positive definite for each $\boldsymbol{u}$; see (2.40). In this case, the relation $J(\boldsymbol{u}, f)\boldsymbol{\eta} = \boldsymbol{0}$ is equivalent to

$$J_0(\boldsymbol{u})\boldsymbol{\eta} = fJ_1(\boldsymbol{u})\boldsymbol{\eta}, \tag{2.42}$$

and hence the problem of finding f and $\boldsymbol{\eta}$ satisfying $J(\boldsymbol{u}, f)\boldsymbol{\eta} = \boldsymbol{0}$, for a given $\boldsymbol{u}$, is a *generalized eigenvalue problem*, for which numerical analysis programs are available. Suppose further that $J_0(\boldsymbol{u})$ and $J_1(\boldsymbol{u})$ are constant matrices independent of $\boldsymbol{u}$, that is, $J_0(\boldsymbol{u}) = J_0$ and $J_1(\boldsymbol{u}) = J_1$. In such a case, the buckling loads $f_{1c} \leq f_{2c} \leq \cdots \leq f_{nc}$ and the associated buckling modes $\boldsymbol{\eta}_{1c}, \boldsymbol{\eta}_{2c}, \ldots, \boldsymbol{\eta}_{nc}$ can be obtained by solving the generalized eigenvalue problem $J_0\boldsymbol{\eta} = fJ_1\boldsymbol{\eta}$. □

2.9 PROBLEMS

Problem 2.1 Obtain the governing equation for each of the following potential systems. In (3), f is a parameter.

(1) $U(x, y) = x^4 + x^2y^2 + y^4$.

(2) $U(x, y, z) = x^4 + 2\sin(x^2)\sin(y^2) + \cos(z^2)$.

(3) $U(x, y, f) = x^4 + y^2 + f(x^2 + y) - y$.

Problem 2.2 Obtain the governing equation and its solution for each of the following potential systems.

$$(1)\quad U(x,y) = x^2 + y^2 - 2x - xy.$$
$$(2)\quad U(x,y,z) = x^2 - y^2 - z^2 + 2(xy + yz + zx) - 4x.$$

Problem 2.3 Obtain the governing equation of a system with potential

$$U(x,y,f) = x^3 - x^2 f + yf - y^2.$$

Plot the solution curve in the x-y plane.

Problem 2.4 For systems with the following potential functions, show that the origin $\mathbf{0}$ is the solution to the governing equation and investigate the stability at the origin.

$$(1)\quad U(x,y) = x^4 + y^4 + 2x^2y^2 + 2x^2 + y^2 + xy.$$
$$(2)\quad U(x,y) = x^4 + x^2 + x^2y^2 + 3xy + y^2.$$
$$(3)\quad U(x,y,z) = x^4 + x^2 + y^2 + z^2 + 2xy + 2yz.$$

Problem 2.5 Consider the system with the potential function

$$U(x,y,f) = f(x^2 + y^2 + x^4) + x^2y^2 + 3xy$$

with a parameter f. Show that the origin $(x,y) = (0,0)$ is the solution to the governing equation for any f. Investigate the dependence of the stability at this point on f.

Problem 2.6 Show the reciprocity and obtain the potential U for each of the following equations.

$$(1)\ \begin{cases} F_x = 4xy^2 + 2x, \\ F_y = 4x^2y + 2y. \end{cases} \qquad (2)\ \begin{cases} F_x = 4x^3 + 2xy^2z^2, \\ F_y = 2x^2yz^2, \\ F_z = 2x^2y^2z + 4z^3. \end{cases}$$

Problem 2.7 Obtain the bifurcation load of the rigid-bar-spring system shown in Fig. 2.10(a). In the rotational spring, a bending moment of $k_\theta\theta$ develops for rotational displacement θ; we set $k_\theta = kL^2$. Employ the rotational displacement u of the rigid bar as the independent variable.

Problem 2.8 Obtain the bifurcation load of the rigid-bar-spring system shown in Fig. 2.10(b). In the rotational spring, a bending moment of $k_\theta\theta$ develops for rotational displacement θ; we set $k_\theta = kL^2$. Employ the rotational displacement u of the rigid bar as the independent variable.

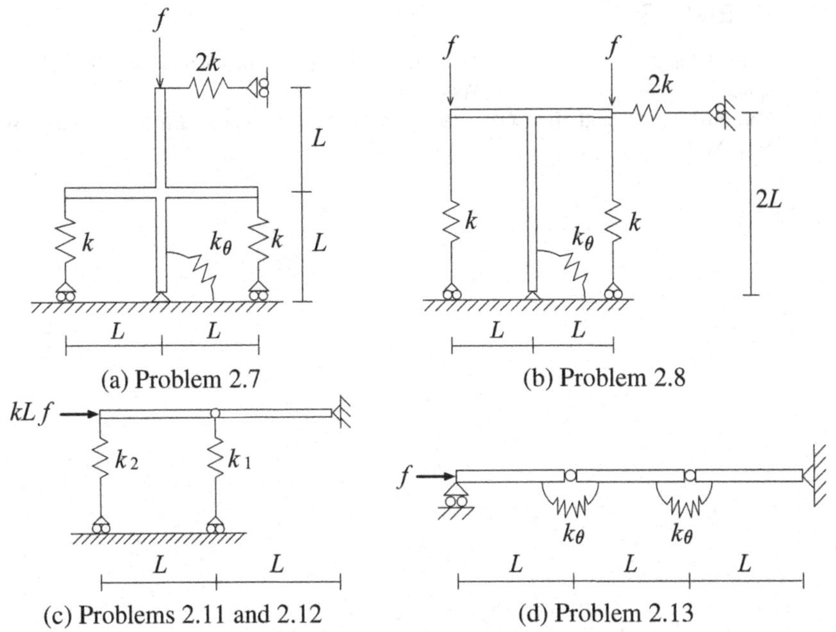

Figure 2.10 Single-degree-of-freedom rigid-bar-spring systems

Problem 2.9 In Section 2.8.1, we have used the rotational displacements to describe the two-degree-of-freedom system in Fig. 2.8. By taking the vertical displacements at the ends of the bars as alternative independent variables, obtain the buckling loads and show that these loads are independent of the choice of the variables to describe the system.

Problem 2.10 Obtain the buckling loads and buckling modes of the n-degree-of-freedom system depicted in Fig. 2.9(a) for (1) $n = 2$ and (2) $n = 3$.

Problem 2.11 Obtain the buckling loads and buckling modes of the two-degree-of-freedom system in Fig. 2.10(c) for $k_1 = 2k$ and $k_2 = k$. Denote by u_1 the rotational displacement of the bar at the right and by u_2 that of the bar at the left; employ these displacements as independent variables.

Problem 2.12 In the two-degree-of-freedom system in Fig. 2.10(c), consider $k_1 = k_2 = k$ as the standard values. Increase either k_1 or k_2 by the same amount of $k\varepsilon$ for a small ε (> 0). In which case does the minimum buckling load increase more? Use rotational displacements as independent variables.

Problem 2.13 Conduct the buckling analysis for the structural system depicted in Fig. 2.10(d). Use rotational displacements as independent variables. In the rotational springs, a bending moment of $k_\theta \theta$ develops for rotational angle θ. The lengths of the three rigid bars are equal to L.

REFERENCES

1. Oden JT and Ripperger EA (1981), *Mechanics of Elastic Structures,* 2nd ed., Hemisphere Publishing Corporation, Washington.
2. Thompson JMT, and Hunt GW (1973), *A General Theory of Elastic Stability*, Wiley, New York.

3 Numerical Analysis I: Path Tracing

3.1 SUMMARY

The bifurcation/buckling analysis in Chapter 2 has dealt mainly with simple cases where the equilibrium paths can be obtained in explicit analytic forms. This chapter introduces numerical analysis methods for tracing equilibrium paths of nonlinear equilibrium equations of a multi-degree-of-freedom system. We present three kinds of procedures: the load control method, the displacement control method, and the arc-length method for path tracing. The procedures presented in this chapter and Chapter 4 are used in numerical bifurcation analysis in the remainder of this book.

The load control method and the displacement control method have simple algorithms; however, they malfunction at the maximal and minimal points of the variable that is controlled. The arc length method[1] is a more sophisticated algorithm and is most powerful among the three methods. We must choose one of these three methods case by case. During the path tracing, the locations of critical points are identified by the eigenanalysis of the Jacobian matrix of the governing equation (Chapter 2). Bifurcating solutions are obtained in Chapter 4 using the so-called branch switching procedure.

The theoretical foundation is given in Section 3.2. The three path tracing methods are presented in Section 3.3. A structural example is given in Section 3.4.

Keywords: • Arc-length method • Critical point • Displacement control method • Load control method • Path tracing • Stability

3.2 THEORETICAL FOUNDATION

We consider a nonlinear *equilibrium equation* (governing equation)

$$\boldsymbol{F}(\boldsymbol{u}, f) = \boldsymbol{0}, \tag{3.1}$$

where $\boldsymbol{F} = (F_1, \ldots, F_n)^\top$ is a vector of sufficiently smooth nonlinear functions, $\boldsymbol{u} = (u_1, \ldots, u_n)^\top$ is an n-dimensional *displacement vector* to describe the state of the system, and f is a load parameter.

The equilibrium equation (3.1) has a total of $n+1$ unknown variables, comprising n displacement variables and one load variable; however, it only has n equations. Accordingly, we need to specify one extra condition to obtain a solution to (3.1).

[1]Details of this method are found, e.g., in Schweizerhof and Wriggers [2], Wagner [4], Crisfield [1], and Shi and Crisfield [3].

DOI: 10.1201/9781003112365-3

According to the imposed extra condition, path tracing methods can be classified as

$$\begin{cases} \textit{load control method}: & \text{the value of } f \text{ is specified,} \\ \textit{displacement control method}: & \text{the value of a displacement } u_i \text{ is specified,} \\ \textit{arc-length method}: & \text{the length of a path (}\textit{arc length}\text{) is specified.} \end{cases}$$

The load control and displacement control methods allow simple formulations; however, these methods tend to encounter a severe difficulty with a maximal/minimal point of the specified variable. The arc-length method is the most powerful path tracing method, although it involves complicated formulations. It is convenient to switch between these three methods appropriately. By repeating the process of specifying some value and obtaining a solution $(\boldsymbol{u}, f)$ to the equilibrium equation, we can arrive at a series of equilibrium points, as illustrated in Fig. 3.1.

A numerical analysis procedure to trace an equilibrium path can be described as follows:

Step 1: Obtain the total potential energy of a system.
Step 2: Obtain the equilibrium equation $\boldsymbol{F}$, the load pattern vector $\partial \boldsymbol{F}/\partial f$, and the tangent stiffness matrix $J = \partial \boldsymbol{F}/\partial \boldsymbol{u}$.
Step 3: Obtain a series of equilibrium points and interpolate them to a smooth curve as an approximation to an equilibrium path.
Step 4: Investigate the stability of equilibrium points, obtain the locations of critical points, and classify the critical points.

Steps 1 and 2 are identical with the procedure already advanced in Chapter 2 "Multi-degree-of-freedom System," except for an addition of the load pattern vector $\partial \boldsymbol{F}/\partial f$ to be used in (3.3) for the classification of critical points, as well as in the displacement control method (Remark 3.2).

In Step 3, if the obtained series of equilibrium points is not smooth or continuous, we may try the following:

- Reduce the interval of the series of equilibrium points.
- Switch to another control method or specify another displacement variable.

In Step 4, we conduct the eigenvalue analysis of the tangent stiffness matrix J and, in turn, to find the location of a critical point on the equilibrium path, as well as

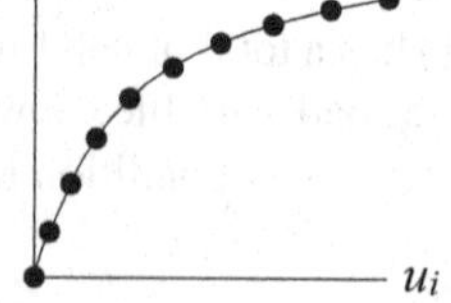

Figure 3.1 A series of equilibrium points and their interpolation

the stability of this path. The stability is determined based on (2.27) in Section 2.6, that is,

$$\begin{cases} \text{eigenvalues of } J \text{ are all positive} & \Rightarrow \quad \text{stable,} \\ \text{at least one eigenvalue of } J \text{ is negative} & \Rightarrow \quad \text{unstable.} \end{cases} \tag{3.2}$$

Critical points are classified as[2]

$$\begin{cases} \boldsymbol{\eta}_{\rm c}^\top \left(\dfrac{\partial \boldsymbol{F}}{\partial f} \right)_{\rm c} = 0 & \Rightarrow \quad \text{bifurcation point,} \\ \boldsymbol{\eta}_{\rm c}^\top \left(\dfrac{\partial \boldsymbol{F}}{\partial f} \right)_{\rm c} \neq 0 & \Rightarrow \quad \text{stationary point of } f, \end{cases} \tag{3.3}$$

where $\boldsymbol{\eta}_{\rm c}$ is the critical eigenvector(s), associated with the zero eigenvalue of the tangent stiffness matrix J. That is, the critical point is a bifurcation point if the vectors $\partial \boldsymbol{F}/\partial f$ and $\boldsymbol{\eta}_{\rm c}$ are orthogonal; the critical point is a stationary point of f if they are not orthogonal. This mathematical classification of critical points is physically reasonable as well.

Remark 3.1 By observing the shape of the fundamental path, we can determine maximal/minimal and inflection points of f, while other critical points are bifurcation points, in general. However, caution must be exercised in tracing a bifurcating path, since a symmetric bifurcation point on the fundamental path is also a maximal/minimal point of f along the bifurcated path. □

In the numerical search of the location of a critical point, we try to detect the change of the sign of an eigenvalue λ_i between two equilibrium points $(\boldsymbol{u}_{\rm A}, f_{\rm A})$ and $(\boldsymbol{u}_{\rm B}, f_{\rm B})$. By denoting the values of λ_i at these points by $\lambda_{i\rm A}$ and $\lambda_{i\rm B}$, we assume that

$$\lambda_{i\rm A}\lambda_{i\rm B} < 0.$$

Then at least one critical point exists between these two points if they are located on the same equilibrium path. As illustrated in Fig. 3.2, the zero-crossing point ($\circ$) of the line segment connecting $(\boldsymbol{u}_{\rm A}, \lambda_{i\rm A})$ and $(\boldsymbol{u}_{\rm B}, \lambda_{i\rm B})$ can be used to approximate the true location $\boldsymbol{u}_{\rm c}$ at ($\bullet$). That is, the location of a critical point can be approximated as

$$\boldsymbol{u}_{\rm c} \approx \frac{\lambda_{i\rm A}\boldsymbol{u}_{\rm B} - \lambda_{i\rm B}\boldsymbol{u}_{\rm A}}{\lambda_{i\rm A} - \lambda_{i\rm B}}, \qquad f_{\rm c} \approx \frac{\lambda_{i\rm A} f_{\rm B} - \lambda_{i\rm B} f_{\rm A}}{\lambda_{i\rm A} - \lambda_{i\rm B}}. \tag{3.4}$$

The accuracy of this approximation can be improved by narrowing the interval between the two points.

---

[2]This classification, to be derived in (5.35) in Remark 5.2 in Section 5.4, is quite convenient from the engineering point of view. Mathematically, however, there is also a critical point that satisfies the condition $\boldsymbol{\eta}_{\rm c}^\top \left(\frac{\partial \boldsymbol{F}}{\partial f} \right)_{\rm c} = 0$ but has no bifurcating solutions (Footnote 2 in Section 5.4.2).

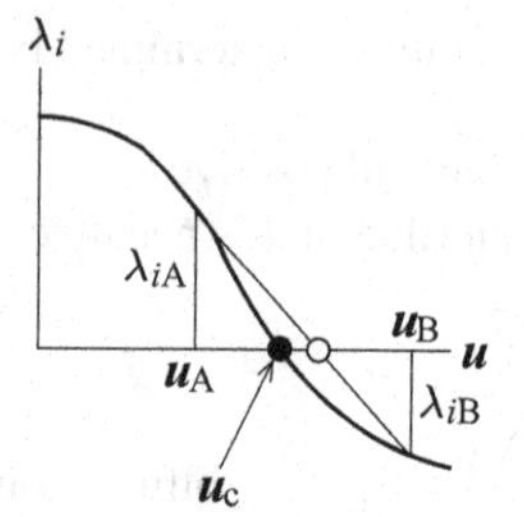

Figure 3.2 The behavior of an eigenvalue λ_i near a critical point

3.3 PATH TRACING METHODS

This section presents three types of path tracing methods: load control method, displacement control method, and arc-length method.

3.3.1 LOAD CONTROL METHOD

In search of an equilibrium point by the load control method, the value of the load f is specified to, say, f_*. Then the equilibrium equation (3.1) becomes

$$\boldsymbol{F}(\boldsymbol{u}, f_*) = \boldsymbol{0}, \tag{3.5}$$

which has n unknown variables and n nonlinear equations.

As a numerical analysis method to solve this nonlinear governing equation, we employ the Newton–Raphson method among several alternatives. For an initial value $(\boldsymbol{u}^{(0)}, f_*)$ of the solution, we would like to obtain a correction vector $\tilde{\boldsymbol{u}}$ of displacements such that $(\boldsymbol{u}^{(0)} + \tilde{\boldsymbol{u}}, f_*)$ is an improved solution to the equilibrium equation (3.5). An incremental form of the equilibrium condition (3.5) at $(\boldsymbol{u}^{(0)} + \tilde{\boldsymbol{u}}, f_*)$ is given by

$$\boldsymbol{F}(\boldsymbol{u}^{(0)} + \tilde{\boldsymbol{u}}, f_*) = \boldsymbol{F}(\boldsymbol{u}^{(0)}, f_*) + J(\boldsymbol{u}^{(0)}, f_*)\,\tilde{\boldsymbol{u}} + (\text{h.o.t.}) = \boldsymbol{0}, \tag{3.6}$$

where (h.o.t.) denotes higher-order terms. We introduce an assumption that the inverse of $J(\boldsymbol{u}^{(0)}, f_*)$ exists, which holds at an ordinary point but fails to hold at a critical point (cf., (2.9)). Under this assumption, we can approximate the correction vector $\tilde{\boldsymbol{u}}$ from (3.6) as

$$\tilde{\boldsymbol{u}} = -J(\boldsymbol{u}^{(0)}, f_*)^{-1}\boldsymbol{F}(\boldsymbol{u}^{(0)}, f_*). \tag{3.7}$$

Then an improved approximation $\boldsymbol{u}^{(1)}$ to the solution $\boldsymbol{u}$ is given by

$$\boldsymbol{u}^{(1)} = \boldsymbol{u}^{(0)} + \tilde{\boldsymbol{u}} = \boldsymbol{u}^{(0)} - J(\boldsymbol{u}^{(0)}, f_*)^{-1}\boldsymbol{F}(\boldsymbol{u}^{(0)}, f_*).$$

We advance the following iterative method, starting with the initial value $\boldsymbol{u}^{(0)}$ and generating a series of improved approximations $\{\boldsymbol{u}^{(\nu)}\}$, as

$$\boldsymbol{u}^{(\nu+1)} = \boldsymbol{u}^{(\nu)} - J(\boldsymbol{u}^{(\nu)}, f_*)^{-1}\boldsymbol{F}(\boldsymbol{u}^{(\nu)}, f_*), \qquad \nu = 0, 1, \ldots. \tag{3.8}$$

The convergence of this iteration may be concluded if, e.g.,

$$|F_i(\boldsymbol{u}^{(\nu)}, f_*)| < \delta, \qquad i = 1, \ldots, n, \tag{3.9}$$

where δ is a convergence criterion, which, for example, should be chosen in view of the rounding error in the numerical computation.

Using the iteration in (3.8) and gradually changing the value of f_*, we can numerically obtain a sequence of equilibrium points $1, 2, 3, 4, \ldots$ in Fig. 3.3(a). Yet, there occurs a problem in the convergence of the iteration in (3.8) near a maximal/minimal point of f, where the tangent stiffness matrix J becomes singular. In particular, the equilibrium path beyond this point cannot be continuously traced by the load control method. For example, in the vicinity of the maximal point A in Fig. 3.3(a), the path tracing captures 5, 6, 7, 8, instead of 4′. The approaching to a maximal/minimal point of f, where the path tracing encounters a problem, is indicated by a rapid increase of the increment of a displacement when the value of f_* is increased/decreased by the same amount. Then it is suggested to switch to another control method.

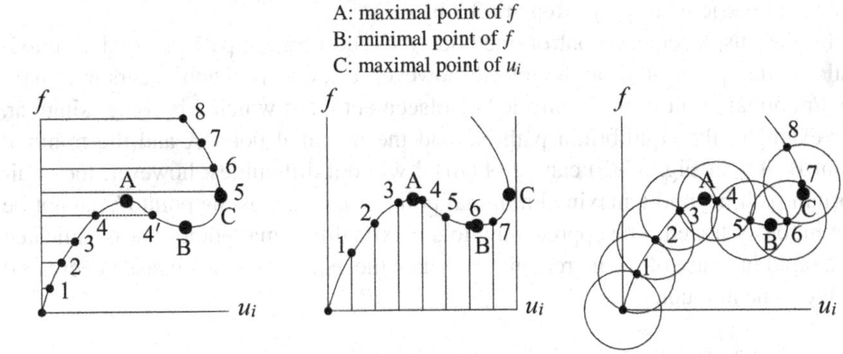

(a) Load control method (b) Displacement control method (c) Arc-length method

Figure 3.3 Path tracing by three kinds of methods

Example 3.1 The load control method is illustrated for the pedagogic system in Example 2.1 in Section 2.3 with

$$\boldsymbol{F} = \begin{pmatrix} x - y - 4x^3 + f \\ y - x - 4y^3 + f \end{pmatrix}, \qquad J = \begin{pmatrix} 1 - 12x^2 & -1 \\ -1 & 1 - 12y^2 \end{pmatrix}. \tag{3.10}$$

For $\boldsymbol{u} = (x, y)^\top$, the iteration in (3.8) can be given explicitly as

$$\begin{pmatrix} x^{(\nu+1)} \\ y^{(\nu+1)} \end{pmatrix}$$
$$= \begin{pmatrix} x^{(\nu)} \\ y^{(\nu)} \end{pmatrix} - \begin{pmatrix} 1 - 12(x^{(\nu)})^2 & -1 \\ -1 & 1 - 12(y^{(\nu)})^2 \end{pmatrix}^{-1} \begin{pmatrix} x^{(\nu)} - y^{(\nu)} - 4(x^{(\nu)})^3 + f_* \\ y^{(\nu)} - x^{(\nu)} - 4(y^{(\nu)})^3 + f_* \end{pmatrix}$$

for $\nu = 0, 1, \ldots$. □

3.3.2 DISPLACEMENT CONTROL METHOD

In the path tracing by the displacement control method, we obtain the solution $(\boldsymbol{u}, f)$ of the equilibrium equation (3.1) for a specified $u_i = u_*$.

By replacing the ith component u_i of $\boldsymbol{u}$ by f, we can introduce a new unknown variable vector as

$$\hat{\boldsymbol{u}} = (u_1, \ldots, u_{i-1}, f, u_{i+1}, \ldots, u_n)^\top.$$

The equilibrium equation (3.1) becomes

$$\hat{\boldsymbol{F}}(\hat{\boldsymbol{u}}) \equiv \boldsymbol{F}(u_1, \ldots, u_{i-1}, u_*, u_{i+1}, \ldots, u_n, f) = \boldsymbol{0}. \tag{3.11}$$

This is an n-dimensional nonlinear equation with n unknown variables, comprising $n-1$ displacements other than u_i and the load parameter f. The solution to (3.11) can be obtained by the Newton–Raphson method, similarly to the load control method. The Jacobian matrix $\hat{J} = \partial\hat{\boldsymbol{F}}/\partial\hat{\boldsymbol{u}}$, in this case, is obtained from the tangent stiffness matrix J by replacing its ith column by the load pattern vector $\partial\boldsymbol{F}/\partial f$. Hence $\hat{J}$ is not a symmetric matrix (cf., Remark 3.2 below).

In the displacement control method, the equilibrium path beyond a maximal/minimal point of f can be traced; however, a severe problem occurs at a maximal/minimal point of the controlled displacement u_i, at which $\hat{J}$ becomes singular. For example, the equilibrium path beyond the maximal point A and the minimal point B of f in Fig. 3.3(b) can be obtained without difficulties; however, the equilibrium path beyond a maximal/minimal point of u_i, such as the point C, cannot be continuously traced. The approaching to a maximal/minimal point of u_i is indicated by a rapid increase of the increment of f when the value of u_* is increased/decreased by the same amount.

Example 3.2 The displacement control method is illustrated for the pedagogic system in Example 3.1. We control the displacement y and set $\hat{\boldsymbol{u}} = (x, f)$. Then $\hat{J}$ can be obtained by replacing the second column of J in (3.10) by $\partial\boldsymbol{F}/\partial f = (1,1)^\top$ as

$$\hat{J} = \begin{pmatrix} 1-12x^2 & 1 \\ -1 & 1 \end{pmatrix}$$

and the iteration for a specified $y = y_*$ is given explicitly as

$$\begin{pmatrix} x^{(\nu+1)} \\ f^{(\nu+1)} \end{pmatrix} = \begin{pmatrix} x^{(\nu)} \\ f^{(\nu)} \end{pmatrix} - \begin{pmatrix} 1-12(x^{(\nu)})^2 & 1 \\ -1 & 1 \end{pmatrix}^{-1} \begin{pmatrix} x^{(\nu)} - y_* - 4(x^{(\nu)})^3 + f^{(\nu)} \\ y_* - x^{(\nu)} - 4{y_*}^3 + f^{(\nu)} \end{pmatrix}$$

for $\nu = 0, 1, \ldots$. □

Remark 3.2 In structural mechanics, the following method is often used in favor of the numerical efficiency in dealing with the symmetric matrix J. Suppose we have an initial approximate solution $(\boldsymbol{u}^{(0)}, f^{(0)})$ to $\boldsymbol{F}(\boldsymbol{u}, f) = \boldsymbol{0}$ with $u_i^{(0)} = u_*$. The correction

$(\tilde{\boldsymbol{u}}, \tilde{f})$ for an improved solution $(\boldsymbol{u}, f) = (\boldsymbol{u}^{(0)} + \tilde{\boldsymbol{u}}, f^{(0)} + \tilde{f})$ is determined from an incremental form of the governing equation (3.1), that is,

$$\boldsymbol{F}(\boldsymbol{u}^{(0)} + \tilde{\boldsymbol{u}}, f^{(0)} + \tilde{f}) = \boldsymbol{F}(\boldsymbol{u}^{(0)}, f^{(0)}) + J(\boldsymbol{u}^{(0)}, f^{(0)})\,\tilde{\boldsymbol{u}} + \frac{\partial \boldsymbol{F}}{\partial f}(\boldsymbol{u}^{(0)}, f^{(0)})\,\tilde{f} + (\text{h.o.t.}). \tag{3.12}$$

At an ordinary point $(\boldsymbol{u}^{(0)}, f^{(0)})$, for which $J(\boldsymbol{u}^{(0)}, f^{(0)})^{-1}$ exists, the equation (3.12) can be solved for $\tilde{\boldsymbol{u}}$ approximately as

$$\tilde{\boldsymbol{u}} \approx \boldsymbol{a}\tilde{f} + \boldsymbol{b} \tag{3.13}$$

with

$$\boldsymbol{a} = (a_1, \ldots, a_n)^\top = -J(\boldsymbol{u}^{(0)}, f^{(0)})^{-1} \frac{\partial \boldsymbol{F}}{\partial f}(\boldsymbol{u}^{(0)}, f^{(0)}),$$
$$\boldsymbol{b} = (b_1, \ldots, b_n)^\top = -J(\boldsymbol{u}^{(0)}, f^{(0)})^{-1} \boldsymbol{F}(\boldsymbol{u}^{(0)}, f^{(0)}).$$

Since the ith coordinate is controlled, we must have $\tilde{u}_i = 0$, whereas $\tilde{u}_i \approx a_i\tilde{f}_i + b_i$ by (3.13). Accordingly, we determine the correction $(\tilde{\boldsymbol{u}}, \tilde{f})$ by $\tilde{u}_i = 0$, $\tilde{f} = -b_i/a_i$, and

$$\tilde{u}_j = a_j\tilde{f} + b_j = -\frac{b_i}{a_i}a_j + b_j \qquad (j = 1, \ldots, n;\ i \neq j),$$

for which (3.13) holds. The correction $(\tilde{\boldsymbol{u}}, \tilde{f})$ obtained in this manner for an improved solution is to be used to define the iteration. □

3.3.3 ARC-LENGTH METHOD

We introduce the simplest formulation of the arc-length method, while more sophisticated variants of the arc-length method are also available (Crisfield [1, Chapter 9]).

In the path tracing by the arc-length method, we specify the arc length (length of the path) between an equilibrium point $(\overline{\boldsymbol{u}}, \overline{f})$ at hand and the equilibrium point $(\boldsymbol{u}, f)$ to be obtained. The arc-length condition is given by

$$||\boldsymbol{u} - \overline{\boldsymbol{u}}||^2 + C(f - \overline{f})^2 = \delta_{\text{a}}{}^2, \tag{3.14}$$

where $||\boldsymbol{u} - \overline{\boldsymbol{u}}||^2 = (\boldsymbol{u} - \overline{\boldsymbol{u}})^\top(\boldsymbol{u} - \overline{\boldsymbol{u}})$, C is a positive constant for scaling, and δ_{a} is an arc length. Note that a proper choice of the values of the scaling constant C and the arc length δ_{a} is vital in successful convergence of the iteration.

We solve a system of $n+1$ equations, comprising the equilibrium equation (3.1) and the arc-length condition (3.14), with $n+1$ unknowns using the iteration

$$\boldsymbol{u}^{(\nu+1)} = \boldsymbol{u}^{(\nu)} + \tilde{\boldsymbol{u}}^{(\nu)}, \qquad f^{(\nu+1)} = f^{(\nu)} + \tilde{f}^{(\nu)}, \qquad \nu = 0, 1, \ldots. \tag{3.15}$$

With the use of the Newton–Raphson method, corrections $(\tilde{\boldsymbol{u}}^{(\nu)}, \tilde{f}^{(\nu)})$ are to be obtained from the following linearized equations of (3.1) and (3.14):

$$\begin{pmatrix} J(\boldsymbol{u}^{(\nu)}, f^{(\nu)}) & \dfrac{\partial \boldsymbol{F}}{\partial f}(\boldsymbol{u}^{(\nu)}, f^{(\nu)}) \\ 2(\boldsymbol{u}^{(\nu)} - \overline{\boldsymbol{u}})^\top & 2C(f^{(\nu)} - \overline{f}) \end{pmatrix} \begin{pmatrix} \tilde{\boldsymbol{u}}^{(\nu)} \\ \tilde{f}^{(\nu)} \end{pmatrix} = \begin{pmatrix} -\boldsymbol{F}(\boldsymbol{u}^{(\nu)}, f^{(\nu)}) \\ \delta_{\text{a}}{}^2 - ||\boldsymbol{u}^{(\nu)} - \overline{\boldsymbol{u}}||^2 - C(f^{(\nu)} - \overline{f})^2 \end{pmatrix}.$$

As illustrated in Fig. 3.3(c), the arc-length method can successfully trace the equilibrium path beyond the maximal point A of f, the minimal one B, and the maximal point C of u_i. However, caution must be exercised on the multiplicity of the solutions to the iteration. An example of such a multiplicity can be seen from Fig. 3.3(c), when we search for point 3 at an arc-length of δ_a from point 2, the point 1 may be obtained erroneously since points 1 and 3 are both located on the circle centered at point 2 with the radius δ_a. To avoid this, an initial value for the Newton–Raphson method is to be chosen appropriately.

3.4 STRUCTURAL EXAMPLE

We conduct the numerical path tracing on the two-bar truss arch depicted in Fig. 3.4(a), whereas the analytical study of the buckling behavior of this arch is conducted in Section 2.7.

Nodes 1 and 2 of this arch are fixed, and node 3 is free to move. The displaced location (x, y) of node 3 is the unknown vector and initial location of the nodes are $(x_1, y_1) = (-1, h)$, $(x_2, y_2) = (1, h)$, and $(x_3, y_3) = (0, 0)$. The arch comprises two identical elastic truss members 1 and 2 both with the modulus of elasticity (Young's modulus) E, the cross-sectional area A, and the initial member length $L = \sqrt{1+h^2}$. The arch is subjected to the vertical load EAf at the crown node 3 (f is a non-dimensional normalized load).

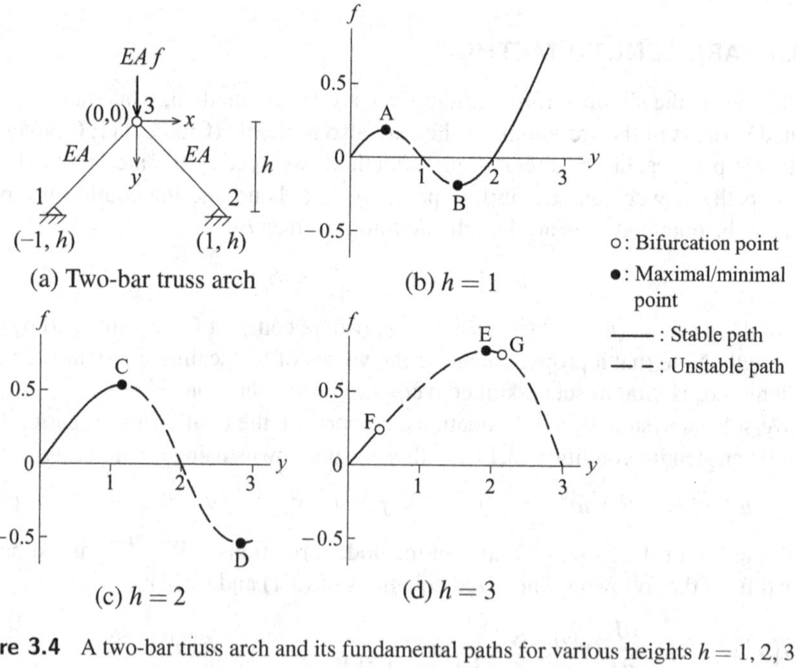

Figure 3.4 A two-bar truss arch and its fundamental paths for various heights $h = 1, 2, 3$

Step 1: The total potential energy is given by

$$U(x,y,f) = \sum_{m=1}^{2} \frac{EA}{2L}(\hat{L}^{(m)} - L)^2 - EAf(y - y_3).$$

Here $\hat{L}^{(m)}$ is the length of the mth member after deformation, being given as

$$\hat{L}^{(m)} = \sqrt{(x - x_m)^2 + (y - y_m)^2}, \quad m = 1, 2.$$

Step 2: The equilibrium equation is obtained as

$$\boldsymbol{F} = \begin{pmatrix} F_x \\ F_y \end{pmatrix} = \begin{pmatrix} \sum_{m=1}^{2} EA\left(\frac{1}{L} - \frac{1}{\hat{L}^{(m)}}\right)(x - x_m) \\ \sum_{m=1}^{2} EA\left(\frac{1}{L} - \frac{1}{\hat{L}^{(m)}}\right)(y - y_m) - EAf \end{pmatrix}, \tag{3.16}$$

where $F_x = \partial U/\partial x$ and $F_y = \partial U/\partial y$. The load pattern vector is a constant vector given by

$$\frac{\partial \boldsymbol{F}}{\partial f} = \begin{pmatrix} \partial F_x/\partial f \\ \partial F_y/\partial f \end{pmatrix} = EA\begin{pmatrix} 0 \\ -1 \end{pmatrix} \tag{3.17}$$

and the tangent stiffness matrix is

$$J(x,y,f) = EA\sum_{m=1}^{2} \begin{pmatrix} \frac{1}{L} - \frac{(y - y_m)^2}{(\hat{L}^{(m)})^3} & \frac{(x - x_m)(y - y_m)}{(\hat{L}^{(m)})^3} \\ \frac{(x - x_m)(y - y_m)}{(\hat{L}^{(m)})^3} & \frac{1}{L} - \frac{(x - x_m)^2}{(\hat{L}^{(m)})^3} \end{pmatrix}. \tag{3.18}$$

Step 3: Prior to the path tracing, it is worth mentioning that, for this simple structure, the fundamental path can be obtained analytically from the equilibrium equations $F_x = 0$ and $F_y = 0$ with (3.16) as

$$x = 0, \quad f = 2\left(\frac{1}{\sqrt{1+h^2}} - \frac{1}{\sqrt{1+(y-h)^2}}\right)(y - h). \tag{3.19}$$

The solution curves of this fundamental path for $h = 1$, 2, 3 are plotted in Figs. 3.4(b)–(d).

For numerical analysis, we employ two kinds of path tracing methods to obtain the fundamental path of this arch. First, the load control method (Section 3.3.1) is employed. Starting with the initial state $(x,y,f) = (0,0,0)$, we obtain a series of equilibrium points shown in Fig. 3.5(a). Here, the value of the load is increased as $f_* = 0, 0.05, 0.10, \ldots$ and the convergence criterion in (3.9) is set with $\delta = 1.0 \times 10^{-6}$. For each height of $h = 1$, 2, and 3, the series of equilibrium points is successfully obtained until point P. Yet, for $h = 1$, for example, point Q at another

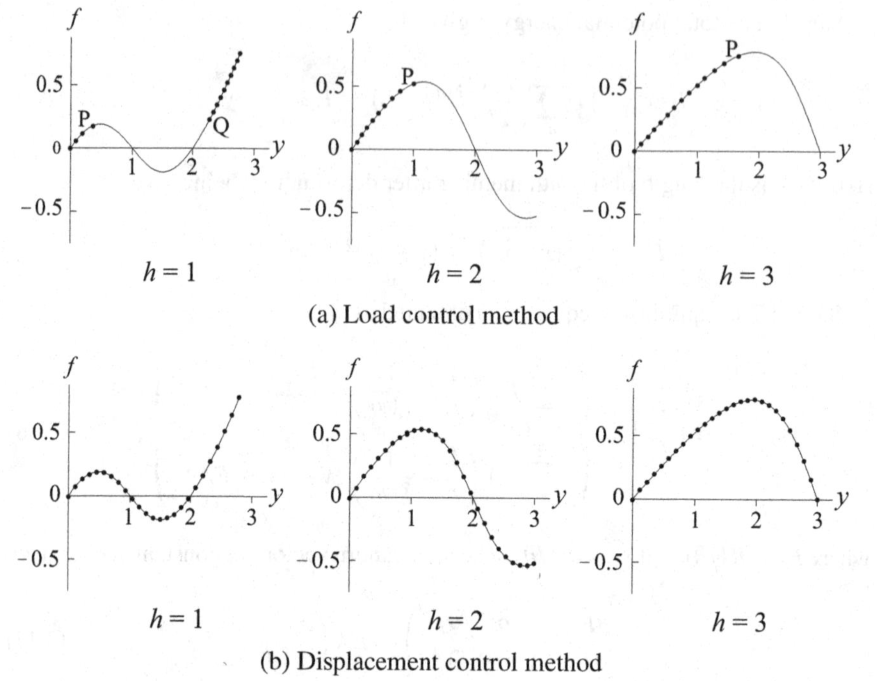

Figure 3.5 The tracing of the fundamental path of the truss arch (•: numerical analysis; solid line: analytical curve)

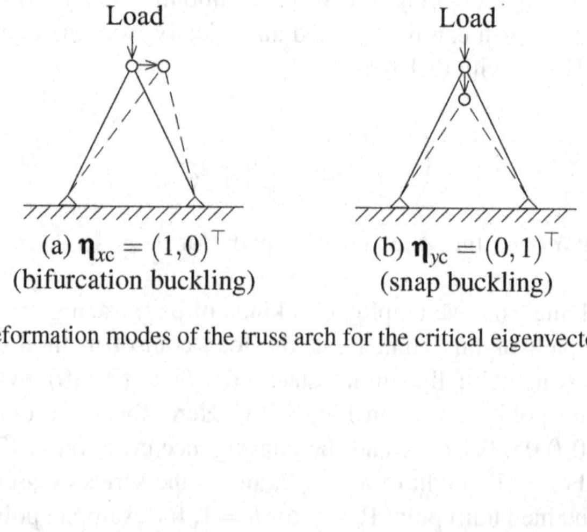

Figure 3.6 Deformation modes of the truss arch for the critical eigenvectors

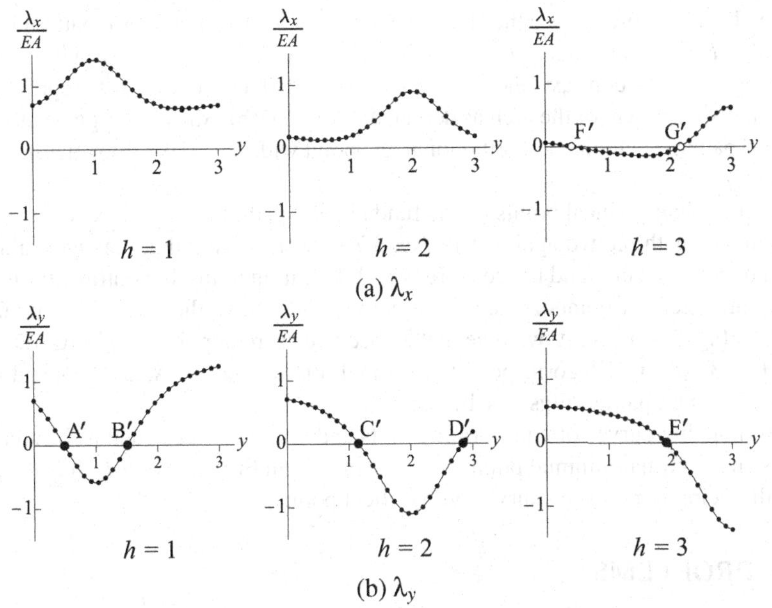

Figure 3.7 Eigenvalues of the tangent stiffness matrix on the fundamental path of the truss arch plotted against y

part of the fundamental path is obtained thereafter. This is due to the malfunction of the load control method near a maximal point of f, at which the assumption of the existence of J^{-1} is violated.

Next, the displacement control method is employed. Starting from the initial state $(x,y,f) = (0,0,0)$, we obtain a series of equilibrium points shown in Fig. 3.5(b) beyond the maximal/minimal points of f, unlike for the load control method. Here, we set $\delta = 1.0 \times 10^{-6}$ and the vertical displacement y is increased from 0 with a constant increment of 0.1.

Step 4: By the eigenvalue analysis of the tangent stiffness matrix, we investigate the stability of equilibrium points, determine the locations of critical points, and classify critical points. This investigation of stability, in general, is to be conducted numerically, but for this simple structure, it can be done analytically as below.

On the fundamental path of (3.19), the tangent stiffness matrix in (3.18) is a diagonal matrix; accordingly, its eigenvalues are given by its diagonal components as

$$\lambda_x = 2EA\left(\frac{1}{\sqrt{1+h^2}} - \frac{(y-h)^2}{[1+(y-h)^2]^{3/2}}\right),$$

$$\lambda_y = 2EA\left(\frac{1}{\sqrt{1+h^2}} - \frac{1}{[1+(y-h)^2]^{3/2}}\right).$$

The eigenvector $\boldsymbol{\eta}_{xc} = (1,0)^\top$ for $\lambda_x = 0$ represents a horizontal sway of the arch as depicted in Fig. 3.6(a). Since the eigenvector $\boldsymbol{\eta}_{xc} = (1,0)^\top$ and the load pattern vector $\partial \boldsymbol{F}/\partial f = EA(0,-1)^\top$ in (3.17) are orthogonal, the point with $\lambda_x = 0$ is a bifurcation point by (3.3). In contrast, the eigenvector $\boldsymbol{\eta}_{yc} = (0,1)^\top$ for $\lambda_y = 0$ represents a vertical displacement of the arch as depicted in Fig. 3.6(b). Since the eigenvector and the load pattern vector are not orthogonal, the point with $\lambda_y = 0$ is a maximal/minimal point of f.

We investigate critical points on the fundamental path. First, we observe λ_x on the fundamental path plotted against y in Fig. 3.7(a). For $h = 1$ and 2, the eigenvalue λ_x does not become zero, and hence there is no bifurcation point. In contrast, for $h = 3$, λ_x becomes zero at points F′ and G′ in correspondence to the bifurcation points F and G in Fig. 3.4(d). Next, we note that λ_y becomes zero at points A′, B′, C′, D′, and E′ in Fig. 3.7(b), which correspond to maximal/minimal points A, B, C, D, and E on the fundamental path in Figs. 3.4(b)–(d).

Interpolating curves of equilibrium points and critical points, such as bifurcation points and maximal/minimal points of f, are plotted in Figs. 3.4(b)–(d). It is the goal of path tracing to plot such curves with critical points.

3.5 PROBLEMS

Problem 3.1 Conduct the path tracing analysis of the two-bar truss arch in Fig. 3.4(a) in Section 3.4 for the height $h = 4$.

Problem 3.2 Consider the two-bar truss arch in Fig. 3.4(a) in Section 3.4. (1) Obtain the range of the height h such that the minimum buckling load becomes $f_c \geq 0.30$. (2) Obtain the range of the height h such that the minimum buckling load is governed by bifurcation. (3) Compare the minimum buckling loads for $h = 2.5$ and $h = 3$.

REFERENCES

1. Crisfield MA (1997), *Non-linear Finite Element Analysis of Solids and Structures*, Vol. 2, Wiley, Chichester.
2. Schweizerhof K, and Wriggers P (1988), Consistent linearization for path following methods in nonlinear FE Analysis, *Computer Methods in Applied Mechanics and Engineering* **59**(3), 261–279.
3. Shi J, and Crisfield MA (1995), Combining arc-length control and line searches in path following, *Commu. Numer. Methods Eng.* **11**(10), 793–803.
4. Wagner W (1991), A path-following algorithm with quadratic predictor, *Comput. & Struct.* **39**(3–4), 339–348.

4 Numerical Analysis II: Branch Switching

4.1 SUMMARY

We have studied a bifurcation/buckling analysis procedure (Chapter 2) and a path tracing procedure (Chapter 3). This chapter introduces a numerical analysis procedure to obtain a bifurcating equilibrium path. This procedure is called "branch switching" as the equilibrium path to be traced is switched from the fundamental path to a bifurcating path. The procedures presented in this chapter and Chapter 3 are used in numerical bifurcation analysis in the remainder of this book. Chapter 5 gives formulas for the direction of a bifurcating path as a theoretical background of the branch switching.

By repeating branch switching[1] and bifurcating path tracing, we can obtain a complete set of equilibrium paths of a nonlinear governing equation. The analysis of secondary and further bifurcations of structures is presented in Chapter 6 and in the literature (e.g., Ikeda and Murota [4, 5]).

The branch switching procedure and the bifurcation analysis procedure for simple bifurcation points are given in Section 4.2. A structural example is presented in Sections 4.3. Basic facts about linear simultaneous equations are presented in Section 4.4 to supplement thc discussion in this chapter.

Keywords: • Asymmetric bifurcation • Bifurcation analysis • Branch switching • Critical point • Linear simultaneous equations • Stability • Symmetric bifurcation

4.2 THEORETICAL FOUNDATION

A procedure to find an equilibrium point on a bifurcating path, which we call a *branch switching*, is presented. In this procedure, an initial value for the Newton–Raphson method is given in the direction of a bifurcating path to search for an equilibrium point on this path. If this point is found, the bifurcating path can be traced by the path tracing procedure given in Chapter 3.

We consider a nonlinear equilibrium equation (governing equation)

$$\boldsymbol{F}(\boldsymbol{u},f)=\boldsymbol{0}, \tag{4.1}$$

where $\boldsymbol{F}=(F_1,\ldots,F_n)^\top$ is a vector of sufficiently smooth nonlinear functions, $\boldsymbol{u}=(u_1,\ldots,u_n)^\top$ is an n-dimensional *displacement vector* to describe the state of the system, and f is a load parameter.

[1]For more accounts of branch switching, see, e.g., De Borst [2], Eriksson [3], Wagner and Wriggers [6], and Crisfield [1].

DOI: 10.1201/9781003112365-4

In the neighborhood of an equilibrium point $(\boldsymbol{u}, f)$ satisfying this equilibrium equation, we consider another equilibrium point $(\boldsymbol{u}+\tilde{\boldsymbol{u}}, f+\tilde{f})$ using incremental variables $(\tilde{\boldsymbol{u}}, \tilde{f})$. Then we can introduce the following incremental equilibrium equation for the incremental variables $(\tilde{\boldsymbol{u}}, \tilde{f})$:

$$\tilde{\boldsymbol{F}}(\tilde{\boldsymbol{u}}, \tilde{f}) = \boldsymbol{0} \tag{4.2}$$

with

$$\tilde{\boldsymbol{F}}(\tilde{\boldsymbol{u}}, \tilde{f}) \equiv \boldsymbol{F}(\boldsymbol{u}+\tilde{\boldsymbol{u}}, f+\tilde{f}) - \boldsymbol{F}(\boldsymbol{u}, f) = J(\boldsymbol{u}, f)\tilde{\boldsymbol{u}} + \frac{\partial \boldsymbol{F}}{\partial f}(\boldsymbol{u}, f)\tilde{f} + \text{(h.o.t.)}, \tag{4.3}$$

where (h.o.t.) denotes higher-order terms. At an ordinary point, where the inverse of $J(\boldsymbol{u}, f)$ exists, (4.2) with (4.3) can be solved for $\tilde{\boldsymbol{u}}$ to arrive asymptotically[2] at

$$\tilde{\boldsymbol{u}} = -J(\boldsymbol{u}, f)^{-1}\frac{\partial \boldsymbol{F}}{\partial f}\tilde{f} + \text{(h.o.t.)},$$

which gives a unique direction of the incremental solution $(\tilde{\boldsymbol{u}}, \tilde{f})$. In contrast, the direction of $(\tilde{\boldsymbol{u}}, \tilde{f})$ cannot be determined in this way for a critical point where J is singular. The direction of a bifurcating solution is studied next.

4.2.1 DIRECTION OF BIFURCATING PATH

We consider the direction of bifurcating solutions for a simple bifurcation point[3] $(\boldsymbol{u}_{\mathrm{c}}, f_{\mathrm{c}})$. Denote the critical eigenvector by $\boldsymbol{\eta}_{\mathrm{c}}$, that is,

$$J(\boldsymbol{u}_{\mathrm{c}}, f_{\mathrm{c}})\boldsymbol{\eta}_{\mathrm{c}} = \boldsymbol{0}. \tag{4.4}$$

Symmetric bifurcation point

For a symmetric bifurcation point, we have $|\tilde{f}| = \mathrm{O}(\|\tilde{\boldsymbol{u}}\|^2)$ from Section 5.4.3, where $\mathrm{O}(\cdot)$ means the order of the term therein. Then (4.2) with (4.3) asymptotically becomes

$$J(\boldsymbol{u}_{\mathrm{c}}, f_{\mathrm{c}})\tilde{\boldsymbol{u}} \approx \boldsymbol{0}.$$

This is an n-dimensional homogeneous linear equation for unknown variable vector $\tilde{\boldsymbol{u}}$ and the solution to this equation is given by a scalar multiple of the eigenvector $\boldsymbol{\eta}_{\mathrm{c}}$ of (4.4), that is, $\tilde{\boldsymbol{u}} \approx C\boldsymbol{\eta}_{\mathrm{c}}$ with some constant C (see Section 4.4). Hence the direction of the bifurcating solution is given by

$$(\tilde{\boldsymbol{u}}, \tilde{f}) \approx (C\boldsymbol{\eta}_{\mathrm{c}}, 0). \tag{4.5}$$

---

[2]The term "asymptotically" means "in a close neighborhood of the bifurcation point."

[3]Recall from (2.12) that $(\boldsymbol{u}_{\mathrm{c}}, f_{\mathrm{c}})$ is a simple critical point if $\operatorname{rank} J(\boldsymbol{u}_{\mathrm{c}}, f_{\mathrm{c}}) = n-1$. We focus on simple bifurcation points in this chapter and Chapter 5, while multiple bifurcation points are studied in Chapter 6.

This expression is useful in path tracing (Section 3.3). Namely, an initial value of the iteration is to be given as

$$(\boldsymbol{u}^{(0)}, f^{(0)}) = (\boldsymbol{u}_c, f_c) + (C\boldsymbol{\eta}_c, 0). \tag{4.6}$$

Note that the constant C is to be specified appropriately in view of the convergence of the iteration.

Example 4.1 Consider a system of nonlinear equations

$$\begin{cases} F_x \equiv fx - x^3 = 0, \\ F_y \equiv -f + y = 0. \end{cases}$$

As depicted in Fig. 4.1, this system has two solution curves

$$\begin{cases} x = 0,\ f = y, & \text{fundamental path,} \\ f = x^2,\ f = y, & \text{bifurcated path,} \end{cases}$$

intersecting at the symmetric bifurcation point $(x_c, y_c, f_c) = (0,0,0)$. The direction of the bifurcated solution curve is $(\tilde{x}, \tilde{y}, \tilde{f}) = (1,0,0)$. The Jacobian matrix

$$J(x,y,f) = \begin{pmatrix} f - 3x^2 & 0 \\ 0 & 1 \end{pmatrix}$$

at the bifurcation point is a singular matrix

$$J(0,0,0) = \begin{pmatrix} 0 & 0 \\ 0 & 1 \end{pmatrix}.$$

The critical eigenvector for the zero eigenvalue is $\boldsymbol{\eta}_c = (1,0)^\top$. The direction of the bifurcated solution curve given by the formula (4.5) with $C = 1$ coincides with $(\tilde{x}, \tilde{y}, \tilde{f}) = (1,0,0)$ above. □

Asymmetric bifurcation point

For an asymmetric bifurcation point, $|\tilde{f}| = \mathrm{O}(\|\tilde{\boldsymbol{u}}\|)$ holds from Section 5.4.2. Then (4.2) with (4.3) asymptotically becomes

$$J(\boldsymbol{u}_c, f_c)\tilde{\boldsymbol{u}} = -\frac{\partial \boldsymbol{F}}{\partial f}(\boldsymbol{u}_c, f_c)\tilde{f} \tag{4.7}$$

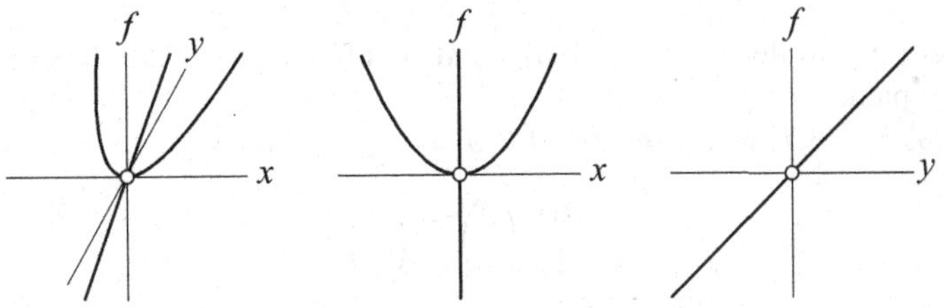

Figure 4.1 A solution curve with a symmetric bifurcation point

in the neighborhood of the bifurcation point $(\boldsymbol{u}_c, f_c)$ (see Section 5.5 for details). This is an n-dimensional inhomogeneous linear equation for the unknown variable vector $\tilde{\boldsymbol{u}}$.

Let $(\tilde{\boldsymbol{u}}, \tilde{f}) = \tilde{f}(\boldsymbol{w}_p, 1)$ denote the tangential direction of the main path. Then $\tilde{\boldsymbol{u}} = \tilde{f}\boldsymbol{w}_p$ is a solution to (4.7). A general solution $\tilde{\boldsymbol{u}}$ to (4.7) is given as the sum of the particular solution $\tilde{f}\boldsymbol{w}_p$ and a scalar multiple of the critical eigenvector $\boldsymbol{\eta}_c$ for the zero eigenvalue.[4] That is, we have

$$\tilde{\boldsymbol{u}} = \tilde{f}(C\boldsymbol{\eta}_c + \boldsymbol{w}_p). \tag{4.8}$$

Here C is a constant to be determined by higher-order terms of (4.3). The direction of the bifurcating solution is determined from (4.8) as

$$(\tilde{\boldsymbol{u}}, \tilde{f}) = \tilde{f}(C\boldsymbol{\eta}_c + \boldsymbol{w}_p, 1). \tag{4.9}$$

Accordingly, the initial value of the iteration for path tracing in Section 3.3 is to be set as

$$(\boldsymbol{u}^{(0)}, f^{(0)}) = (\boldsymbol{u}_c, f_c) + \tilde{f}(C\boldsymbol{\eta}_c + \boldsymbol{w}_p, 1). \tag{4.10}$$

The constant C and the incremental load $\tilde{f}$ are to be specified by trial and errors in the numerical analysis.

Example 4.2 Consider a system of nonlinear equations

$$\begin{cases} F_x \equiv 3x^2 - 2xf = 0, \\ F_y \equiv -f + 2y = 0. \end{cases}$$

This system has two solution curves

$$\begin{cases} x = 0, \ f = 2y, & \text{fundamental path,} \\ f = 3x/2, \ f = 2y, & \text{bifurcated path,} \end{cases}$$

intersecting at the asymmetric bifurcation point $(x_c, y_c, f_c) = (0,0,0)$. The direction of the bifurcated solution is $(\tilde{x}, \tilde{y}, \tilde{f}) = (2/3, 1/2, 1)$.

The Jacobian matrix

$$J(x,y,f) = \begin{pmatrix} 6x - 2f & 0 \\ 0 & 2 \end{pmatrix}$$

at this point becomes a singular matrix $J(0,0,0) = \begin{pmatrix} 0 & 0 \\ 0 & 2 \end{pmatrix}$. The critical eigenvector for the zero eigenvalue is $\boldsymbol{\eta}_c = (1,0)^\top$ and is different from the direction of the bifurcating path.

The equation (4.7) with $(\partial \boldsymbol{F}/\partial f)(0,0,0) = (0,-1)^\top$ reads

$$\begin{pmatrix} 0 & 0 \\ 0 & 2 \end{pmatrix} \begin{pmatrix} \tilde{x} \\ \tilde{y} \end{pmatrix} = \begin{pmatrix} 0 \\ 1 \end{pmatrix} \tilde{f}.$$

---

[4]This is due to the general results for linear equations in Section 4.4.

The particular solution $(\tilde{x}_p, \tilde{y}_p, \tilde{f}_p) = (0, 1/2, 1)$ gives the direction of the main path. The direction of the bifurcating path is given by a superposition of some scalar multiple of the eigenvector $\boldsymbol{\eta}_c = (1,0)^\top$ and the direction of the main path. Indeed, this is the case as

$$\begin{pmatrix} \tilde{x} \\ \tilde{y} \\ \tilde{f} \end{pmatrix} = \left[\frac{2}{3} \begin{pmatrix} 1 \\ 0 \\ 0 \end{pmatrix} + \begin{pmatrix} 0 \\ 1/2 \\ 1 \end{pmatrix} \right] \tilde{f} = \begin{pmatrix} 2/3 \\ 1/2 \\ 1 \end{pmatrix} \tilde{f},$$

which coincides with the formula (4.9) with the choice of $C = 2/3$. □

4.2.2 BIFURCATION ANALYSIS PROCEDURE

A numerical procedure for bifurcation analysis can be described as follows:

Step 1: Obtain the fundamental path.

Step 2: Obtain the locations of critical points on the path. Classify the critical points into simple and multiple critical points. Classify simple critical points into stationary points of f and bifurcation points. Classify bifurcation points into symmetric and asymmetric points.

Step 3: Search for a bifurcating path and investigate the stability of the bifurcating path in view of its post-bifurcation shape (Remark 4.2 below).

Step 4: Repeat Steps 2 and 3 for each bifurcating path.

Steps 1 and 2 have already been introduced in Chapter 3. In Step 2, we can classify a bifurcation point as a symmetric bifurcation point if $\boldsymbol{\eta}_c$ and $-\boldsymbol{\eta}_c$ are associated with the same physical behavior.

Step 3 is a procedure that is newly introduced in this chapter. In this step, we search for an equilibrium point on a bifurcating path in the direction of the bifurcating solution (Section 4.2.1). We search herein for a bifurcating solution by appropriately choosing an incremental value of the specified variable and the constant C in (4.6) for a symmetric bifurcation point and C and $\tilde{f}$ in (4.10) for an asymmetric bifurcation point.[5] Recall that the direction of the bifurcating path varies with symmetric and asymmetric bifurcation points.

Remark 4.1 In obtaining a bifurcating path, it suffices to obtain the path in the direction of $\boldsymbol{\eta}_c$, if $\boldsymbol{\eta}_c$ and $-\boldsymbol{\eta}_c$ are associated with the same physical behavior; the path in the direction of $-\boldsymbol{\eta}_c$ can be obtained by a simple geometrical operation (see Section 6.2). In contrast, if $\boldsymbol{\eta}_c$ and $-\boldsymbol{\eta}_c$ are associated with different physical behaviors, the paths in both directions must be obtained. □

[5] When the displacement control method in Section 3.3.2 is employed, it is usually pertinent to control the displacement component that has the largest absolute value in the components of $\boldsymbol{\eta}_c$ in (4.4).

Remark 4.2 When we trace a fundamental path by increasing f and encounter a symmetric bifurcation point, the stability of this bifurcation point can be investigated with reference to the slope of the bifurcating path just after bifurcation. That is,

$$\begin{cases} \text{bifurcating path has a positive slope} & \Rightarrow \text{ stable,} \\ \text{bifurcating path has a negative slope} & \Rightarrow \text{ unstable} \end{cases} \tag{4.11}$$

(see Remark 5.3 in Section 5.4.3). When the bifurcation point is unstable, the load at this point gives the maximum load of the system. When the bifurcation point is stable, the stability of equilibrium points on the bifurcating path is to be investigated in search of the maximum load. □

4.3 BAR-SPRING SYSTEM

The bifurcation analysis procedure in the previous section is applied to the bar-spring system with two degrees of freedom in Fig. 4.2(a). The two rigid bars are supported by springs with two different spring properties:

$$k_1 = k_2 = \begin{cases} k, & \text{linear-symmetric spring,} \\ k\left(1 + \dfrac{3}{2}\dfrac{x}{L}\right), & \text{nonlinear-asymmetric spring,} \end{cases} \tag{4.12}$$

where x is a vertical displacement of the spring. This system has a symmetric bifurcation point for the linear–symmetric spring and an asymmetric bifurcation point for the nonlinear-asymmetric spring. The rotations u_1 and u_2 of the rigid bars are chosen as the unknown variables (Fig. 4.2(b)).

4.3.1 LINEAR-SYMMETRIC SPRING

The bifurcation analysis procedure of Section 4.2.2 is illustrated for the bar-spring system with the linear-symmetric spring property. This system has a symmetric bifurcation point. For details of structural analysis, refer to Section 2.8.1.

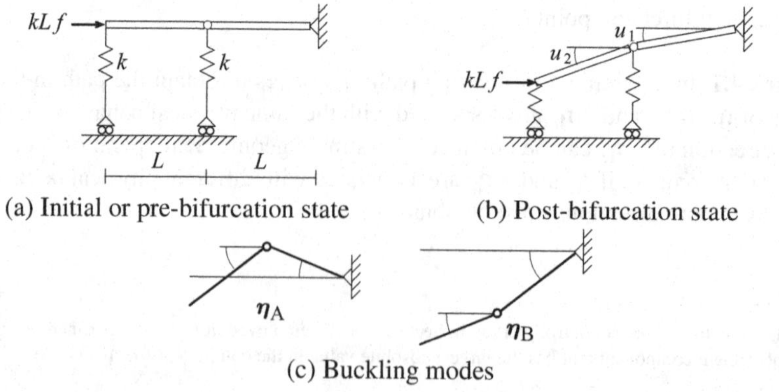

Figure 4.2 A bar-spring system with two degrees of freedom and its buckling modes

Step 1: The fundamental path is the trivial solution $\boldsymbol{u} = (u_1, u_2)^\top = \boldsymbol{0}$.
Step 2: On the fundamental path $\boldsymbol{u} = \boldsymbol{0}$, the tangent stiffness matrix becomes

$$J(\boldsymbol{0}, f) = kL^2 \begin{pmatrix} 2-f & 1 \\ 1 & 1-f \end{pmatrix}. \tag{4.13}$$

This $J(\boldsymbol{0}, f)$ is the same as the matrix in (2.38) in Section 2.8.1, and therefore the bifurcation loads and the bifurcation modes (critical eigenvectors of the zero eigenvalue) are given, respectively, as

$$f_A = \frac{3-\sqrt{5}}{2} \approx 0.382, \qquad f_B = \frac{3+\sqrt{5}}{2} \approx 2.618, \tag{4.14}$$

$$\boldsymbol{\eta}_A = \begin{pmatrix} -0.5257 \\ 0.8507 \end{pmatrix}, \qquad \boldsymbol{\eta}_B = \begin{pmatrix} 0.8507 \\ 0.5257 \end{pmatrix}. \tag{4.15}$$

These bifurcation modes are depicted in Fig. 4.2(c). The corresponding critical points are bifurcation points A and B (Fig. 4.3(a)).

Since this structural system has upside-down symmetry, the positive direction of the bifurcation mode is identical with the negative one. Hence bifurcation points A and B are both symmetric bifurcation points. Then by Remark 4.1, we need to obtain only the solution in the direction of the eigenvector $\boldsymbol{\eta}_c$ since

$$(\boldsymbol{u}, f) \text{ is a solution} \Rightarrow (-\boldsymbol{u}, f) \text{ is also a solution.} \tag{4.16}$$

Step 3: An initial value $(\boldsymbol{u}^{(0)}, f^{(0)})$ to find a bifurcating equilibrium point is given by (4.6) with $\boldsymbol{u}_c = (0,0)^\top$ and $f_c = f_A, f_B$ as

$$(\boldsymbol{u}^{(0)}, f^{(0)}) = (\boldsymbol{u}_c, f_c) + (C\boldsymbol{\eta}_c, 0)$$
$$\approx \begin{cases} (C(-0.5257, 0.8507)^\top, 0.382), & \text{bifurcation point A,} \\ (C(0.8507, 0.5257)^\top, 2.618), & \text{bifurcation point B.} \end{cases}$$

With the use of this initial value ($C = 0.01$) for the iteration, we obtain the bifurcating paths depicted in Fig. 4.3. In path tracing by the displacement control method in

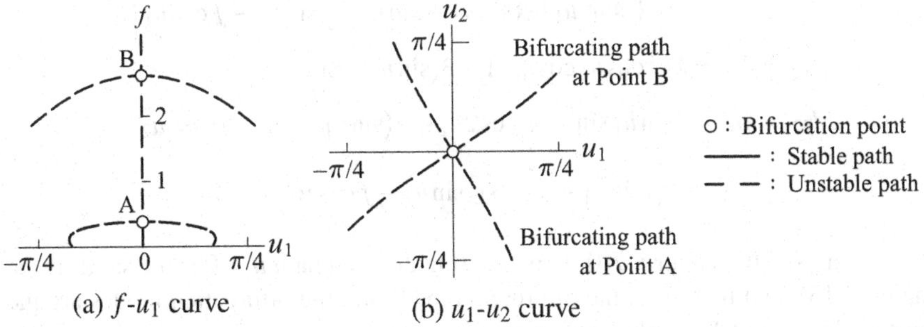

Figure 4.3 Equilibrium paths of the bar-spring system with the linear-symmetric spring (symmetric bifurcation point)

Section 3.3.2, the component of the critical eigenvector with a larger absolute value is controlled, that is, u_2 for the bifurcation point A and u_1 for the bifurcation point B. Both points A and B are unstable bifurcation points since the slope of the bifurcating path is negative for each of these two points (cf., (4.11) in Remark 4.2).

Step 4: Since there is no bifurcation point on the bifurcating path, there is no secondary bifurcation.

4.3.2 NONLINEAR-ASYMMETRIC SPRING

The bifurcation analysis procedure of Section 4.2.2 is illustrated for the bar-spring system in Fig. 4.2(a) with the nonlinear-asymmetric spring property $k_1 = k_2 = k(1 + 3x/(2L))$. This system has an asymmetric bifurcation point.

Step 1: The total potential energy is

$$U(\boldsymbol{u}, f) = \frac{kL^2}{2}\big[\sin^2 u_1 + \sin^3 u_1 + (\sin u_1 + \sin u_2)^2 + (\sin u_1 + \sin u_2)^3 - 2(2 - \cos u_1 - \cos u_2) f\,\big],$$

where $\boldsymbol{u} = (u_1, u_2)^\top$. The equilibrium equation is

$$\frac{\partial U}{\partial u_1} = kL^2\Big[\sin 2u_1 + \cos u_1 \sin u_2 + \frac{3}{2}\{\sin^2 u_1 + (\sin u_1 + \sin u_2)^2\}\cos u_1 - f \sin u_1\Big] = 0,$$

$$\frac{\partial U}{\partial u_2} = kL^2\Big[\sin u_1 \cos u_2 + \frac{1}{2}\sin 2u_2 + \frac{3}{2}(\sin u_1 + \sin u_2)^2 \cos u_2 - f \sin u_2\Big] = 0.$$

The fundamental path is the trivial solution $\boldsymbol{u} = (u_1, u_2)^\top = \boldsymbol{0}$.

Step 2: The partial differentiation of the equilibrium equation leads to the tangent stiffness matrix $J(\boldsymbol{u}, f) = (J_{ij} \mid i, j = 1, 2)$ with the components:

$$J_{11} = kL^2\Big[2\cos 2u_1 - \sin u_1 \sin u_2 + 3(2\sin u_1 + \sin u_2)\cos^2 u_1 - \frac{3}{2}\{\sin^2 u_1 + (\sin u_1 + \sin u_2)^2\}\sin u_1 - f \cos u_1\Big],$$

$$J_{12} = J_{21} = kL^2 \cos u_1 \cos u_2 \,[1 + 3(\sin u_1 + \sin u_2)],$$

$$J_{22} = kL^2\Big[-\sin u_1 \sin u_2 + \cos 2u_2 + 3(\sin u_1 + \sin u_2)\cos^2 u_2 - \frac{3}{2}(\sin u_1 + \sin u_2)^2 \sin u_2 - f \cos u_2\Big].$$

The tangent stiffness matrix $J(\boldsymbol{0}, f)$ on the fundamental path $\boldsymbol{u} = \boldsymbol{0}$ is the same as the one in (4.13), and therefore, the bifurcation loads and the bifurcation modes are the same as those in (4.14) and (4.15), respectively.

By the asymmetry of the spring property, this structural system does not have upside down symmetry. The positive direction of the bifurcation mode and the negative

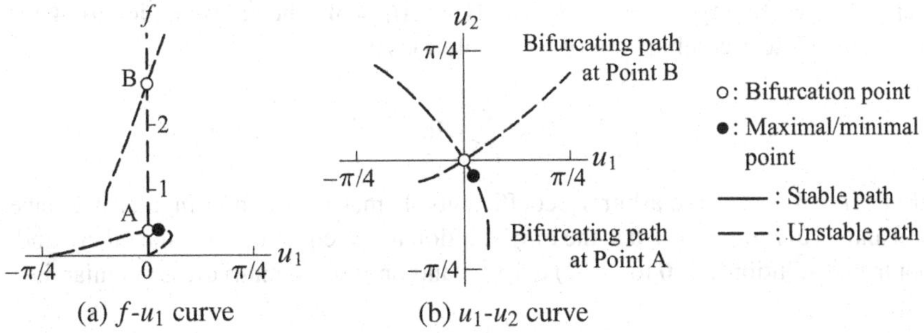

Figure 4.4 Equilibrium paths of the bar-spring system with the nonlinear-asymmetric spring (asymmetric bifurcation point)

one are physically not identical. Hence the bifurcation points A and B are asymmetric bifurcation points.

Step 3: An initial value $(\boldsymbol{u}^{(0)}, f^{(0)})$ to find a bifurcating equilibrium point is given by (4.10) with $\boldsymbol{u}_{\rm c} = (0,0)^\top$ and $f_{\rm c} = f_{\rm A}$, $f_{\rm B}$ as

$$
\begin{aligned}
(\boldsymbol{u}^{(0)}, f^{(0)}) &= (\boldsymbol{u}_{\rm c}, f_{\rm c}) + \tilde{f}(C\boldsymbol{\eta}_{\rm c} + \boldsymbol{w}_{\rm p}, 1) \\
&= \begin{cases} (C\tilde{f}(-0.5257, 0.8507)^\top,\ 0.382 + \tilde{f}), & \text{bifurcation point A,} \\ (C\tilde{f}(0.8507, 0.5257)^\top,\ 2.618 + \tilde{f}), & \text{bifurcation point B.} \end{cases}
\end{aligned}
$$

Here we have chosen $\boldsymbol{w}_{\rm p} = \boldsymbol{0}$ because the fundamental path is given by the trivial solution $\tilde{\boldsymbol{u}} = \boldsymbol{0}$. With the use of this initial value ($C = \pm 1.0$, $\tilde{f} = 10^{-2}$), we can obtain the bifurcating path depicted in Fig. 4.4. In path tracing by the displacement control method (Section 3.3.2), the component of the critical eigenvector with a larger absolute value is controlled, that is, u_2 for the bifurcation point A and u_1 for the bifurcation point B. Since point A is an asymmetric bifurcation point, we need to obtain the two bifurcating half branches:[6] one in the direction of $\boldsymbol{\eta}_{\rm c}$ and the other in $-\boldsymbol{\eta}_{\rm c}$ (Remark 4.1). Such is also the case for the bifurcation point B.

Step 4: Since there is no bifurcation point on the bifurcating path, there is no secondary bifurcation.

4.4 APPENDIX: LINEAR SIMULTANEOUS EQUATIONS

We briefly introduce basic facts about linear simultaneous equations that are used in Section 4.2.1.

First, we consider *homogeneous linear simultaneous equations*

$$A\boldsymbol{x} = \boldsymbol{0} \tag{4.17}$$

for an n-dimensional unknown vector $\boldsymbol{x}$ with an $n \times n$ matrix A. Assume that the rank of A is equal to $n-k$, that is, $\text{rank}A = n-k$ $(0 \le k \le n)$. Then there are k linearly

[6] These two half branches are connected at the bifurcation point A to form a bifurcating path.

independent vectors $\boldsymbol{\eta}_i$ $(i = 1, \ldots, k)$ satisfying $A\boldsymbol{\eta}_i = \mathbf{0}$. Then any solution to (4.17) is given by a linear combination of these solutions as

$$\boldsymbol{x} = \sum_{i=1}^{k} c_i \boldsymbol{\eta}_i.$$

Here c_i $(i = 1, \ldots, k)$ are arbitrary coefficients. If matrix A is nonsingular, we have $k = 0$ and, accordingly, $\boldsymbol{x} = \mathbf{0}$ is the only solution to the equations. On the other hand, a nontrivial solution $\boldsymbol{x} \neq \mathbf{0}$ to (4.17) exists if and only if the matrix A is singular, i.e.,

$$\det A = 0. \tag{4.18}$$

Next, we consider *inhomogeneous linear simultaneous equations*

$$A\boldsymbol{x} = \boldsymbol{b}.$$

If a solution $\boldsymbol{x}_{\mathrm{p}}$ satisfying this equation exists, all solutions to this equation can be expressed as the sum of this particular solution $\boldsymbol{x}_{\mathrm{p}}$ and a linear combination of the solutions $\boldsymbol{\eta}_i$ $(i = 1, \ldots, k)$ to the homogeneous equations as

$$\boldsymbol{x} = \sum_{i=1}^{k} c_i \boldsymbol{\eta}_i + \boldsymbol{x}_{\mathrm{p}}.$$

Example 4.3 For inhomogeneous linear equations

$$\begin{pmatrix} 1 & 2 & -1 \\ -3 & 2 & 1 \\ -4 & 0 & 2 \end{pmatrix} \begin{pmatrix} x \\ y \\ z \end{pmatrix} = \begin{pmatrix} 1 \\ 1 \\ 0 \end{pmatrix},$$

a row transformation is applied to arrive at

$$\begin{pmatrix} 1 & 0 & -1/2 \\ 0 & 1 & -1/4 \\ 0 & 0 & 0 \end{pmatrix} \begin{pmatrix} x \\ y \\ z \end{pmatrix} = \begin{pmatrix} 0 \\ 1/2 \\ 0 \end{pmatrix}.$$

We have $n = 3$, $k = 1$, and $\mathrm{rank}A = 2$. Choosing a solution $(x, y, z) = (1/2, 1/4, 1)$ to the homogeneous equations and a particular solution $(x_{\mathrm{p}}, y_{\mathrm{p}}, z_{\mathrm{p}}) = (0, 1/2, 0)$, we arrive at the solution $(x, y, z) = c(1/2, 1/4, 1) + (0, 1/2, 0)$ parametrized by c. □

4.5 PROBLEMS

Problem 4.1 Solve the following linear equations.

$$\text{(1)} \quad \begin{pmatrix} 1 & 3 & 2 \\ 2 & 1 & 2 \\ 1 & -2 & 0 \end{pmatrix} \begin{pmatrix} x \\ y \\ z \end{pmatrix} = \begin{pmatrix} 6 \\ 5 \\ -1 \end{pmatrix}.$$

$$\text{(2)} \quad \begin{pmatrix} 1 & -1 & 1 \\ 1 & -1 & 1 \\ 1 & -1 & 1 \end{pmatrix} \begin{pmatrix} x \\ y \\ z \end{pmatrix} = \begin{pmatrix} 1 \\ 1 \\ 1 \end{pmatrix}.$$

Problem 4.2 Conduct the bifurcation analysis of the two-bar truss arch in Fig. 3.4(a) in Section 3.4 for the height $h = 3$.

Problem 4.3 Conduct the bifurcation analysis of the bar-spring system in Fig. 4.2 for linear-symmetric springs with $k_1 = 3k$ and $k_2 = 2k$.

REFERENCES

1. Crisfield MA (1997), *Non-linear Finite Element Analysis of Solids and Structures*, Vol. 2, Wiley, Chichester.
2. De Borst R (1987), Computation of post-bifurcation and post-failure behavior of strain-softening solids, *Comput. & Struct.* **25**(2), 211–224.
3. Eriksson A (1987), Using eigenvector projections to improve convergence in non-linear finite element equilibrium iterations, *Int. J. Numer. Methods Eng.* **24**(3), 497–512.
4. Ikeda K, and Murota K (2019), *Imperfect Bifurcation in Structures and Materials: Engineering Use of Group-Theoretic Bifurcation Theory*, 3rd ed., Springer, New York.
5. Ikeda K, Murota K, and Fujii H (1991), Bifurcation hierarchy of symmetric structures, *Int. J. Solids Struct.* **27**(12), 1551–1573.
6. Wagner W, and Wriggers P (1988), A simple method for the calculation of postcritical branches, *Eng. Comput.* **5**, 103–109.

5 Bifurcation Theory I: Basics

5.1 SUMMARY

The bifurcation equation is introduced as a basic tool to describe bifurcation behavior in the neighborhood of a critical point. Simple critical points are classified in view of the expanded form of the bifurcation equation, and the qualitative (bifurcation) behaviors in the neighborhood of these points are investigated. Chapter 2 and this chapter give the theoretical foundation of this book. Whereas Chapter 2 focuses on the critical point (buckling load and buckling mode), this chapter deals with post-buckling behavior. Some results of this chapter have already been used in the search for a bifurcating path (Chapter 4). We focus on simple critical points, while multiple bifurcation points are studied in Chapter 6.

It is, in general, very difficult to elucidate the qualitative bifurcation behavior of a nonlinear multi-degree-of-freedom system. The bifurcation equation, which has much fewer degrees of freedom than the original equation, has been used in nonlinear mathematics to overcome such a difficulty. The bifurcation equation is derived as follows.

- Coordinate transformations of unknown variables and equations are conducted.
- Transformed variables are divided into "active coordinates" related to the bifurcation modes and other "passive coordinates."
- The passive coordinates are eliminated from the governing equation to arrive at the so-called bifurcation equation with reduced degrees of freedom.

This process to derive the bifurcation equation is called the elimination of passive coordinates (Thompson and Hunt [5]) or the Liapunov–Schmidt–Koiter reduction (Koiter [3]; Sattinger [4]). In particular, the bifurcation equation for a simple critical point, which we study in this chapter, is a scalar equation in a single unknown variable. Moreover, by considering only a few governing terms in the power series expansion of the bifurcation equation, we can set forth general forms of this equation for simple critical points. Such a strategy to investigate the bifurcation behavior in the neighborhood of a bifurcation point, which we call an *asymptotic theory*, plays an important role in bifurcation theory (Thompson and Hunt [5]; Koiter [3]; Ikeda and Murota [2]). We would intentionally avoid rigorous mathematical formalism, such as the implicit function theorem, for which the readers may refer to the book of Ikeda and Murota [2].

A simple example of a bifurcation equation is presented in Section 5.2. The derivation of the bifurcation equation is advanced in Section 5.3. The classification of simple critical points is conducted in Section 5.4. The direction of bifurcating paths is studied in Section 5.5. Structural examples of three kinds of bifurcations are introduced in Section 5.6.

DOI: 10.1201/9781003112365-5

Keywords: • Asymptotic analysis • Bifurcation equation • Liapunov–Schmidt–Koiter reduction • Passive coordinates • Simple critical point • Stability

5.2 SIMPLE EXAMPLE OF BIFURCATION EQUATION

To illustrate the bifurcation equation, we consider the propped cantilever shown in Fig. 5.1(a). The propped cantilever comprises a truss member that is simply supported at a rigid foundation and supported by horizontal and vertical springs. The equilibrium equation for this cantilever is

$$\boldsymbol{F}(\boldsymbol{u},f) = \begin{pmatrix} F_x \\ F_y \end{pmatrix} = EA \begin{pmatrix} \left(\dfrac{1}{L} - \dfrac{1}{\hat{L}}\right)(x - x_1) + F_{sx} \\ \left(\dfrac{1}{L} - \dfrac{1}{\hat{L}}\right)(y - y_1) + F_{sy} - f \end{pmatrix} = \begin{pmatrix} 0 \\ 0 \end{pmatrix}, \tag{5.1}$$

where $\boldsymbol{u} = (x,y)^\top$ is the location of node 2 after displacement and f is the non-dimensional vertical load normalized with respect to the cross-sectional rigidity EA of the truss member (E denotes Young's modulus and A the cross-sectional area); (x_1, y_1) represents the location of the fixed node 1; L and $\hat{L}$ denote the length of the member before and after displacement, respectively; and F_{sx} and F_{sy} are the horizontal and vertical normalized non-dimensional forces exerted by the springs, respectively. The lengths L and $\hat{L}$ are given by

$$L = [(x_2 - x_1)^2 + (y_2 - y_1)^2]^{1/2}, \qquad \hat{L} = [(x - x_1)^2 + (y - y_1)^2]^{1/2}, \tag{5.2}$$

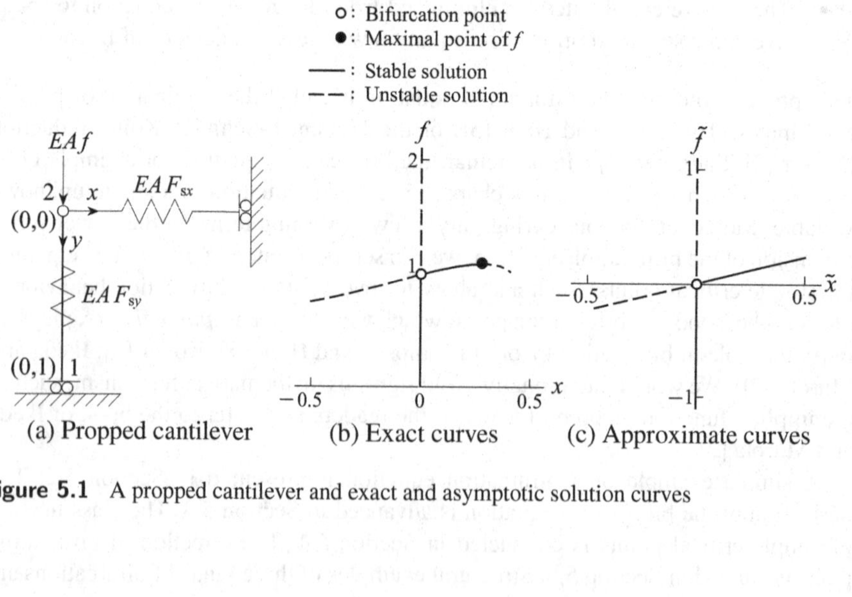

Figure 5.1 A propped cantilever and exact and asymptotic solution curves

where (x_2, y_2) denotes the location of node 2 before displacement, and the normalized forces F_{sx} and F_{sy} are defined by

$$F_{sx} = \frac{x - x_2}{L} + \left(\frac{x - x_2}{L}\right)^2, \qquad F_{sy} = \frac{y - y_2}{L}. \tag{5.3}$$

We have $(x_1, y_1) = (0, 1)$ and $(x_2, y_2) = (0, 0)$, and the initial member length L is given by $L = [(0-0)^2 + (1-0)^2]^{1/2} = 1$. Then, we have

$$F_{sx} = x + x^2, \qquad F_{sy} = y. \tag{5.4}$$

The horizontal force $|F_{sx}|$ is larger for positive x than for negative x with the same $|x|$.

In the following, "direct" and "asymptotic" analyses are compared. In the direct analysis, the solutions to the governing equation are obtained directly in explicit forms, which is possible for this simple example. In the asymptotic analysis, the governing equation is expanded into a power series. By focusing on the leading terms, we derive a single equation—the bifurcation equation—that describes the important aspects of the local bifurcation behavior.

Direct Analysis

The use of (5.2) and (5.4) in (5.1) leads to the equilibrium equation as

$$\boldsymbol{F}(\boldsymbol{u}, f) = \begin{pmatrix} F_x \\ F_y \end{pmatrix} = EA \begin{pmatrix} \left(1 - \dfrac{1}{[x^2 + (y-1)^2]^{1/2}}\right) x + x + x^2 \\ \left(1 - \dfrac{1}{[x^2 + (y-1)^2]^{1/2}}\right) (y-1) + y - f \end{pmatrix} = \begin{pmatrix} 0 \\ 0 \end{pmatrix}. \tag{5.5}$$

The solutions (x, y, f) of (5.5) form curves that are expressed as

$$\begin{aligned} &x = 0, \quad f = 2y, \quad y < 1, && \text{fundamental path,} \\ &y = 1 - [(x+2)^{-2} - x^2]^{1/2}, \quad f = 1 + x[(x+2)^{-2} - x^2]^{1/2}, && -2 < x \le \sqrt{2} - 1, \\ & && \text{bifurcated path.} \end{aligned} \tag{5.6}$$

These solution curves are plotted in Fig. 5.1(b) as an f versus x relation. Two paths intersect at the bifurcation point $(x_c, y_c, f_c) = (0, 1/2, 1)$. This is an asymmetric bifurcation point.

The location of this point on the fundamental path can be determined as a point where the Jacobian matrix $J(x, y, f)$ with $x = 0$ becomes singular. Note that

$$J(0, y, f) = \left.\begin{pmatrix} F_{x,x} & F_{x,y} \\ F_{y,x} & F_{y,y} \end{pmatrix}\right|_{x=0} = EA \begin{pmatrix} 2 - \dfrac{1}{|y-1|} & 0 \\ 0 & 2 \end{pmatrix},$$

where $F_{x,y}$, for example, denotes the derivative of F_x with respect to y. The Jacobian matrix at the bifurcation point $(x_c, y_c, f_c) = (0, 1/2, 1)$ is given by

$$J_c = EA \begin{pmatrix} 0 & 0 \\ 0 & 2 \end{pmatrix}, \tag{5.7}$$

of which $\boldsymbol{\eta}_1 = (1,0)^\top$ is the critical eigenvector; physically, $\boldsymbol{\eta}_1$ represents the tumbling of node 2 in the x-direction.

Asymptotic Analysis

The actual bifurcation analysis is complicated even for a system with a few degrees of freedom. Such complexity can be circumvented by asymptotic analysis as described below.

We investigate the properties of the solution locally around the critical point $(x_c, y_c, f_c) = (0, 1/2, 1)$. Define the incremental variables $(\tilde{x}, \tilde{y}, \tilde{f})$ as

$$(\tilde{x}, \tilde{y}, \tilde{f}) = (x, y, f) - (0, 1/2, 1). \tag{5.8}$$

Expanding the equilibrium equation (5.1) around $(x_c, y_c, f_c) = (0, 1/2, 1)$ yields a set of incremental equations. The Taylor expansion of the first equation $F_x = F_x(x, y, f)$ in (5.1) gives

$$\begin{aligned} &F_x(x,y,f) - F_x(0,1/2,1) \\ &\quad = F_x(\tilde{x}, 1/2+\tilde{y}, 1+\tilde{f}) - F_x(0,1/2,1) \\ &\quad = (F_{x,x})_c\, \tilde{x} + (F_{x,y})_c\, \tilde{y} + (F_{x,f})_c\, \tilde{f} \\ &\qquad + \left[\frac{1}{2}(F_{x,xx})_c\, \tilde{x}^2 + (F_{x,xy})_c\, \tilde{x}\,\tilde{y} + \frac{1}{2}(F_{x,yy})_c\, \tilde{y}^2 \right. \\ &\qquad \left. + (F_{x,xf})_c\, \tilde{x}\tilde{f} + (F_{x,yf})_c\, \tilde{y}\tilde{f} + \frac{1}{2}(F_{x,ff})_c\, \tilde{f}^2 \right] + (\text{h.o.t.}) \\ &\quad = EA(\tilde{x}^2 - 4\tilde{x}\,\tilde{y}) + (\text{h.o.t.}) = 0, \end{aligned} \tag{5.9}$$

where $(\cdot)_c$ denotes a value evaluated at the point (x_c, y_c, f_c) and (h.o.t.) denotes higher-order terms. Similarly, the expansion of the second equation $F_y = F_y(x, y, f)$ in (5.1) gives

$$F_y(x,y,f) - F_y(0,1/2,1) = EA(2\tilde{y} - \tilde{f}) + (\text{h.o.t.}) = 0. \tag{5.10}$$

Equation (5.10) can be solved for $\tilde{y}$ as

$$\tilde{y} = \tilde{f}/2 + (\text{h.o.t.}). \tag{5.11}$$

Substitution of this equation into (5.9) yields

$$\hat{F}(\tilde{x}, \tilde{f}) \equiv \tilde{x}^2 - 2\tilde{x}\tilde{f} + (\text{h.o.t.}) = 0, \tag{5.12}$$

which denotes an incremental relation between $\tilde{x}$ (associated with the bifurcation mode) and $\tilde{f}$. As (5.12) begins with a quadratic term in $\tilde{x}$, two solutions $\tilde{x}$ exist, for a given $\tilde{f}$, in the neighborhood of the bifurcation point. Thus, bifurcation can be interpreted as the emergence of multiple solutions due to the vanishing of lower-order terms.

Equation (5.12) is called the *bifurcation equation*. It has been derived from the full system of equations by eliminating the variable $\tilde{y}$, which is not associated with the critical eigenvector of the Jacobian matrix J_c in (5.7) at the critical point. Such a reduction process to the bifurcation equation is called the *Liapunov–Schmidt–Koiter reduction*, the general treatment of which is the topic of the next section.

By omitting higher-order terms (h.o.t.) in (5.12), we obtain an equation

$$\tilde{x}^2 - 2\tilde{x}\tilde{f} = 0$$

that approximates (5.12) in the neighborhood of $(\tilde{x}, \tilde{f}) = (0,0)$. Figure 5.1(c) shows the curves expressed by this equation. These approximate curves simulate the exact curves in Fig. 5.1(b) quite well in the neighborhood of the bifurcation point but less accurately away from it due to the omission of higher-order terms in the bifurcation equation (5.12).

The bifurcation equation (5.12), despite its simplicity, retains important information about bifurcation behavior. It is a fundamental strategy of bifurcation theory to extract information about bifurcation behavior by examining the asymptotic form of the bifurcation equation, which is much simpler and easier to handle than the original equation.

5.3 DERIVATION OF BIFURCATION EQUATION

We march on to derive the bifurcation equation at simple critical points in a general setting, following the three steps described in Section 5.1. The simple example in Section 5.2 has illustrated this derivation.

We consider a nonlinear *equilibrium equation* (governing equation)

$$\boldsymbol{F}(\boldsymbol{u}, f) = \boldsymbol{0}, \tag{5.13}$$

where $\boldsymbol{F} = (F_1, \ldots, F_n)^\top$ is a vector of sufficiently smooth nonlinear functions, $\boldsymbol{u} = (u_1, \ldots, u_n)^\top$ is an n-dimensional *displacement vector* to describe the state of the system, and f is a load parameter. The system is assumed to have a potential and, therefore, the Jacobian matrix (tangent stiffness matrix) is a symmetric matrix.[1]

A simple critical point $(\boldsymbol{u}_c, f_c)$ satisfies the governing equation (5.13) and the criticality (buckling) condition

$$\det J(\boldsymbol{u}_c, f_c) = 0 \tag{5.14}$$

[1] See Ikeda and Murota [2] for the derivation of the bifurcation equation for a non-potential system.

(cf., (2.10) in Section 2.4), where $J(\boldsymbol{u}, f)$ denotes the Jacobian matrix. Since the Jacobian matrix $J_c \equiv J(\boldsymbol{u}_c, f_c)$ has a zero eigenvalue at a simple critical point, we can choose the eigenvalues λ_i and eigenvectors $\boldsymbol{\eta}_i$ of J_c $(i = 1, \ldots, n)$ so that

$$\begin{cases} \lambda_1 = 0, & J_c \boldsymbol{\eta}_1 = \boldsymbol{0}, \\ \lambda_i \neq 0, & J_c \boldsymbol{\eta}_i = \lambda_i \boldsymbol{\eta}_i, \quad i = 2, \ldots, n. \end{cases} \tag{5.15}$$

Since J_c is a symmetric matrix, the set of eigenvectors $\boldsymbol{\eta}_i$ $(i = 1, \ldots, n)$ can be chosen to be orthonormal, that is, $\boldsymbol{\eta}_i^\top \boldsymbol{\eta}_j = \delta_{ij}$ $(i, j = 1, \ldots, n)$. Here, δ_{ij} is the *Kronecker delta*, defined as $\delta_{ij} = 1$ for $i = j$ and $\delta_{ij} = 0$ for $i \neq j$. Then the matrix

$$H = [\boldsymbol{\eta}_1, \ldots, \boldsymbol{\eta}_n] \tag{5.16}$$

is an orthogonal matrix, that is $H^\top H = I$ (I denotes the identity matrix). To distinguish the eigenvector $\boldsymbol{\eta}_1$ from others, we introduce the notation $\boldsymbol{\eta}_c$ for $\boldsymbol{\eta}_1$, and call it the *critical eigenvector*. That is, we have

$$J_c \boldsymbol{\eta}_c = \boldsymbol{0}. \tag{5.17}$$

We consider an equilibrium point $(\boldsymbol{u}, f)$, i.e., a solution to the governing equation (5.13) in the neighborhood of the simple critical point $(\boldsymbol{u}_c, f_c)$. Using the orthogonal transformation matrix H in (5.16), the incremental variables $(\boldsymbol{w}, \tilde{f})$ between these two points are defined by

$$\boldsymbol{u} - \boldsymbol{u}_c = H\boldsymbol{w} = \sum_{i=1}^{n} \boldsymbol{\eta}_i w_i, \qquad \tilde{f} = f - f_c, \tag{5.18}$$

where $\boldsymbol{w} = (w_1, \ldots, w_n)^\top$.

By taking the inner product of the eigenvector $\boldsymbol{\eta}_i$ and the governing equation (5.13) in the incremental variables $(\boldsymbol{w}, \tilde{f})$, we obtain an incremental governing equation

$$\tilde{F}_i(\boldsymbol{w}, \tilde{f}) \equiv \boldsymbol{\eta}_i^\top \left[\boldsymbol{F}(\boldsymbol{u}_c + H\boldsymbol{w}, f_c + \tilde{f}) - \boldsymbol{F}(\boldsymbol{u}_c, f_c)\right] = 0, \qquad i = 1, \ldots, n \tag{5.19}$$

(we have $\boldsymbol{F}(\boldsymbol{u}_c, f_c) = \boldsymbol{0}$).

In the derivation of the incremental governing equation (5.19) from the governing equation, the following two facts are important.

- By (5.15), the first unknown variable w_1 has different property than other unknown variables w_i $(i = 2, \ldots, n)$ since $\boldsymbol{\eta}_1$ corresponds to the zero eigenvalue. The variable w_1 is called the *active coordinate* and w_i $(i = 2, \ldots, n)$ are called *passive coordinates*.
- Using the transformation matrix H, we can diagonalize the Jacobian matrix J_c as

$$H^\top J_c H = \begin{pmatrix} \lambda_1 & & \\ & \ddots & \\ & & \lambda_n \end{pmatrix},$$

that is,

$$\boldsymbol{\eta}_i^\top J_c \boldsymbol{\eta}_j = \delta_{ij} \lambda_j, \qquad i, j = 1, \ldots, n. \tag{5.20}$$

In the neighborhood of the critical point $(\boldsymbol{u}_{\rm c}, f_{\rm c})$, the incremental governing equation (5.19) is expanded into a power series as

$$\begin{aligned}\tilde{F}_i(\boldsymbol{w},\tilde{f}) &= \boldsymbol{\eta}_i^\top \left[\boldsymbol{F}(\boldsymbol{u}_{\rm c}+H\boldsymbol{w}, f_{\rm c}+\tilde{f}) - \boldsymbol{F}(\boldsymbol{u}_{\rm c}, f_{\rm c})\right] \\ &= \boldsymbol{\eta}_i^\top \left[\left(\frac{\partial \boldsymbol{F}}{\partial \boldsymbol{u}}\right)_{\rm c} H\boldsymbol{w} + \left(\frac{\partial \boldsymbol{F}}{\partial f}\right)_{\rm c} \tilde{f}\right] + (\text{term of } w_1^2) + (\text{h.o.t.}) \\ &= \lambda_i w_i + a_i \tilde{f} + b_i w_1^2 + (\text{h.o.t.}), \quad i = 1,\dots,n. \end{aligned} \tag{5.21}$$

(It is to be noted that higher order terms (h.o.t.) in this equation contain quadratic terms w_i^2 $(i \neq 1)$ other than w_1^2.) Here, $(\cdot)_{\rm c}$ denotes the evaluation at the critical point $(\boldsymbol{u}_{\rm c}, f_{\rm c})$, and the coefficient a_i is defined by

$$a_i = \boldsymbol{\eta}_i^\top \left(\frac{\partial \boldsymbol{F}}{\partial f}\right)_{\rm c} \tag{5.22}$$

and, from (5.20), we have

$$\boldsymbol{\eta}_i^\top \left(\frac{\partial \boldsymbol{F}}{\partial \boldsymbol{u}}\right)_{\rm c} H\boldsymbol{w} = \boldsymbol{\eta}_i^\top J_{\rm c} \sum_{j=1}^{n} \boldsymbol{\eta}_j w_j = \sum_{j=1}^{n} \delta_{ij} \lambda_j w_j = \lambda_i w_i. \tag{5.23}$$

Since $\lambda_1 = 0$ in (5.15), the first equation $(i = 1)$ of (5.21) becomes

$$\tilde{F}_1(\boldsymbol{w},\tilde{f}) = a_1 \tilde{f} + b_1 w_1^2 + (\text{h.o.t.}), \tag{5.24}$$

which is qualitatively different from the remaining equations with $i = 2,\dots,n$. The leading term (lowest-order term) of w_1 in (5.24) is w_1^2. This is why the term of w_1^2 is explicitly given in (5.21).

The remaining equations $\tilde{F}_i(\boldsymbol{w},\tilde{f}) = 0$ for $i = 2,\dots,n$ in (5.21) can be solved for $w_2,\dots,w_n$ as

$$w_i = -\frac{1}{\lambda_i}(a_i \tilde{f} + b_i w_1^2) + (\text{h.o.t.}), \quad i = 2,\dots,n. \tag{5.25}$$

Note that higher-order terms (h.o.t.) in this equation are functions only in (w_1, f); accordingly, the displacement vector $\boldsymbol{w}$ can be expressed as a smooth function $\boldsymbol{w} = \boldsymbol{\varphi}(w_1)$ in variable w_1. Note that there is only one "active" variable $w \equiv w_1$, while other "passive" variables $w_2,\dots,w_n$ are uniquely determined for a given w. By eliminating w_i $(i = 2,\dots,n)$ in the higher-order terms of $\tilde{F}_1$ in (5.24) using (5.25), we can derive the *bifurcation equation*

$$\hat{F}(w,\tilde{f}) \equiv \tilde{F}_1(\boldsymbol{\varphi}(w),\tilde{f}) = 0. \tag{5.26}$$

This process of deriving the bifurcation equation is called the *elimination of the passive coordinates* or the *Liapunov–Schmidt–Koiter reduction*. For a solution w to the bifurcation equation (5.26), the vector

$$\boldsymbol{u} = \boldsymbol{u}_{\rm c} + \sum_{i=1}^{n} \varphi_i(w) \boldsymbol{\eta}_i \tag{5.27}$$

gives a solution to $\boldsymbol{F}(\boldsymbol{u}, f) = \boldsymbol{0}$, where $\boldsymbol{\varphi}(w) = (\varphi_1(w),\dots,\varphi_n(w))$.

The criticality condition for the bifurcation equation is equivalent to the criticality condition of the original system (Ikeda and Murota [2]). That is, by denoting the Jacobian of the bifurcation equation as $\hat{J} = \partial\hat{F}/\partial w$, we have

$$\hat{J}(w,\tilde{f}) = 0 \iff \det J(\boldsymbol{u}, f) = 0. \tag{5.28}$$

We can expand the bifurcation equation (5.26) into a power series of w and $\tilde{f}$ involving an appropriate number of terms as

$$\hat{F}(w,\tilde{f}) = \sum_{i\geq 0}\sum_{j\geq 0} A_{ij} w^i \tilde{f}^j, \tag{5.29}$$

where

$$A_{ij} = \frac{1}{i!\,j!}\frac{\partial^{i+j}\hat{F}}{\partial w^i \partial \tilde{f}^j}(0,0), \quad i,j = 0,1,2,\ldots. \tag{5.30}$$

Note that $(w,\tilde{f}) = (0,0)$ corresponds to the critical point and satisfies the bifurcation equation (5.26) and the criticality condition (5.28). That is, we have

$$A_{00} = 0, \qquad A_{10} = 0. \tag{5.31}$$

As has been demonstrated for the simple example in Section 5.2, when the absolute values of incremental variables w and $\tilde{f}$ are sufficiently small, which is the case in a neighborhood of the critical point, the bifurcation equation can be approximated by considering only the leading terms. For example, if A_{20} is nonzero, the term of $A_{30}w^3$ can be omitted as $|A_{30}w^3|$ becomes much smaller than $|A_{20}w^2|$. It is a basic strategy in bifurcation theory to investigate the asymptotic behavior of the solutions by omitting higher-order terms in the bifurcation equation.

For particular structures, it is often difficult to explicitly derive the expansion coefficients. However, it is a major viewpoint of this chapter, as well as the philosophy of bifurcation theory, to investigate if these coefficients are zero or nonzero instead of actually evaluating the coefficients.

Remark 5.1 In this chapter, the reduction to the bifurcation equation is conducted only for critical points. Nonetheless, such reduction is applicable to an *ordinary point* with $A_{10} \neq 0$. In the neighborhood of an ordinary point, (5.26) becomes

$$\hat{F}(w,\tilde{f}) = A_{10}w + A_{01}\tilde{f} + (\text{h.o.t.}) = 0 \tag{5.32}$$

and the Jacobian becomes asymptotically as

$$\hat{J}(w,\tilde{f}) \approx A_{10}\ (\neq 0). \tag{5.33}$$

By (5.32), the variable w is uniquely determined from $\tilde{f}$ as $w \approx -A_{01}/A_{10}\tilde{f}$. From (5.33), the stability does not change near an ordinary point, and the system is stable near this point for $A_{10} > 0$ and unstable for $A_{10} < 0$. □

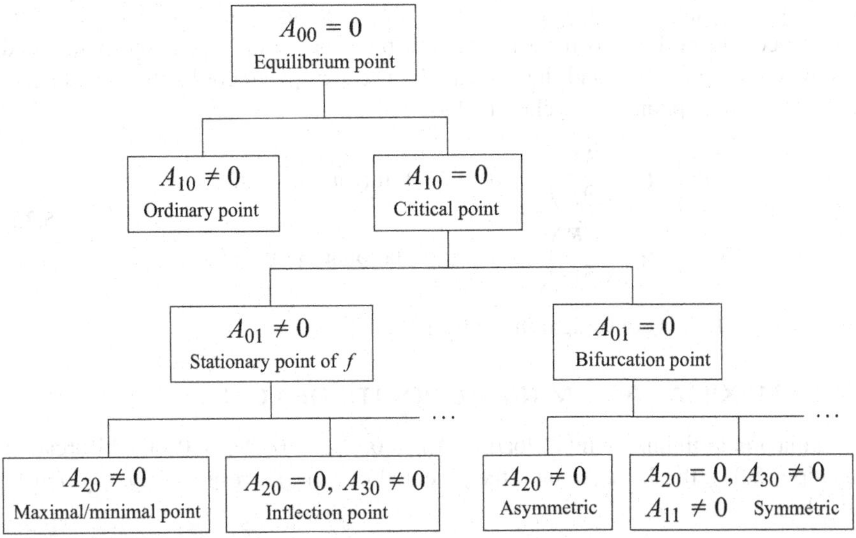

Figure 5.2 Classification of equilibrium points

5.4 CLASSIFICATION OF SIMPLE CRITICAL POINTS

With reference to the vanishing and non-vanishing of the coefficients A_{ij} $(i,j = 0,1,\ldots)$ of the expanded form (5.29) of the bifurcation equation (5.26) at a critical point with $(w,\tilde{f}) = (0,0)$, we classify simple critical points and investigate the behavior around the critical points. See Ikeda and Murota [2] and Fujii and Yamaguti [1] for a more mathematical description of this classification.

For a critical point, we have $A_{10} = 0$ by (5.31), and hence (5.29) becomes

$$\hat{F}(w,\tilde{f}) = (A_{20}w^2 + A_{30}w^3 + \cdots) + (A_{01}\tilde{f} + A_{02}\tilde{f}^2 + \cdots) + (A_{11}w\tilde{f} + A_{21}w^2\tilde{f} + A_{12}w\tilde{f}^2 + \cdots). \tag{5.34}$$

The asymptotic property of this function can be described by the leading terms of each of the three sets of terms in the parentheses. Since the function $\hat{F}(w,\tilde{f})$ is quadratic in w, the equation $\hat{F}(w,\tilde{f}) = 0$ does not determine w uniquely for each $\tilde{f}$. According to the vanishing/non-vanishing of the value of A_{01}, critical points are classified into *stationary point* of f and *bifurcation point*, as shown in Fig. 5.2. According to the vanishing/non-vanishing of the value of A_{20}, a bifurcation point $(A_{01} = 0)$ can be classified into an *asymmetric bifurcation point* and a *symmetric bifurcation point*. The critical points shown by $(\cdots)$ in Fig. 5.2 are rare and are out of the scope of this book.

Remark 5.2 The coefficient A_{01} in (5.34) has the concrete form of

$$A_{01} = \boldsymbol{\eta}_{\rm c}^{\top} \left(\frac{\partial \boldsymbol{F}}{\partial f} \right)_{\rm c},$$

where $\boldsymbol{\eta}_c = \boldsymbol{\eta}_1$ and $(\partial \boldsymbol{F}/\partial f)_c$ is the load pattern vector. From this relation, we see that the condition $A_{01} = 0$ for a bifurcation point is the orthogonality of the load pattern vector $(\partial \boldsymbol{F}/\partial f)_c$ and the critical eigenvector $\boldsymbol{\eta}_c$. Accordingly, we can also say that a critical point can be classified as

$$\begin{cases} \boldsymbol{\eta}_c^\top \left(\dfrac{\partial \boldsymbol{F}}{\partial f}\right)_c = 0 & \Rightarrow \text{ bifurcation point,} \\ \boldsymbol{\eta}_c^\top \left(\dfrac{\partial \boldsymbol{F}}{\partial f}\right)_c \neq 0 & \Rightarrow \text{ stationary point of } f. \end{cases} \tag{5.35}$$

This classification is convenient in applications. □

5.4.1 MAXIMAL AND MINIMAL POINTS OF LOAD

For a maximal/minimal point of load f $(A_{10} = 0,\ A_{01} \neq 0,\ A_{20} \neq 0)$, the bifurcation equation (5.34), the Jacobian, and the potential of $\hat{F}(w, \tilde{f})$, respectively, asymptotically become

$$\hat{F}(w, \tilde{f}) \approx A_{20}w^2 + A_{01}\tilde{f}, \tag{5.36}$$

$$\hat{J}(w, \tilde{f}) \approx 2A_{20}w, \tag{5.37}$$

$$\hat{U}(w, \tilde{f}) \approx \frac{1}{3}A_{20}w^3 + A_{01}w\tilde{f}. \tag{5.38}$$

From $\hat{F}(w, \tilde{f}) = 0$ for $\hat{F}$ in (5.36), we obtain $\tilde{f} \approx -(A_{20}/A_{01})w^2$. According to the sign of A_{20}/A_{01}, the maximal/minimal point of f is classified as

$$\begin{cases} A_{20}/A_{01} > 0 & \Rightarrow \text{ maximal point,} \\ A_{20}/A_{01} < 0 & \Rightarrow \text{ minimal point.} \end{cases}$$

For $\tilde{f} = 0$, corresponding to the maximal/minimal point, the potential becomes

$$\hat{U}(w, 0) \approx \frac{1}{3}A_{20}w^3,$$

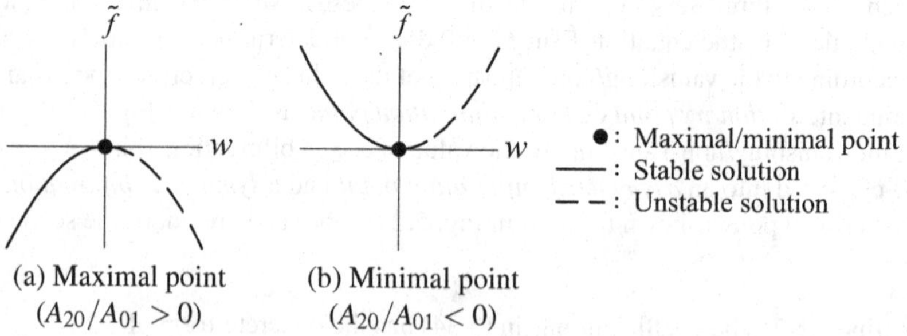

Figure 5.3 Asymptotic behavior in the neighborhood of a maximal/minimal point of f (stability is described for the case of $A_{20} < 0$)

and has an inflection point at $w = 0$. Accordingly, the maximal/minimal point of f, located at $w = 0$, is unstable. Then, also from (5.37), the stability in the neighborhood of this point is classified as

$$\begin{cases} A_{20}w > 0 & \Rightarrow \text{ stable (ordinary point)}, \\ A_{20}w = 0 & \Rightarrow \text{ unstable (maximal/minimal point of } f), \\ A_{20}w < 0 & \Rightarrow \text{ unstable (ordinary point)}. \end{cases} \tag{5.39}$$

Thus the stability of the system changes at the maximal/minimal point of f. Asymptotic behavior in the neighborhood of this point is depicted in Fig. 5.3.

5.4.2 ASYMMETRIC BIFURCATION POINT

For an asymmetric bifurcation point ($A_{10} = A_{01} = 0, A_{20} \neq 0$), we have

$$\hat{F}(w,\tilde{f}) \approx A_{20}w^2 + A_{11}w\tilde{f} + A_{02}\tilde{f}^2, \tag{5.40}$$

$$\hat{J}(w,\tilde{f}) \approx 2A_{20}w + A_{11}\tilde{f}, \tag{5.41}$$

$$\hat{U}(w,\tilde{f}) \approx \frac{1}{3}A_{20}w^3 + \frac{1}{2}A_{11}w^2\tilde{f} + A_{02}w\tilde{f}^2. \tag{5.42}$$

For $\tilde{f} = 0$, corresponding to the bifurcation point, the potential becomes

$$\hat{U}(w,0) \approx \frac{1}{3}A_{20}w^3$$

and has an inflection point at $w = 0$. Accordingly, the asymmetric bifurcation point, located at $w = 0$, is unstable.

The behavior of the system is investigated for two different cases: the case with the trivial solution ($A_{02} = 0$) and that without it ($A_{02} \neq 0$). The propped cantilever with an asymmetric spring in Section 5.6.3 corresponds to the case with the trivial solution.

Case with trivial solution ($A_{02} = 0$)

In the case of $A_{02} = 0$, (5.40) becomes

$$\hat{F}(w,\tilde{f}) \approx A_{20}w^2 + A_{11}w\tilde{f}$$

and $\hat{F}(w,\tilde{f}) = 0$ has two solution curves:

$$\begin{cases} w = 0, & \text{trivial solution (main path)}, \\ w \approx -(A_{11}/A_{20})\tilde{f}, & \text{bifurcating path}, \end{cases}$$

which are plotted in Fig. 5.4(a). The bifurcating path is asymmetric in the sense that it is not invariant to the reflection $w \longmapsto -w$.

From (5.41), the Jacobian becomes

$$\begin{cases} \hat{J}(0,\tilde{f}) \approx A_{11}\tilde{f}, & \text{trivial solution}, \\ \hat{J}(-(A_{11}/A_{20})\tilde{f},\tilde{f}) \approx -A_{11}\tilde{f}, & \text{bifurcating path}. \end{cases} \tag{5.43}$$

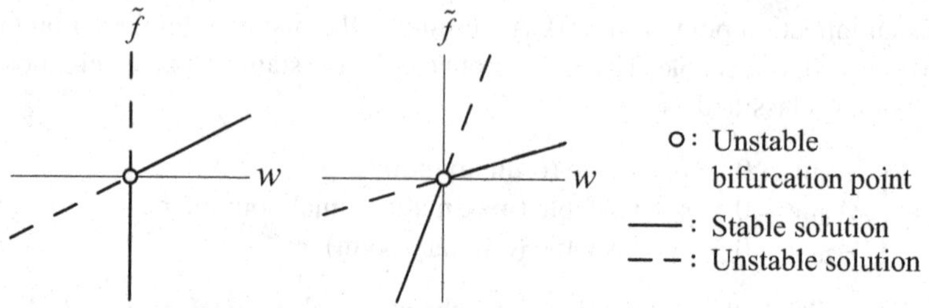

(a) With trivial solution ($A_{02} = 0$) (b) Without trivial solution ($A_{02} > 0$)

Figure 5.4 Asymptotic behavior in the neighborhood of a (nondegenerate) asymmetric bifurcation point ($A_{11} < 0$ and $A_{20} > 0$)

In the case of an engineering interest where the trivial solution $w = 0$ is stable until reaching the bifurcation point, we have $\hat{J}(0, \tilde{f}) > 0$ for $\tilde{f} < 0$ in (5.43), that is,

$$A_{11} < 0. \tag{5.44}$$

Stable solutions for this case are shown by solid lines and unstable ones by dashed lines in Fig. 5.4(a). The trivial solution is unstable above the bifurcation point ($\tilde{f} > 0$) and stable below this point ($\tilde{f} < 0$). In contrast, the bifurcated solution is unstable below ($\tilde{f} < 0$) and stable above the point ($\tilde{f} > 0$). This phenomenon is often called an *exchange of stability*.

Case without trivial solution ($A_{02} \neq 0$)

We consider the case of $A_{02} \neq 0$ without the trivial solution. When the condition of nondegeneracy[2]

$$A_{11}{}^2 - 4A_{02}A_{20} > 0 \tag{5.45}$$

holds, $\hat{F}(w, \tilde{f}) = 0$ for $\hat{F}$ in (5.40) has two solutions

$$w \approx \frac{-A_{11} \pm \sqrt{A_{11}{}^2 - 4A_{02}A_{20}}}{2A_{20}} \tilde{f} \tag{5.46}$$

and the point $(w, \tilde{f}) = (0, 0)$ is thus a bifurcation point at which these two solution curves intersect (Fig. 5.4(b)).

[2] When $A_{11}{}^2 - 4A_{02}A_{20} < 0$, the equation $\hat{F}(w, \tilde{f}) = 0$ has no solution other than $(w, \tilde{f}) = (0, 0)$, and this critical point is an isolated point.

5.4.3 SYMMETRIC BIFURCATION POINT

For a symmetric bifurcation point ($A_{10} = A_{01} = A_{20} = 0$, $A_{30} \neq 0$, $A_{11} \neq 0$), we have

$$\hat{F}(w,\tilde{f}) \approx A_{30}w^3 + A_{11}w\tilde{f} + A_{02}\tilde{f}^2, \tag{5.47}$$

$$\hat{J}(w,\tilde{f}) \approx 3A_{30}w^2 + A_{11}\tilde{f}, \tag{5.48}$$

$$\hat{U}(w,\tilde{f}) \approx \frac{1}{4}A_{30}w^4 + \frac{1}{2}A_{11}w^2\tilde{f} + A_{02}w\tilde{f}^2. \tag{5.49}$$

For $\tilde{f} = 0$, corresponding to the bifurcation point, the potential reduces to

$$\hat{U}(w,0) \approx \frac{1}{4}A_{30}w^4. \tag{5.50}$$

Accordingly, the stability of this bifurcation point is classified as

$$\begin{cases} A_{30} < 0 \Rightarrow & \text{unstable}, \\ A_{30} > 0 \Rightarrow & \text{stable}. \end{cases} \tag{5.51}$$

Thus the stability depends on the sign of the coefficient A_{30}.

Similarly to the asymmetric bifurcation point, we consider below the case with the trivial solution ($A_{02} = 0$) and that without it ($A_{02} \neq 0$) separately. The propped cantilever in Section 5.2 corresponds to the case with the trivial solution.

Case with trivial solution ($A_{02} = 0$)

For the case of $A_{02} = 0$, (5.47) becomes

$$\hat{F}(w,\tilde{f}) \approx A_{30}w^3 + A_{11}w\tilde{f}, \tag{5.52}$$

and the bifurcation equation $\hat{F}(w,\tilde{f}) = 0$ has the trivial solution $w = 0$ and a bifurcating path (Fig. 5.5):

$$\begin{cases} w = 0, & \text{trivial solution (main path)}, \\ \tilde{f} \approx -\dfrac{A_{30}}{A_{11}}w^2, & \text{bifurcating path}. \end{cases} \tag{5.53}$$

The Jacobian is obtained by substituting (5.53) into (5.48) as

$$\begin{cases} \hat{J}(0,\tilde{f}) \approx A_{11}\tilde{f}, & \text{trivial solution}, \\ \hat{J}(w,\tilde{f}) \approx 2A_{30}w^2, & \text{bifurcating path}. \end{cases} \tag{5.54}$$

The bifurcated path is unstable for $A_{30} < 0$ and stable for $A_{30} > 0$, similarly to the stability of the bifurcation point in (5.51).

In the case of an engineering interest where the trivial solution $w = 0$ is stable until reaching the bifurcation point, we have $\hat{J}(0,\tilde{f}) > 0$ for $\tilde{f} < 0$. Then from (5.54), we have

$$A_{11} < 0. \tag{5.55}$$

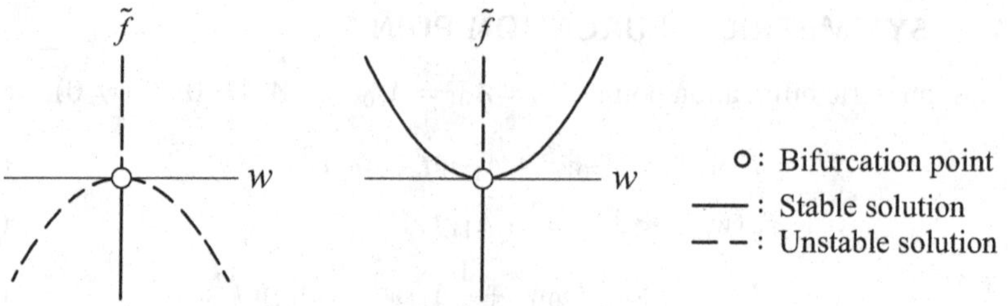

(a) Unstable bifurcation point ($A_{30} < 0$)

(b) Stable bifurcation point ($A_{30} > 0$)

Figure 5.5 Asymptotic behavior in the neighborhood of a symmetric bifurcation point with the trivial solution (stability is described for $A_{11} < 0$)

Figure 5.5 shows solution curves and stability of these curves when $A_{02} = 0$. The fundamental path beyond the bifurcation point ($\tilde{f} > 0$) is unstable both for unstable and stable bifurcation points. For an unstable bifurcation point ($A_{30} < 0$ in Fig. 5.5(a)), when the load f is increased from a small value ($\tilde{f} < 0$), the system becomes unstable upon reaching the bifurcation point ($\tilde{f} = 0$), en route to a failure. For a stable bifurcation point ($A_{30} > 0$ in Fig. 5.5(b)), beyond the bifurcation point ($\tilde{f} > 0$), the bifurcated solution is stable. When the load f is increased from a small value ($\tilde{f} < 0$), there is a continuation of stable equilibria, stable trivial solution ($\tilde{f} < 0$) followed by the stable bifurcating path ($\tilde{f} > 0$). As a consequence, the system does not undergo a failure even at the onset of the bifurcation.

Case without trivial solution ($A_{02} \neq 0$)

In the case of $A_{02} \neq 0$, the bifurcation equation $\hat{F}(w, \tilde{f}) = 0$ for $\hat{F}$ in (5.47) has two solution curves:[3]

$$\begin{cases} \tilde{f} = -\dfrac{A_{11}}{A_{02}} w + \mathrm{O}(w^2), & \text{main path,} \\ \tilde{f} = -\dfrac{A_{30}}{A_{11}} w^2 + \mathrm{O}(w^3), & \text{bifurcating path.} \end{cases} \tag{5.56}$$

The main path has a nonzero slope at the bifurcation point (Fig. 5.6)) and is a nontrivial solution.

The Jacobian is obtained by substituting (5.56) into (5.48) to arrive at asymptotically the same form as (5.54) for the case with the trivial solution. Accordingly, the stability of the system is qualitatively identical with the case of the trivial solution.

Remark 5.3 We consider the case of an engineering interest where the main path, with or without the trivial solution, is stable until reaching the bifurcation point ($A_{11} < 0$ from (5.55)). Then, from (5.53) and (5.56), we have $A_{30} > 0$ or $A_{30} < 0$

[3]See the answer to Problem 5.6 for the proof of (5.56).

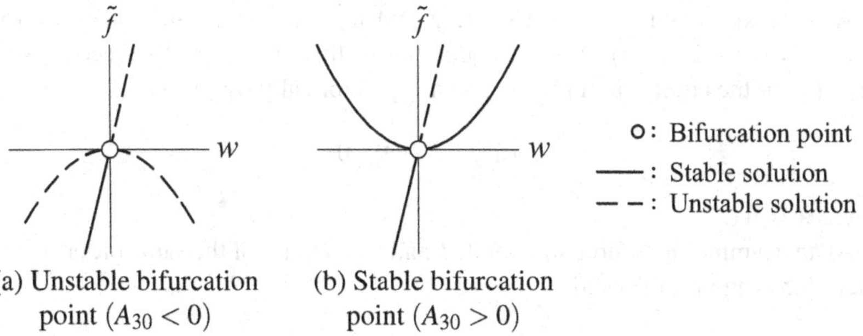

Figure 5.6 Asymptotic behavior in the neighborhood of a symmetric bifurcation point without the trivial solution (stability is described for $A_{11} < 0$)

according to whether f increases or decreases on a bifurcating path just after bifurcation. By (5.51), bifurcation point is stable when $A_{30} > 0$ and is unstable when $A_{30} < 0$. Hence we can identify the stability of the bifurcation point, as well as the bifurcating path, by

$$\begin{cases} f \text{ increases along the bifurcating path} & \Rightarrow \quad \text{stable}, \\ f \text{ decreases along the bifurcating path} & \Rightarrow \quad \text{unstable}, \end{cases}$$

without resort to the concrete form of the coefficient A_{30} (cf., Remark 4.2 in Section 4.2.2). □

5.5 DIRECTION OF BIFURCATING PATHS

We investigate the direction of equilibrium paths in the neighborhood of a critical point as a theoretical supplement to Section 4.2.1. Recall from (5.18) and (5.25) that

$$\tilde{\boldsymbol{u}} = \boldsymbol{u} - \boldsymbol{u}_{\mathrm{c}} = H\boldsymbol{w} = \sum_{i=1}^{n} \boldsymbol{\eta}_i w_i, \tag{5.57}$$

$$w_i = -\frac{1}{\lambda_i}(a_i\tilde{f} + b_i w_1^2) + (\text{h.o.t.}), \quad i = 2, \ldots, n. \tag{5.58}$$

At an ordinary point, (5.58) holds also for $i = 1$ (cf., Remark 5.1). Because w_1 and $\tilde{f}$ are of the same order in the neighborhood of an ordinary point, the term of ${w_1}^2$ can be omitted. Then (5.57) asymptotically becomes

$$\tilde{\boldsymbol{u}} = \boldsymbol{u} - \boldsymbol{u}_{\mathrm{c}} = \sum_{i=1}^{n} \boldsymbol{\eta}_i w_i \approx -\sum_{i=1}^{n} \boldsymbol{\eta}_i \frac{a_i}{\lambda_i}\tilde{f}.$$

Accordingly, the tangential direction of the equilibrium path is given as

$$(\tilde{\boldsymbol{u}}, \tilde{f}) \approx \left(-\sum_{i=1}^{n} \boldsymbol{\eta}_i \frac{a_i}{\lambda_i}\tilde{f}, \tilde{f} \right). \tag{5.59}$$

At a maximal/minimal point of f, $\tilde{f}$ and w_1^2 are of the same order by (5.36), whereas w_i $(i=2,\ldots,n)$ are higher-order terms than w_1 by (5.58). Accordingly, the direction of the equilibrium path is given asymptotically by (5.57) as

$$(\tilde{\boldsymbol{u}},\tilde{f}) \approx (w\boldsymbol{\eta}_{\rm c},0),$$

where $w=w_1$.

At an asymmetric bifurcation point, $\tilde{f}$ and $w=w_1$ are of the same order by (5.40). Then (5.57) and (5.58) yield

$$\tilde{\boldsymbol{u}} = \boldsymbol{u}-\boldsymbol{u}_{\rm c} \approx w\boldsymbol{\eta}_{\rm c} - \sum_{i=2}^{n} \boldsymbol{\eta}_i \frac{a_i}{\lambda_i}\tilde{f}.$$

Therefore, the direction of the bifurcating solution is given by

$$(\tilde{\boldsymbol{u}},\tilde{f}) \approx \left(w\boldsymbol{\eta}_{\rm c} - \sum_{i=2}^{n} \boldsymbol{\eta}_i \frac{a_i}{\lambda_i}\tilde{f}, \tilde{f} \right).$$

At a symmetric bifurcation point, $\tilde{f}$ and w_1^2 are of the same order on a bifurcating path by (5.53) and (5.56). In addition, w_i $(i=2,\ldots,n)$ are of the same order as $\tilde{f}$ by (5.58), and hence these terms can be omitted asymptotically relative to the term of $w=w_1$. Accordingly, the direction of the bifurcating solution is given by

$$(\tilde{\boldsymbol{u}},\tilde{f}) \approx (w\boldsymbol{\eta}_{\rm c},0).$$

5.6 STRUCTURAL EXAMPLES OF THREE KINDS OF BIFURCATIONS

Using structural examples, we illustrate the bifurcation behaviors associated with three kinds of simple bifurcation points: unstable-symmetric, stable-symmetric, and asymmetric points. As we have seen in Sections 5.4.2 and 5.4.3, simple bifurcation points can be classified into these three kinds by the vanishing/non-vanishing of the coefficient A_{20} and the sign of A_{30} in the expression (5.29) of the bifurcation equation.

For this purpose, we refer to the propped cantilever,[4] comprising rigid bar supported by a spring, depicted in Fig. 5.7. The bottom of this bar is fixed to the ground by a hinge that allows only rotation, and its top is supported horizontally by a spring and is subjected to the vertical load kLf, where f represents the load normalized by kL using the length L of the member and spring constant k.

This is the same structure as the one studied in Section 1.5, but we introduce here nonlinear spring properties to demonstrate that such properties entail different post-bifurcation behaviors. We consider three kinds of springs, which develop a force

---

[4]This propped cantilever is taken from Thompson and Hunt [2].

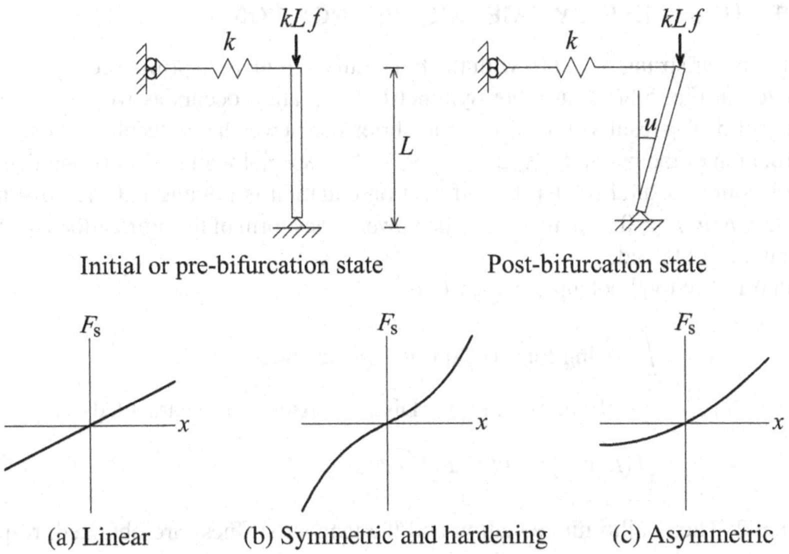

Figure 5.7 A propped cantilever and spring properties

against a horizontal displacement x as follows (see Figs. 5.7 (a)–(c)):

$$F_s(x) = \begin{cases} kx, & \text{linear spring,} \\ k\left(x + 2\dfrac{x^3}{L^2}\right), & \text{symmetric-hardening spring,} \\ k\left(x + \dfrac{3}{2}\dfrac{x^2}{L}\right), & \text{asymmetric spring.} \end{cases} \tag{5.60}$$

These three kinds of material properties correspond to the three kinds of bifurcation points.

When the load f is small, the bar retains its upright state, supported by the spring. When the load f is increased to some extent, the bar cannot resist the load any further and starts to rotate, undergoing the bifurcation/buckling, as shown at the top-right of Fig. 5.7. Note that the displacement of the bar is horizontal, but the load is vertical. It is a characteristic of bifurcation behavior that the direction of the displacement is perpendicular to the direction of the load.

To be made clear in the analyses below, this propped cantilever, irrespective of the spring property, displays the same upright state and encounters the bifurcation at the same load level. Yet, the types of bifurcation points and post-bifurcation behaviors vary with the spring properties and exhibit three different behaviors: "unstable–symmetric bifurcation," "stable–symmetric bifurcation," and "asymmetric bifurcation." The symmetry of the spring force is related to the symmetry of the post-bifurcation behaviors.

5.6.1 UNSTABLE–SYMMETRIC BIFURCATION

For the linear spring of (5.60) with the bilaterally symmetric spring force $F_s(x) = kx$ depicted in Fig. 5.8(a), unstable–symmetric bifurcation occurs as we have seen in Section 1.5. The results of Section 1.5 are briefly reviewed here. Recall that u denotes the rotation of the bar and $-\pi/2 \leq u \leq \pi/2$. As a special feature of this one-degree-of-freedom structural model, the bifurcation equation is nothing but the governing equation $F(u, f) = 0$, and moreover, the power series form of the bifurcation equation need not be obtained.

Step 1: The total potential energy U is

$$\begin{aligned} U &= \int (\text{Spring force})\, \mathrm{d}(\text{Spring displacement}) \\ &\quad - (\text{External force}) \times (\text{Distance external force traveled}) \\ &= \frac{1}{2}k(L\sin u)^2 - kLf \cdot L(1 - \cos u). \end{aligned} \tag{5.61}$$

Step 2: The equilibrium equation and the tangent stiffness are obtained, respectively, as

$$F(u, f) \equiv \frac{\partial U}{\partial u} = kL^2 \sin u(\cos u - f) = 0, \tag{5.62}$$

$$J(u, f) \equiv \frac{\partial F}{\partial u} = kL^2(\cos 2u - f\cos u). \tag{5.63}$$

Step 3: From the equilibrium equation (5.62), we obtain two kinds of equilibrium paths

$$\begin{cases} u = 0, & \text{fundamental path (trivial solution),} \\ f = \cos u, & \text{bifurcated path.} \end{cases}$$

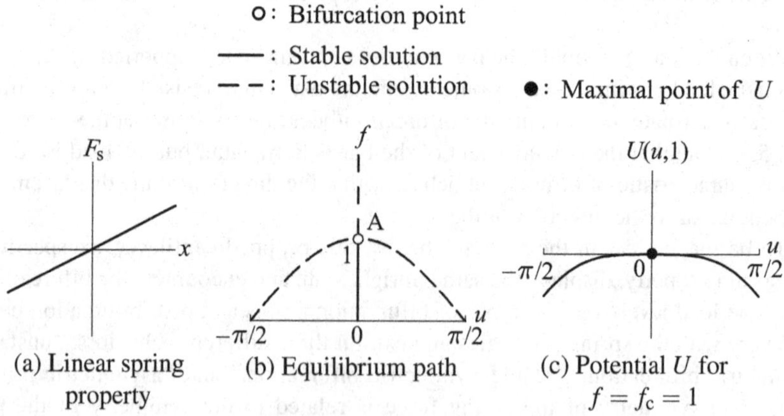

Figure 5.8 Bifurcation properties of the propped cantilever with the linear spring

As shown in Fig. 5.8(b), the bifurcation point is located at A with $(u,f) = (0,1)$ at which the two paths intersect. The bifurcated paths are bilaterally symmetric. A solution u is unique for each $f \geq 1$, whereas there are three solutions u for each $f < 1$. Thus the multiplicity of the solution u for each f is engendered by bifurcation. The coefficients A_{20} and A_{30} at the bifurcation point $(u_c, f_c) = (0,1)$ can be obtained explicitly as

$$A_{20} \equiv \frac{\partial^2 F}{\partial u^2}(0,1) = kL^2(-2\sin 2u + f\sin u)\big|_{(u,f)=(0,1)} = 0,$$

$$A_{30} \equiv \frac{\partial^3 F}{\partial u^3}(0,1) = kL^2(-4\cos 2u + f\cos u)\big|_{(u,f)=(0,1)} = -3kL^2 < 0.$$

Accordingly, this point is an *unstable–symmetric bifurcation point* (cf., Section 5.4.3).

Step 4: On the fundamental path $u = 0$, the tangent stiffness in (5.63) becomes

$$J(0,f) = kL^2(1-f).$$

Since $J(0,1) = 0$ holds, $(u_c, f_c) = (0,1)$ is a critical point, which is a bifurcation point. Then an equilibrium point on this fundamental path is classified as

$$\begin{cases} f < 1 & \Rightarrow \text{ stable ordinary point,} \\ f = 1 & \Rightarrow \text{ critical point (bifurcation point),} \\ f > 1 & \Rightarrow \text{ unstable ordinary point.} \end{cases}$$

Accordingly, the fundamental path is stable when the load is small $(f < 1)$.

On the bifurcated path $f = \cos u$, the tangent stiffness in (5.63) becomes

$$J(u,\cos u) = kL^2(\cos 2u - \cos^2 u) = -kL^2 \sin^2 u < 0 \qquad (u \neq 0).$$

Accordingly, this bifurcated path is unstable.

At the bifurcation point $(u_c, f_c) = (0,1)$, the total potential energy function U with constant $f = f_c = 1$ is obtained from (5.61) as

$$U(u,1) = 2kL^2\left(\sin^2\frac{u}{2}\cos^2\frac{u}{2} - \sin^2\frac{u}{2}\right) = -2kL^2\sin^4\frac{u}{2}.$$

This function takes a local maximum at the displacement u_c for the bifurcation point; accordingly, this bifurcation point is unstable (see the potential distribution in Fig. 5.8(c)).

5.6.2 STABLE–SYMMETRIC BIFURCATION

Stable–symmetric bifurcation takes place for the symmetric-hardening spring of (5.60) with the spring property (Fig. 5.9(a)):

$$F_s(x) = k\left(x + 2\frac{x^3}{L^2}\right).$$

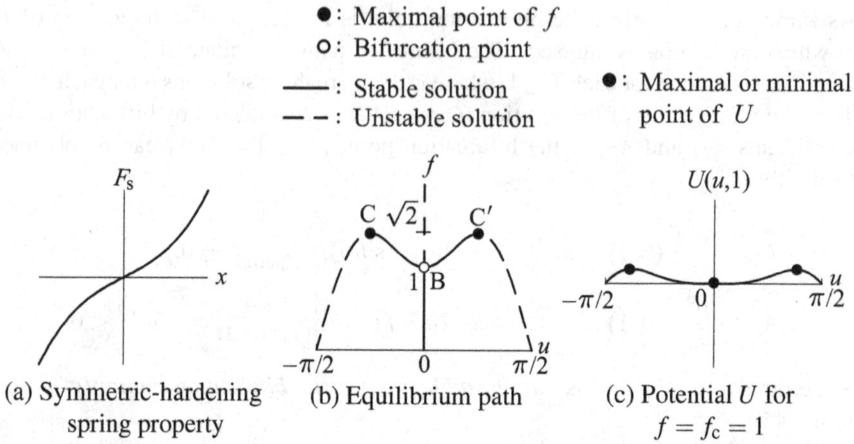

Figure 5.9 Bifurcation properties of the propped cantilever with the symmetric-hardening spring

The nonlinear (cubic) term $2x^3/L^2$ of this spring property stabilizes the post-bifurcation state, unlike the linear spring, leading to stable–symmetric bifurcation as shown below.

Step 1: The total potential energy U is evaluated to

$$U = \frac{k}{2}\left[(L\sin u)^2 + \frac{(L\sin u)^4}{L^2}\right] - kLf \cdot L(1-\cos u). \tag{5.64}$$

Step 2: The equilibrium equation and the tangent stiffness, respectively, become

$$\begin{aligned} F(u,f) &\equiv \frac{\partial U}{\partial u} = kL^2 \sin u \left[\cos u(1+2\sin^2 u) - f\right] \\ &= kL^2\left(\sin 2u - \frac{1}{4}\sin 4u - f\sin u\right) = 0, \end{aligned} \tag{5.65}$$

$$J(u,f) \equiv \frac{\partial F}{\partial u} = kL^2\left(2\cos 2u - \cos 4u - f\cos u\right). \tag{5.66}$$

Step 3: From the equilibrium equation (5.65), we can obtain two kinds of equilibrium paths

$$\begin{cases} u = 0, & \text{fundamental path (trivial solution),} \\ f = \cos u(1+2\sin^2 u), & \text{bifurcated path.} \end{cases}$$

As shown in Fig. 5.9(b), these paths intersect at the bifurcation point B at $(u,f) = (0,1)$ and the post-bifurcation behavior is bilaterally symmetric.

The coefficients A_{20} and A_{30} at the bifurcation point $(u_c, f_c) = (0,1)$ can be obtained explicitly from (5.66) as

$$A_{20} \equiv \frac{\partial^2 F}{\partial u^2}(0,1) = kL^2(-4\sin 2u + 4\sin 4u + f\sin u)\big|_{(u,f)=(0,1)} = 0,$$

$$A_{30} \equiv \frac{\partial^3 F}{\partial u^3}(0,1) = kL^2(-8\cos 2u + 16\cos 4u + f\cos u)\big|_{(u,f)=(0,1)} = 9kL^2 > 0.$$

Accordingly, this point is a *stable–symmetric bifurcation point* (cf., Section 5.4.3).

Step 4: On the fundamental path $u = 0$, the tangent stiffness in (5.66) is evaluated to

$$J(0, f) = kL^2(1 - f).$$

The fundamental path, accordingly, is stable for $f < 1$ and unstable for $f > 1$.

On the bifurcated path $f = \cos u(1 + 2\sin^2 u)$, the tangent stiffness in (5.66) becomes

$$J(u, \cos u(1 + 2\sin^2 u)) = 3kL^2 \sin^2 u \cos 2u.$$

The equilibrium point on this bifurcated path is classified as

$$\begin{cases} 0 < |u| < \pi/4 & \Rightarrow \text{ stable ordinary point,} \\ u = \pm\pi/4 & \Rightarrow \text{ critical point (maximal point of } f\text{),} \\ \pi/4 < |u| \leq \pi/2 & \Rightarrow \text{ unstable ordinary point.} \end{cases}$$

As the load f increases from the initial state $(u, f) = (0,0)$, the system remains stable until reaching the bifurcation load $f = f_c = 1$ and the bar remains upright. At $f_c = 1$, bifurcation takes place and the bar starts to tilt leftward or rightward. Thereafter the load increases stably as the tilting proceeds until reaching the maximal load $f = \sqrt{2}$ at the points C and C′ with $u = \pm\pi/4$.

The total potential energy $U(u, 1)$ for the bifurcation load $f = f_c = 1$ is depicted in Fig. 5.9(c) (see (5.64)). This bifurcation point is stable because $U(u, 1)$ takes a local minimum at $u = u_c = 0$ for the bifurcation point.

5.6.3 ASYMMETRIC BIFURCATION

For the asymmetric spring of (5.60) that develops a bilaterally asymmetric force $F_s(x) = k(x + 3x^2/(2L))$ (Fig. 5.10(a)), we encounter an asymmetric bifurcation as explained below.

Step 1: The total potential energy U becomes

$$U = \frac{k}{2}\left[(L\sin u)^2 + \frac{(L\sin u)^3}{L}\right] - kLf \cdot L(1 - \cos u). \tag{5.67}$$

Step 2: The equilibrium equation and the tangent stiffness, respectively, become

$$F(u, f) \equiv \frac{\partial U}{\partial u} = kL^2 \sin u \left(\cos u + \frac{3}{4}\sin 2u - f\right) = 0, \tag{5.68}$$

$$J(u, f) \equiv \frac{\partial F}{\partial u} = kL^2\left[\cos 2u - \frac{3}{8}(\sin u - 3\sin 3u) - f\cos u\right]. \tag{5.69}$$

Step 3: The equilibrium equation (5.68) has two kinds of equilibrium paths (Fig. 5.10(b)):

$$\begin{cases} u=0, & \text{fundamental path (trivial path)},\\ f=\cos u+\dfrac{3}{4}\sin 2u, & \text{bifurcated path.}\end{cases}$$

As shown in Fig. 5.10(b), these paths intersect at the bifurcation point C at $(u,f)=(0,1)$ and the post-bifurcation behavior is bilaterally asymmetric.

The coefficient A_{20} at the bifurcation point $(u_c,f_c)=(0,1)$ can be obtained explicitly as

$$A_{20}\equiv kL^2\left(-2\sin 2u-\frac{3}{8}(\cos u-9\cos 3u)+f\sin u\right)\bigg|_{(u,f)=(0,1)}=3kL^2\neq 0.$$

Accordingly, this point is an *asymmetric bifurcation point* (cf., Sections 5.4.2).

Step 4: On the fundamental path $u=0$, we have $J(0,f)=kL^2(1-f)$. Accordingly, this path is stable for $f<1$ and unstable for $f>1$.

On the bifurcated path $f=\cos u+(3/4)\sin 2u$, we have

$$J\left(u,\cos u+\frac{3}{4}\sin 2u\right)=kL^2\sin u\left(\frac{3}{2}\cos 2u-\sin u\right).$$

Based on the sign of the function in the parentheses on the right-hand side of this equation, the stability of the bifurcated path can be classified as depicted in Fig. 5.10(b).

As the load f increases from the initial state $(u,f)=(0,0)$, the system remains stable until reaching the bifurcation point D at $f=f_c=1$ and the bar remains upright. The fundamental path becomes unstable above this point. On a half branch

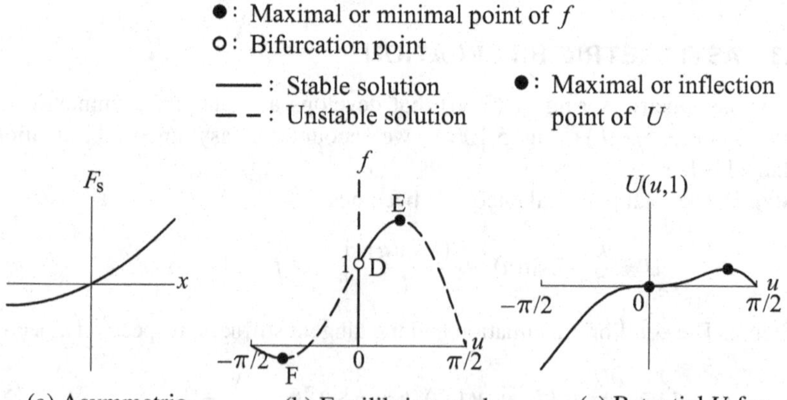

(a) Asymmetric spring property (b) Equilibrium path (c) Potential U for $f=f_c=1$

Figure 5.10 Bifurcation properties of the propped cantilever with the asymmetric spring

emanating from the bifurcation point in the direction of $u > 0$, the bar tilts rightward and the structural system is stable until reaching the maximal point E. On the other half branch ($u < 0$), the bar tilts leftward and the system is unstable until reaching the minimal point F of load. As a consequence of this bilateral asymmetry of the spring force, the bifurcation behavior is also asymmetric. At this asymmetric bifurcation point, the stability changes at the bifurcation point along the fundamental path and also along the bifurcated path. This type of bifurcation point is called an *asymmetric bifurcation point.*

As shown in in Fig. 5.10(c), the total potential energy $U(u,1)$ for the bifurcation load $f = f_c = 1$ has an inflection point at $u = u_c = 0$, corresponding to the bifurcation point. Accordingly, this bifurcation point is unstable. The maximal point E is also unstable.

5.7 PROBLEMS

Problem 5.1 Obtain the asymptotic form of the bifurcation equation (5.26) for an inflection point of f and investigate the stability.

Problem 5.2 (1) Show that the equilibrium equation

$$F(u,f) = (\sin 2u - 1 + \cos u)\left(\cos 2u + \frac{3}{4}\sin 2u - f\right) = 0$$

has an asymmetric bifurcation point. (2) Investigate the asymptotic behavior, including stability, of the solution curves in the neighborhood of this bifurcation point.

Problem 5.3 Consider the propped cantilever at the top of Fig. 5.7 in Section 5.6 and the symmetric-hardening spring in (5.60). Investigate the asymptotic behavior, including stability, in the neighborhood of the stable bifurcation point.

Problem 5.4 Consider the propped cantilever at the top of Fig. 5.7 in Section 5.6 and a symmetric-hardening spring that develops a force of $k(x + \alpha x^3/L^2)$ against displacement x. Obtain the range of α such that the bifurcation point is stable.

Problem 5.5 Consider the equilibrium equation

$$\begin{pmatrix} \left(1 - [x^2 + (1-y)^2]^{-1/2}\right) x + x \\ \left(1 - [x^2 + (1-y)^2]^{-1/2}\right)(y-1) + 2y - f \end{pmatrix} = \begin{pmatrix} 0 \\ 0 \end{pmatrix}$$

in (x,y) with a bifurcation parameter f, where $0 < y < 1$. Obtain the asymptotic form of the bifurcation equation.

Problem 5.6 Derive (5.56).

Problem 5.7 Show that the potential in (5.64) takes a local minimum at the bifurcation point.

Problem 5.8 Answer the following questions for the rigid-bar-spring system depicted in Fig. 5.11. In the two rotational springs, a bending moment of $k_\theta \theta$ develops for rotational displacement θ; we set $k_\theta = \alpha k L^2$. Employ the rotational displacement u of the rigid bars as the independent variable (both bars have the same u by symmetry). (1) Obtain the bifurcation load, the trivial solution, and the bifurcating solution for $\alpha = 1$. (2) Investigate the stability of the bifurcation point and the bifurcating solution just after bifurcation for $\alpha = 1$. (3) Find the range of α such that the bifurcation load f_c satisfies $f_c \geq 10kL$.

Problem 5.9 Derive the bifurcation equation for the two-bar truss arch in Fig. 2.5 in Section 2.7.

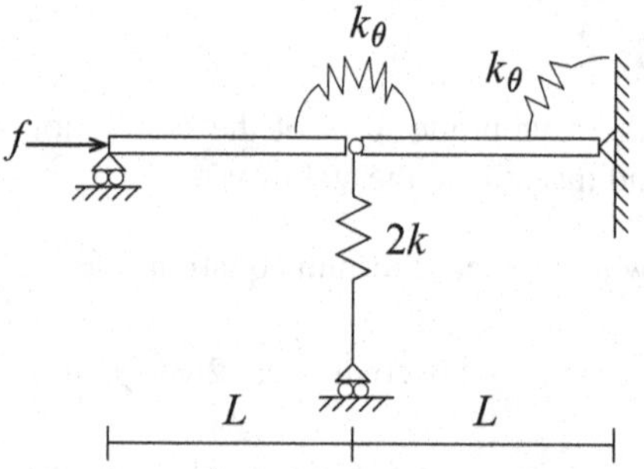

Figure 5.11 A single-degree-of-freedom rigid-bar-spring system

REFERENCES

1. Fujii H, and Yamaguti M (1980), Structure of singularities and its numerical realization in nonlinear elasticity, *J. Math. Kyoto Univ.* **20**(3), 489–590.
2. Ikeda K, and Murota K (2019), *Imperfect Bifurcation in Structures and Materials: Engineering Use of Group-Theoretic Bifurcation Theory*, 3rd ed., Springer, New York.
3. Koiter WT (1945), On the Stability of Elastic Equilibrium, *Dissertation,* Delft Univ. Tech. (English translation: NASA Tech. Trans. F **10: 833**, 1967).
4. Sattinger DH (1979), *Group Theoretic Methods in Bifurcation Theory*, Lecture Notes in Mathematics **762**, Springer, Berlin.
5. Thompson JMT, and Hunt GW (1973), *A General Theory of Elastic Stability*, Wiley, New York.

6 Bifurcation Theory II: Symmetric Structures

6.1 SUMMARY

The mechanism of bifurcation for simple critical points has been studied in Chapter 5 by investigating the vanishing and non-vanishing of the terms in the bifurcation equation of a power series form. This chapter makes clear that such vanishing and non-vanishing occur systematically as a consequence of the symmetry of a system. The relation between symmetry and bifurcation is explained with reference to group-theoretic bifurcation theory. It is useful to learn this theory in understanding the bifurcation mechanism of individual structures, such as domes and shells with regular-polygonal (dihedral-group) symmetry.

Many bifurcation phenomena are associated with the breaking of symmetry and the mathematical mechanism of these phenomena is elucidated by group-theoretic bifurcation theory.[1] Such symmetry breaking is observed for domes and shells (Yamaki [78]) and honeycomb structures (Gibson and Ashby [12]; Saiki, Ikeda, and Murota [56]), as well as in flow patterns (Koschmieder [41]). Figure 6.1 sketches a hemispherical dome in (a) and its loss of axisymmetry (rotational symmetry about the vertical axis) by bifurcation in (b).

This chapter introduces fundamental concepts of group-theoretic bifurcation theory with reference to simple structures with simple symmetries. Group is used to describe the symmetry of a structure and, in turn, to describe the symmetry of the governing equation in terms of equivariance. It is noted again that the symmetry (equivariance) of a system results in a systematic vanishing and non-vanishing of the terms in the bifurcation equation. Such a bifurcation equation gives rise to bifurcating solutions with reduced symmetries, which are labeled by subgroups of the group describing the symmetry of the given system. In particular, we study the bifurcation of a system with regular-polygonal (dihedral-group) symmetry, which has a double bifurcation point with two critical eigenvectors.

Keywords: • Bifurcation • Dihedral group • Dome • Equivariance • Group • Group-theoretic bifurcation theory • Symmetry

6.2 BIFURCATION DUE TO REFLECTION SYMMETRY

We explain how the symmetry of a system gives rise to bifurcation behavior with reference to two simple examples with reflection symmetry.

[1] For more mathematical accounts on group-theoretic bifurcation theory, see, e.g., Sattinger [57], Fujii and Yamaguti [10], Golubitsky and Schaeffer [15], Golubitsky, Stewart, and Schaeffer [17], Ikeda, Murota, and Fujii [33], and Ikeda and Murota [32].

DOI: 10.1201/9781003112365-6

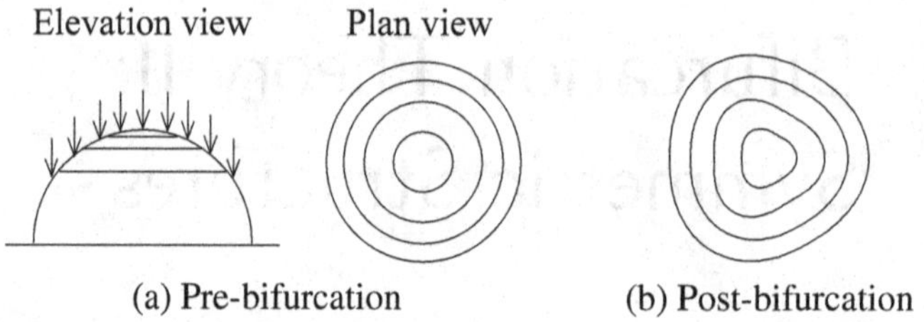

Figure 6.1 A hemispherical dome undergoing symmetry-breaking bifurcation

6.2.1 PROPPED CANTILEVER

Consider the propped cantilever in Fig. 6.2, a rigid bar of length 1 pinned to the ground and supported by a linear spring of spring constant k. It has a single displacement variable u that denotes the tilted angle of the rigid bar; the bar stands upright when $u = 0$. The total potential energy is expressed as

$$U(u,f) = \frac{1}{2}k\sin^2 u - f(1-\cos u), \tag{6.1}$$

and the equilibrium equation is given by

$$F(u,f) \equiv \frac{\partial U}{\partial u} = \sin u\,(k\cos u - f) = 0. \tag{6.2}$$

This system is *invariant* to the *reflection*

$$u \mapsto -u. \tag{6.3}$$

Indeed, the potential function (6.1) is invariant to this reflection, that is,

$$U(u,f) = U(-u,f). \tag{6.4}$$

As a consequence of this *invariance*, the equilibrium equation satisfies the condition

$$-F(u,f) = F(-u,f). \tag{6.5}$$

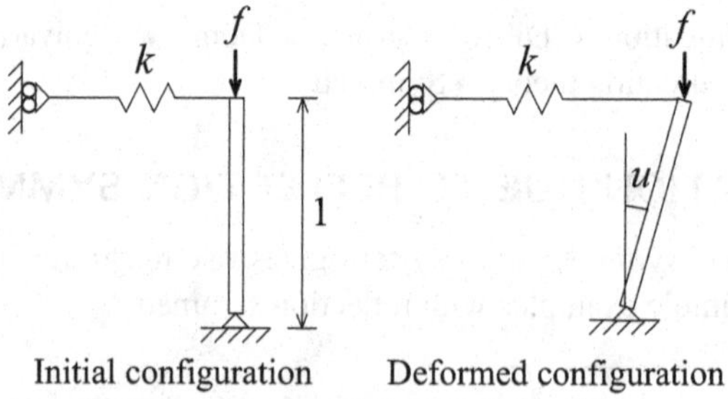

Figure 6.2 A propped cantilever

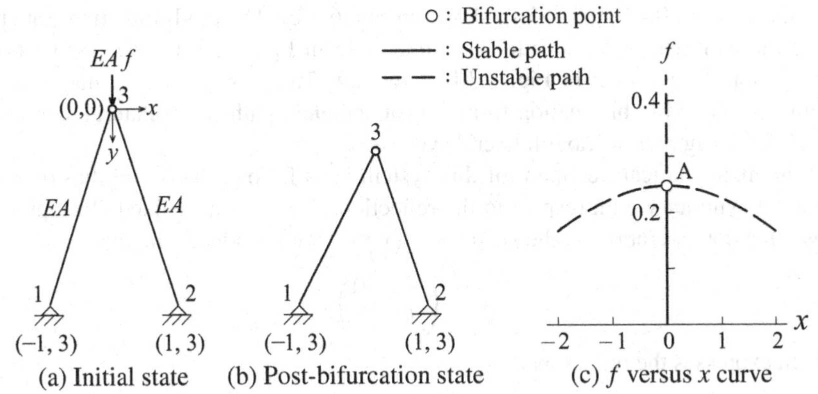

Figure 6.3 A two-bar truss arch and its bifurcation behavior

Equation (6.5) shows that $F(u,f)$ is an odd function in u. Hence we can write

$$F(u,f) = uF_1(u^2,f)$$

for some function F_1. Therefore, $u = 0$ is a trivial solution of (6.2) (i.e., $F(0,f) = 0$ for all f) and a bifurcating solution can arise from $F_1(u^2,f) = 0$. In this way, symmetry causes bifurcation.

6.2.2 TWO-BAR TRUSS ARCH WITH BILATERAL SYMMETRY

As a second example, we consider the truss arch in Fig. 6.3 that has bilateral (reflection) symmetry. Node 3 is free to move, and nodes 1 and 2 are fixed to the ground. The equilibrium equation of this arch is given by (3.16) as[2]

$$\boldsymbol{F}(\boldsymbol{u},f) = \begin{pmatrix} \sum_{m=1}^{2} EA\left(\frac{1}{L} - \frac{1}{\hat{L}^{(m)}}\right)(x - x_m) \\ \sum_{m=1}^{2} EA\left(\frac{1}{L} - \frac{1}{\hat{L}^{(m)}}\right)(y - y_m) - EAf \end{pmatrix} = \begin{pmatrix} 0 \\ 0 \end{pmatrix}, \qquad (6.6)$$

where $\boldsymbol{u} = (x,y)^\top$ denotes the displaced location of node 3, and $(x_1,y_1) = (-1,3)$ and $(x_2,y_2) = (1,3)$ denote the fixed locations of nodes 1 and 2, respectively. The two members have the same modulus of elasticity E, cross-sectional area A, and the initial length $L = \sqrt{10}$. Their member lengths after deformation are

$$\hat{L}^{(1)} = \sqrt{(x+1)^2 + (y-3)^2}, \qquad \hat{L}^{(2)} = \sqrt{(x-1)^2 + (y-3)^2}.$$

The fundamental path corresponds to the solutions with $x = 0$. On this fundamental path, we focus on the first critical point A, a simple bifurcation point at

[2] See Section 2.7 or 3.4 for details of the analysis of this truss arch.

$(x_c, y_c, f_c) = (0, 0.4473, 0.2478)$, shown in Fig. 6.3(c). The pre-bifurcation state has bilateral symmetry (similarly to the initial state in Fig. 6.3(a)), whereas the post-bifurcation state shown in Fig. 6.3(b) does not. The truss arch thus undergoes a symmetry-breaking bifurcation from the fundamental path with bilateral symmetry to a bifurcating path without bilateral symmetry.

The mathematical treatment of this system is as follows. This arch has obvious bilateral symmetry with respect to the reflection $x \mapsto -x$ with y fixed. To represent this reflection operation on the vector $\boldsymbol{u} = (x, y)^\top$, we introduce a matrix

$$T = \begin{pmatrix} -1 & 0 \\ 0 & 1 \end{pmatrix},$$

which expresses the reflection as

$$T \begin{pmatrix} x \\ y \end{pmatrix} = \begin{pmatrix} -x \\ y \end{pmatrix}.$$

The equilibrium equation $\boldsymbol{F}(\boldsymbol{u}, f)$ in (6.6) of this truss arch satisfies the condition

$$T\boldsymbol{F}(\boldsymbol{u}, f) = \boldsymbol{F}(T\boldsymbol{u}, f), \tag{6.7}$$

which follows from

$$T\boldsymbol{F}(\boldsymbol{u}, f) = T \begin{pmatrix} F_1(x, y, f) \\ F_2(x, y, f) \end{pmatrix} = \begin{pmatrix} -F_1(x, y, f) \\ F_2(x, y, f) \end{pmatrix},$$

$$\boldsymbol{F}(T\boldsymbol{u}, f) = \boldsymbol{F}(-x, y, f) = \begin{pmatrix} F_1(-x, y, f) \\ F_2(-x, y, f) \end{pmatrix},$$

in which

$$\begin{aligned} &F_1(x, y, f) \\ &= EA\left\{\left(\frac{1}{\sqrt{10}} - \frac{1}{\sqrt{(x+1)^2 + (y-3)^3}}\right)(x+1)\right. \\ &\qquad \left. + \left(\frac{1}{\sqrt{10}} - \frac{1}{\sqrt{(x-1)^2 + (y-3)^3}}\right)(x-1)\right\} \\ &= EA\left(\frac{2x}{\sqrt{10}} - \frac{x+1}{\sqrt{(x+1)^2 + (y-3)^3}} - \frac{x-1}{\sqrt{(x-1)^2 + (y-3)^3}}\right), \\ &F_2(x, y, f) \\ &= EA\left\{\left(\frac{1}{\sqrt{10}} - \frac{1}{\sqrt{(x+1)^2 + (y-3)^3}}\right)(y-3)\right. \\ &\qquad \left. + \left(\frac{1}{\sqrt{10}} - \frac{1}{\sqrt{(x-1)^2 + (y-3)^3}}\right)(y-3) - f\right\} \\ &= EA\left(\frac{2}{\sqrt{10}} - \frac{1}{\sqrt{(x+1)^2 + (y-3)^3}} - \frac{1}{\sqrt{(x-1)^2 + (y-3)^3}}\right)(y-3) - f, \end{aligned}$$

and hence

$$F_1(-x,y,f) = -F_1(x,y,f), \qquad F_2(-x,y,f) = F_2(x,y,f).$$

The condition (6.7) indicates the geometrical objectivity that the transformation $(x,y) \mapsto (-x,y)$ has an identical effect as the transformation $(F_1,F_2) \mapsto (-F_1,F_2)$.

At the simple critical point A, we can derive the bifurcation equation $\hat{F}(w,\tilde{f}) = 0$ as described in Section 5.3. Here w is a scalar variable associated with the critical eigenvector $\boldsymbol{\eta}_{\rm c}$ and $\tilde{f} = f - f_{\rm c}$. As a consequence of the symmetry condition (6.7) for the equilibrium equation, the bifurcation equation is equipped with the property

$$\hat{F}(-w,\tilde{f}) = -\hat{F}(w,\tilde{f}), \tag{6.8}$$

which shows that the function $\hat{F}$ is an odd function in w. Equation (6.8) indicates that if $(w,\tilde{f})$ is a solution to $\hat{F} = 0$, $(-w,\tilde{f})$ is also a solution, thereby indicating the existence of multiple solutions if $w \neq 0$.

By the condition (6.8) that $\hat{F}$ is an odd function in w, the power series expansion of the bifurcation equation $\hat{F} = 0$ takes the following form:

$$\hat{F}(w,\tilde{f}) \approx w(A_{10} + A_{30}w^2 + A_{11}\tilde{f} + \text{(h.o.t.)}) = 0, \tag{6.9}$$

where (h.o.t.) denotes higher order terms (of even orders in w). Since $A_{10} = 0$ holds at a critical point, the bifurcation equation (6.9) takes the same form as (5.52). Accordingly, the critical point is a symmetric bifurcation point. Then the bifurcating solution

$$\tilde{f} = -\frac{A_{30}}{A_{11}}w^2 + \text{(h.o.t.)} \tag{6.10}$$

is an even function in w (higher order terms are also even functions in w) and has bilateral symmetry.

The solution $(\boldsymbol{u}, f)$ in the original variable is determined from $(w,\tilde{f})$ as

$$\boldsymbol{u} = \boldsymbol{u}_{\rm c} + w_1\boldsymbol{\eta}_1 + w_2\boldsymbol{\eta}_2, \qquad f = f_{\rm c} + \tilde{f},$$

with $\boldsymbol{\eta}_1 = \boldsymbol{\eta}_{\rm c}$ and $(w_1,w_2) = (\varphi_1(w),\varphi_2(w))$ in the notation of (5.27) in Section 5.3. It follows from the general theory that $\boldsymbol{\varphi} = (\varphi_1,\varphi_2)$ is also an odd function, that is, $\boldsymbol{\varphi}(-w) = -\boldsymbol{\varphi}(w)$. Therefore, if $\boldsymbol{u} = \boldsymbol{u}_{\rm c} + \tilde{\boldsymbol{u}}$ is a bifurcating solution to $\boldsymbol{F}(\boldsymbol{u},f) = \boldsymbol{0}$, then $\boldsymbol{u}' = \boldsymbol{u}_{\rm c} - \tilde{\boldsymbol{u}}$ is also a bifurcating solution.

6.3 BASICS OF GROUPS

The concept of a group is introduced along with associated geometric transformations. This concept is useful to express the symmetry of a structure in mathematical terms.

6.3.1 GROUPS AND SUBGROUPS

A set G is called a *group* if, for any pair of elements g and h of G, an element of G called the *product* of g and h is specified, and if the following (i) through (iii) are satisfied, where the product of g and h is denoted as gh.

(i) The *associative law* holds:

$$(g\,h)\,k = g\,(h\,k), \qquad g,h,k \in G. \tag{6.11}$$

(ii) There exists an element $e \in G$ (called the *identity element*) such that

$$e\,g = g\,e = g, \qquad g \in G. \tag{6.12}$$

(iii) For any $g \in G$, there exists $h \in G$ (called the *inverse* of g) such that

$$g\,h = h\,g = e. \tag{6.13}$$

It is required in (ii) that there should exist a single element e for which the identity (6.12) is true for all elements g of G. It can be shown that the identity element e in (ii) is uniquely determined. The inverse of g in (iii) is unique for each g and denoted as g^{-1}.

It is often convenient to represent the product operation in the form of a table, called the *multiplication table*, in which the rows and columns are indexed by the elements of G and the product gh of g and h is given as the entry of the table with row-index g and column-index h. See Table 6.1 for concrete examples.

A nonempty subset of a group G is called a *subgroup* of G if it forms a group with respect to the same product operation defined in G. A subgroup of G is called a *proper subgroup* if it is distinct from G.

6.3.2 DIHEDRAL AND CYCLIC GROUPS

The *cyclic group* of degree n, conventionally denoted as C_n, is a group consisting of powers of a single element r with $r^n = e$. That is,

$$\mathrm{C}_n = \{e, r, r^2, \ldots, r^{n-1}\} \tag{6.14}$$

Table 6.1
Multiplication tables for groups

(a) $\mathrm{C}_3 = \{e, r, r^2\}$

	e	r	r^2
e	e	r	r^2
r	r	r^2	e
r^2	r^2	e	r

(b) $\mathrm{D}_3 = \{e, r, r^2, s, sr, sr^2\}$

	e	r	r^2	s	sr	sr^2
e	e	r	r^2	s	sr	sr^2
r	r	r^2	e	sr^2	s	sr
r^2	r^2	e	r	sr	sr^2	s
s	s	sr	sr^2	e	r	r^2
sr	sr	sr^2	s	r^2	e	r
sr^2	sr^2	s	sr	r	r^2	e

(c) $\mathrm{D}_1 = \{e, s\}$

	e	s
e	e	s
s	s	e

with the product operation given by $r^i r^j = r^{i+j}$. For $n = 1$, in particular, we have $\mathrm{C}_1 = \{e\}$. The multiplication table of C_3 is given in Table 6.1(a).

The *dihedral group* of degree $n \geq 2$ is a group with $2n$ elements given as

$$\mathrm{D}_n = \{e, r, \ldots, r^{n-1}, s, sr, \ldots, sr^{n-1}\}, \tag{6.15}$$

where r and s are assumed to satisfy identities

$$r^n = s^2 = (sr)^2 = e \tag{6.16}$$

as well as $r^i r^j = r^{i+j}$. This means, for example, the product of $g = sr$ and $h = sr^2$ is calculated as $gh = (sr)(sr^2) = (sr)^2 r = e\,r = r$. The multiplication table of D_3 is given in Table 6.1(b). By convention we use the notation D_n also for $n = 1$, with the understanding that $\mathrm{D}_1 = \{e, s\}$, for which the multiplication table is given in Table 6.1(c).

We use a bracket $\langle\cdot\rangle$ to denote the *group generated* by the listed element(s). For example, $\mathrm{D}_1 = \langle s \rangle$ and $\mathrm{D}_n = \langle r, s \rangle$ for $n \geq 2$.

Subgroups of D_n comprise dihedral and cyclic groups whose degree m divides n:

$$\mathrm{D}_m = \langle r^{n/m}, s \rangle = \{r^{in/m}, sr^{in/m} \mid i = 0, 1, \ldots, m-1\}, \tag{6.17}$$

$$\mathrm{C}_m = \langle r^{n/m} \rangle = \{r^{in/m} \mid i = 0, 1, \ldots, m-1\}, \tag{6.18}$$

where $1 \leq m \leq n$ and $\mathrm{C}_1 = \{e\}$. These subgroups express partial symmetries of the symmetry represented by D_n. Cyclic groups C_m represent rotation-symmetric patterns; the group C_1, in particular, represents a completely asymmetric pattern. Dihedral groups D_m indicate reflection symmetric patterns. See Section 6.3.3 (Fig. 6.5, in particular) for a concrete example.

Remark 6.1 To be precise, there are several subgroups of D_n that are isomorphic to D_m. Namely, for $k = 1, \ldots, n/m$,

$$\mathrm{D}_m^{k,n} = \langle r^{n/m}, sr^{k-1} \rangle \tag{6.19}$$

is also a subgroup of D_n, where $\mathrm{D}_m^{1,n} = \mathrm{D}_m$. The subgroups $\mathrm{D}_m^{k,n}$ indicate reflection symmetric patterns, where the index k distinguishes the direction of the reflection line. In this book, to simplify the description, we identify $\mathrm{D}_m^{k,n}$ with D_m without distinguishing the difference in the direction of the reflection line. □

6.3.3 TRUSS DOME WITH REGULAR-TRIANGULAR SYMMETRY

As a structural example with symmetry, we consider the regular-triangular truss dome in Fig. 6.4. This dome is subjected to the vertical (z-directional) load f at each free node, and the truss members have the same cross-sectional rigidity EA. This dome has symmetric geometrical configuration, rigidity distribution, and external force, which are invariant to the dihedral group $\mathrm{D}_3 = \{e, r, r^2, s, sr, sr^2\}$ of degree 3. That is, this dome is D_3-symmetric.

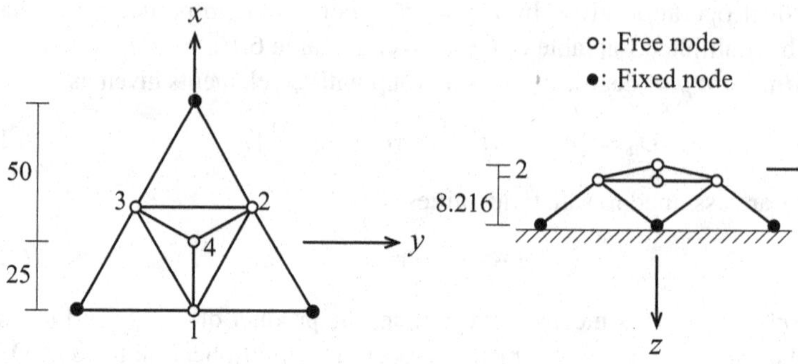

Figure 6.4 A regular-triangular truss dome

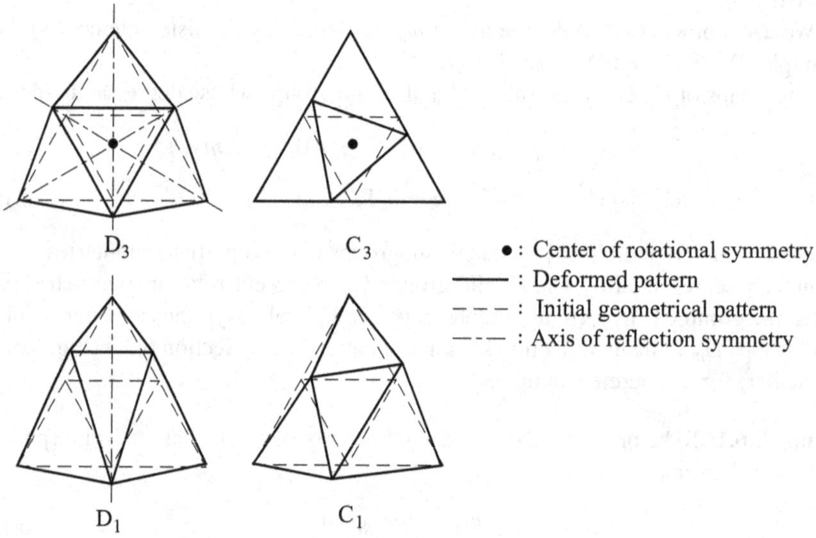

Figure 6.5 Deformation patterns of the regular-triangular free nodes of the truss dome and the associated subgroups of D_3

Deformation patterns of this truss dome are classified by subgroups of group D_3. Pre-bifurcation deformation patterns are regular-triangular and D_3-symmetric. The symmetry of bifurcating patterns is labeled by subgroups of D_3, such as

$$C_3 = \{e, r, r^2\}, \qquad D_1 = \{e, s\}, \qquad C_1 = \{e\}. \tag{6.20}$$

Here, C_3 denotes the $2\pi/3$-rotation symmetry, D_1 is the bilateral symmetry, and C_1 represents the asymmetry. Typical deformation patterns labeled by the subgroups in (6.20) are depicted in Fig. 6.5.

6.4 GROUP-THEORETIC BIFURCATION THEORY

In Section 6.2, we have seen, in terms of two simple structural examples, that bifurcation occurs as a consequence of reflection symmetry. In this section, we present an outline of the group-theoretic bifurcation theory that clarifies the mathematical mechanism underlying the bifurcation phenomena under general symmetry. We follow the formulation and notation of Ikeda and Murota [32], to which the reader is referred for details.

6.4.1 GENERAL THEORY

Following Section 5.3, we consider a nonlinear *equilibrium equation* (governing equation)

$$\boldsymbol{F}(\boldsymbol{u}, f) = \boldsymbol{0}, \tag{6.21}$$

where $\boldsymbol{F} = (F_1, \ldots, F_n)^\top$ is a vector of sufficiently smooth nonlinear functions, $\boldsymbol{u} = (u_1, \ldots, u_n)^\top$ is an n-dimensional vector to describe the state of the system, and f is a load parameter. The system is assumed to have a potential, and hence the Jacobian matrix (tangent stiffness matrix) is a symmetric matrix.

To express the symmetry of the system in the governing equation (6.21), we employ a group G comprising a set of geometrical transformations (expressing, e.g., reflections and rotations). We assume that, for each element g of G, an $n \times n$ nonsingular matrix $T(g)$ is given and that the collection of those matrices satisfies the following relations

$$T(g)T(h) = T(gh), \qquad T(e) = I, \qquad T(g^{-1}) = T(g)^{-1} \tag{6.22}$$

for all $g, h \in G$, where e denotes the identity element of G and g^{-1} is the inverse of g. The three conditions in (6.22) correspond, respectively, to the conditions (i) to (iii) defining a group in Section 6.3.1. Such collection of matrices $T(g)$ associated with the elements g of G is called a *matrix representation* of group G, and the matrices $T(g)$ are *representation matrices*. The representation matrix $T(g)$ is intended to be an expression of a geometrical transformation g on the vector $\boldsymbol{u}$ or the equation $\boldsymbol{F}$. A vector $\boldsymbol{u}$ is called *G-symmetric* if $T(g)\boldsymbol{u} = \boldsymbol{u}$ holds for all $g \in G$.

The symmetry of a system can be formulated in terms of the equilibrium equation that the condition

$$T(g)\boldsymbol{F}(\boldsymbol{u}, f) = \boldsymbol{F}(T(g)\boldsymbol{u}, f) \tag{6.23}$$

holds for all $g \in G$. If this is the case, the equation $\boldsymbol{F}(\boldsymbol{u}, f)$ is called *equivariant* to the group G. The equivariance (6.23) is a mathematical formulation of the geometrical symmetry that transforming $\boldsymbol{u}$ by $T(g)$ turns out to be identical, in effect, with transforming $\boldsymbol{F}$ by $T(g)$. We can see from (6.23) that if $\boldsymbol{u}$ is a solution to $\boldsymbol{F}(\boldsymbol{u}, f) = \boldsymbol{0}$, then $T(g)\boldsymbol{u}$ is also a solution for any $g \in G$. This implies that there are multiple solutions if $\boldsymbol{u}$ is not G-symmetric. It is emphasized that the group-equivariance of the equation is a different concept than the G-symmetry of a particular solution $\boldsymbol{u}$.

Example 6.1 Our analysis of the two-bar truss arch in Section 6.2.2 fits in the formulation above. The bilateral symmetry of the arch is described by the dihedral group of degree 1, that is, $G = \mathrm{D}_1 = \{e, s\}$, where s denotes the reflection $x \mapsto -x$. We consider a matrix representation defined by

$$T(e) = \begin{pmatrix} 1 & 0 \\ 0 & 1 \end{pmatrix}, \quad T(s) = \begin{pmatrix} -1 & 0 \\ 0 & 1 \end{pmatrix}. \tag{6.24}$$

This is indeed a matrix representation as it satisfies the relations in (6.22); for example, we have $T(g)T(h) = T(gh)$ for $g = h = s$ since $T(s)T(s) = I$ and $T(s^2) = T(e) = I$. The equivariance condition (6.23) for $g = s$ coincides with (6.7), whereas the condition (6.23) for $g = e$ is trivially satisfied since $T(e)$ is the identify matrix. □

In Section 6.2.2 we have dealt with a simple bifurcation point, but a multiple critical point often appears for a symmetric system in general. For a critical point $(\boldsymbol{u}_\mathrm{c}, f_\mathrm{c})$ with multiplicity, say, M, we can choose the eigenvalues λ_i and eigenvectors $\boldsymbol{\eta}_i$ $(i = 1, \ldots, n)$ of the Jacobian matrix $J_\mathrm{c} = J(\boldsymbol{u}_\mathrm{c}, f_\mathrm{c})$ such that

$$\begin{cases} \lambda_i = 0, & J_\mathrm{c}\boldsymbol{\eta}_i = \boldsymbol{0}, \qquad i = 1, \ldots, M, \\ \lambda_i \neq 0, & J_\mathrm{c}\boldsymbol{\eta}_i = \lambda_i \boldsymbol{\eta}_i, \quad i = M+1, \ldots, n. \end{cases}$$

Then we can derive a bifurcation equation

$$\hat{\boldsymbol{F}}(\boldsymbol{w}, \tilde{f}) = \boldsymbol{0}, \tag{6.25}$$

which is an M-dimensional equation in an M-dimensional vector $\boldsymbol{w} = (w_1, \ldots, w_M)^\top$. When the governing equation has group equivariance (6.23), the bifurcation equation inherits the equivariance in the form of

$$\hat{T}(g)\hat{\boldsymbol{F}}(\boldsymbol{w}, f) = \hat{\boldsymbol{F}}(\hat{T}(g)\boldsymbol{w}, f) \tag{6.26}$$

for all $g \in G$, where $\hat{T}$ is another matrix representation of G consisting of $M \times M$ nonsingular matrices $\hat{T}(g)$ associated with $g \in G$.

Example 6.2 For the simple bifurcation point of the two-bar truss arch (see Section 6.2.2), we have $M = 1$ and the bifurcation equation $\hat{F}(w, \tilde{f}) = 0$ in a scalar variable w. The equivariance (6.26) for $g = s$ with $\hat{T}(s) = -1$ coincides with the condition $\hat{F}(-w, \tilde{f}) = -\hat{F}(w, \tilde{f})$ in (6.8), showing that $\hat{F}$ is an odd function in w. We have seen in Section 6.2.2 that the oddness of $\hat{F}$ is the reason for the existence of a bifurcating solution. □

Just as the bilateral symmetry imposes oddness as in Example 6.2, the equivariance condition (6.26) imposes a strong restriction on the form of the function $\hat{\boldsymbol{F}}(\boldsymbol{w}, f)$. This means that the bifurcation equation of a symmetric system takes on a special form, which gives rise to multiple solutions. Furthermore, a detailed analysis of the bifurcation equation reveals the symmetry of the bifurcating solutions. It

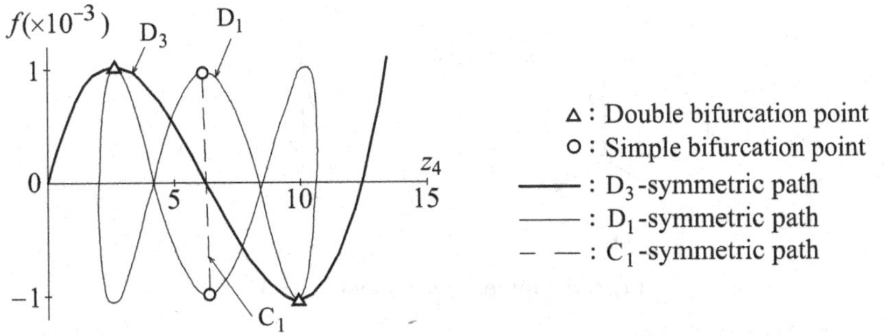

Figure 6.6 Load versus displacement curves of the regular-triangular truss dome in Fig. 6.4 (Ikeda, Murota, and Fujii [33])

is a general phenomenon that a bifurcated solution is less symmetric than the pre-bifurcation solution. In other words, the symmetry of a bifurcated solution $\boldsymbol{u}$ is represented by a subgroup of G. The subgroup representing the symmetry of a solution $\boldsymbol{u}$ is defined by $\{g \in G \mid T(g)\boldsymbol{u} = \boldsymbol{u}\}$.

Accordingly, *recursive bifurcation* (repeated occurrence of bifurcations) is characterized by a hierarchy of subgroups:

$$G \to G_1 \to G_2 \to \cdots. \tag{6.27}$$

Here, $G_i \to G_{i+1}$ means a bifurcation of G_{i+1}-symmetric solutions from G_i-symmetric solutions (G_{i+1} is a *proper subgroup* of G_i). The hierarchy in (6.27) means *bifurcation hierarchy* associated with the recursive reduction of symmetry from G-symmetry, G_1-symmetry, G_2-symmetry, and so on. It is a major achievement of group-theoretic bifurcation theory that we can make a list of possible hierarchies of subgroups valid for all systems with G-symmetry, independent of individual systems.

Example 6.3 As a structural example of the hierarchy, we recall the regular-triangular truss dome in Fig. 6.4. Figure 6.6 plots the equilibrium path of this truss dome undergoing recursive bifurcation, where $\triangle$ denotes a double bifurcation point. The bifurcation phenomenon of this dome corresponds to the bifurcation hierarchy $D_3 \to D_1 \to C_1$. □

6.4.2 BIFURCATION OF A REGULAR-TRIANGULAR SYSTEM

As another example of a D_3-symmetric structure, we consider the axisymmetric truss tent, which is rotationally symmetric about the z-axis, as shown in Fig. 6.7(a). This tent comprises three identical truss members with the same length $L = \sqrt{5}$ and the same cross-sectional rigidity EA, and is subjected to a vertical load EAf at the crown node. The displacement variables are defined by the x-, y-, and z-directional displacements of the crown node as $\boldsymbol{u} = (x, y, z)$.

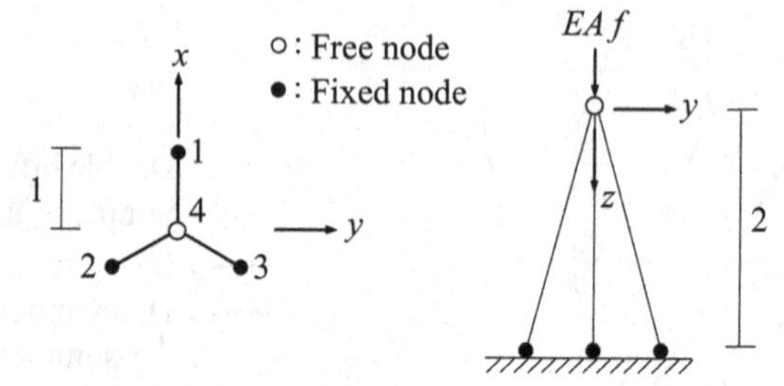

(a) Axisymmetric truss tent ($n = 3$)

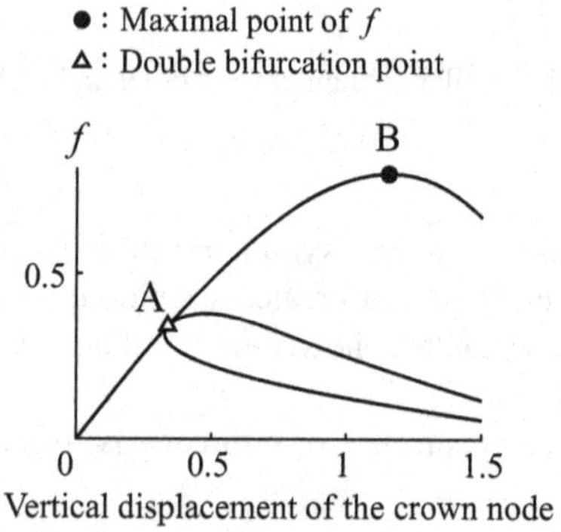

(b) Load versus displacement curves

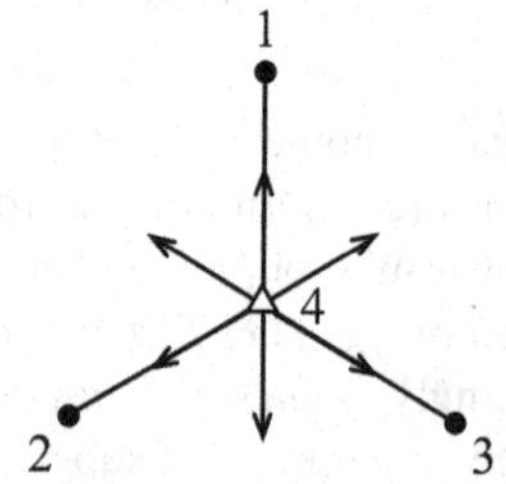

(c) Directions of bifurcating paths

Figure 6.7 An axisymmetric truss tent ($n = 3$) and the bifurcation behavior at the double bifurcation point A (Ikeda et al. [36])

The function $\boldsymbol{F}(\boldsymbol{u}, f)$ for the governing equation is given by

$$\boldsymbol{F}(\boldsymbol{u}, f) = \begin{pmatrix} F_1 \\ F_2 \\ F_3 \end{pmatrix} = \sum_{m=1}^{3} EA \left(\frac{1}{L} - \frac{1}{\hat{L}^{(m)}} \right) \begin{pmatrix} x - x_m \\ y - y_m \\ z - z_m \end{pmatrix} - \begin{pmatrix} 0 \\ 0 \\ EAf \end{pmatrix}, \tag{6.28}$$

where (x_m, y_m, z_m) $(m = 1, 2, 3)$ are the location of the fixed nodes, which are given by

$$(x_1, y_1, z_1) = (1, 0, 2),$$
$$(x_2, y_2, z_2) = (-1/2, -\sqrt{3}/2,\ 2),$$
$$(x_3, y_3, z_3) = (-1/2,\ \sqrt{3}/2,\ 2),$$

and

$$\hat{L}^{(m)} = \sqrt{(x - x_m)^2 + (y - y_m)^2 + (z - z_m)^2}, \qquad m = 1, 2, 3.$$

This tent, which is symmetric in geometrical configuration, in stiffness distribution, and in loading, remains invariant under two geometrical transformations: the counterclockwise[3] rotation r about the z-axis by an angle $2\pi/3$ and the reflection $s: y \mapsto -y$ with respect to the xz-coordinate plane. This geometric invariance is mathematically expressed as the invariance with respect to the dihedral group of degree three

$$\mathrm{D}_3 = \langle r, s \rangle = \{e,\ r,\ r^2,\ s,\ sr,\ sr^2\}$$

for the matrix representation T defined by

$$T(r) = \begin{pmatrix} \cos(2\pi/3) & -\sin(2\pi/3) & 0 \\ \sin(2\pi/3) & \cos(2\pi/3) & 0 \\ 0 & 0 & 1 \end{pmatrix} = \begin{pmatrix} a & -b & 0 \\ b & a & 0 \\ 0 & 0 & 1 \end{pmatrix},\ T(s) = \begin{pmatrix} 1 & 0 & 0 \\ 0 & -1 & 0 \\ 0 & 0 & 1 \end{pmatrix}, \tag{6.29}$$

where $a = -1/2$ and $b = \sqrt{3}/2$.

The function $\boldsymbol{F}(\boldsymbol{u}, f)$ in (6.28) is equivariant to D_3 in the sense of (6.23). With the short-hand notation $\boldsymbol{F}(\boldsymbol{u}, f) = (F_1(x,y,z), F_2(x,y,z), F_3(x,y,z))^\top$ omitting f, the equivariance condition (6.23) for $g = s$ reads:

$$F_1(x,y,z) = F_1(x,-y,z), \tag{6.30}$$
$$-F_2(x,y,z) = F_2(x,-y,z), \tag{6.31}$$
$$F_3(x,y,z) = F_3(x,-y,z), \tag{6.32}$$

whereas the condition (6.23) for $g = r$ reads:

$$aF_1(x,y,z) - bF_2(x,y,z) = F_1(ax - by, bx + ay, z), \tag{6.33}$$
$$bF_1(x,y,z) + aF_2(x,y,z) = F_2(ax - by, bx + ay, z), \tag{6.34}$$
$$F_3(x,y,z) = F_3(ax - by, bx + ay, z). \tag{6.35}$$

These relations can be verified by straightforward calculations using the concrete form of $\boldsymbol{F}(\boldsymbol{u}, f)$ in (6.28) and the representation matrices $T(r)$ and $T(s)$ in (6.29).

Since D_3 is generated by r and s, the relations (6.30)–(6.35) obtained from (6.23) for $g = r$ and s imply the equivariance (6.23) for all element g of D_3. It is emphasized that the content of the rather abstract expression (6.23) for D_3-symmetry is given by the concrete equations (6.30)–(6.35).

As the loading parameter f is increased from 0, the tent encounters a double bifurcation point A before reaching a maximal point B of f, as plotted in Fig. 6.7(b). The maximal point B is a simple critical point and the critical eigenvector for the zero eigenvalue of the Jacobian matrix is $\boldsymbol{\eta}_\mathrm{B} = (0,0,1)^\top$.

The double bifurcation point A has two critical eigenvectors $\boldsymbol{\eta}_1 = (1,0,0)^\top$ and $\boldsymbol{\eta}_2 = (0,1,0)^\top$ for the zero eigenvalue. A superposition of these two eigenvectors

$$\boldsymbol{\eta}_\mathrm{A}(\alpha) = \cos\alpha \begin{pmatrix} 1 \\ 0 \\ 0 \end{pmatrix} + \sin\alpha \begin{pmatrix} 0 \\ 1 \\ 0 \end{pmatrix} = \begin{pmatrix} \cos\alpha \\ \sin\alpha \\ 0 \end{pmatrix} \quad (0 \le \alpha < 2\pi) \tag{6.36}$$

[3] This counterclockwise rotation appears to be clockwise in Fig. 6.7 since the z-axis is directed downward.

is also a critical eigenvector, where α expresses the direction of the eigenvector.

For this double bifurcation point, we obtain a bifurcation equation $\hat{\boldsymbol{F}}(\boldsymbol{w},\tilde{f}) = \boldsymbol{0}$ in two-dimensional vector $\boldsymbol{w} = (w_1, w_2)^\top$, as in (6.25). The function $\hat{\boldsymbol{F}}(\boldsymbol{w},\tilde{f})$ is subject to the inherited equivariance (6.26) for $\hat{T}$ defined by

$$\hat{T}(r) = \begin{pmatrix} \cos(2\pi/3) & -\sin(2\pi/3) \\ \sin(2\pi/3) & \cos(2\pi/3) \end{pmatrix} = \begin{pmatrix} a & -b \\ b & a \end{pmatrix}, \quad \hat{T}(s) = \begin{pmatrix} 1 & 0 \\ 0 & -1 \end{pmatrix}.$$

That is, the function $\hat{\boldsymbol{F}}(\boldsymbol{w},\tilde{f}) = (\hat{F}_1(w_1,w_2), \hat{F}_2(w_1,w_2))^\top$ satisfies the following relations:

$$\hat{F}_1(w_1,w_2) = \hat{F}_1(w_1,-w_2), \tag{6.37}$$

$$-\hat{F}_2(w_1,w_2) = \hat{F}_2(w_1,-w_2), \tag{6.38}$$

$$a\hat{F}_1(w_1,w_2) - b\hat{F}_2(w_1,w_2) = \hat{F}_1(aw_1 - bw_2, bw_1 + aw_2), \tag{6.39}$$

$$b\hat{F}_1(w_1,w_2) + a\hat{F}_2(w_1,w_2) = \hat{F}_2(aw_1 - bw_2, bw_1 + aw_2). \tag{6.40}$$

By investigating the properties of the power series expansion of $\hat{\boldsymbol{F}}(\boldsymbol{w},\tilde{f})$ imposed by the conditions (6.37)–(6.40), we can show that the bifurcating solutions exist in the special directions of $\boldsymbol{\eta}_{\mathrm{A}}(\alpha)$ with $\alpha = \pi j/3$ for $j = 0, 1, \ldots, 5$. These directions correspond to three bifurcating paths, i.e., six half branches (two half branches connected at the bifurcation point form a bifurcating path). This is the symmetry-breaking bifurcation associated with

$$\mathrm{D}_3 \to \mathrm{D}_1.$$

The half branches for $\alpha = 0$, $2\pi/3$, and $4\pi/3$ represent the tumbling of the crown towards a member and other half branches for $\alpha = \pi$, $5\pi/3$, and $\pi/3$ represent the tumbling of the crown towards the middle of two members. The two half branches for $\alpha = 0$ and π correspond to physically different behaviors.

The above argument can be extended to a general axisymmetric truss tent described by D_n for $n \geq 3$. The expression (6.36) is valid for any n. By the group-theoretic bifurcation analysis, it can be shown that the bifurcating solutions exist in the special directions of

$$\alpha = \frac{\pi j}{n}, \qquad j = 0, 1, \ldots, 2n-1. \tag{6.41}$$

The two half branches correspond to the same physical behavior for n even and to different behaviors for n odd.

The bifurcation mechanism for the dihedral group D_n is completely clarified in the literature (Sattinger [58]; Golubitsky, Stewart, and Schaeffer [17]; Ikeda and Murota [32]). To sum up, the multiplicity of a bifurcation point is, generically, equal to one or two, and the symmetries of the bifurcating solutions are given by

$$\begin{cases} \mathrm{C}_n,\ \mathrm{D}_{n/2} = \langle r^2, s\rangle,\ \mathrm{D}'_{n/2} = \langle r^2, rs\rangle,\ \mathrm{D}_m, & n\text{: even}, \\ \mathrm{C}_n,\ \mathrm{D}_m, & n\text{: odd}. \end{cases} \tag{6.42}$$

Here m is a divisor of n satisfying $1 \leq m < n/2$. The bifurcating solutions with the symmetries C_n, $D_{n/2}$, and $D'_{n/2}$ branch at a simple bifurcation point when n is even. The bifurcating solutions with the symmetries D_m branch at a double bifurcation point.

By conducting a similar group-theoretic analysis for the bifurcating branches, we can find the symmetries of the secondary bifurcating solutions. In this manner, the bifurcation mechanism of a hierarchical bifurcation of D_n-symmetric system can be obtained.

6.4.3 TRUSS DOME WITH REGULAR-HEXAGONAL SYMMETRY

As an example of a D_6-symmetric structure, we consider the regular-hexagonal truss dome (Hangai truss dome) depicted in Fig. 6.8(a). This dome has 24 members with the same cross-sectional rigidity EA and is subjected to a proportional loading of z-directional load $0.5f$ at the crown node and z-directional load f at the free nodes surrounding the crown node. The displacements of the seven free nodes are represented by a vector $\boldsymbol{u}$ of dimension $n = 21$. The D_6-symmetry of this truss dome is expressed by the equivariance condition (6.23) using 21×21 matrices $T(g)$ defined for $g \in D_6$.

The deformation patterns of this dome are described by subgroups of D_6 as

$$
\begin{cases}
D_6: & \text{regular hexagonal symmetry,} \\
D_3: & \text{three-axis symmetry,} \\
D_2: & \text{two-axis symmetry,} \\
D_1: & \text{one-axis symmetry,} \\
C_6: & \pi/3\text{-rotation symmetry,} \\
C_3: & 2\pi/3\text{-rotation symmetry,} \\
C_2: & \pi\text{-rotation symmetry,} \\
C_1: & \text{asymmetry.}
\end{cases}
$$

Figure 6.9 depicts some of these patterns.

The rule[4] of the hierarchical bifurcation of the D_6-symmetric system is depicted in Fig. 6.10. Here, the solid arrow is related to a simple bifurcation point and the dashed arrow to a double bifurcation point. This figure, for example, shows a possible occurrence of a hierarchical bifurcation associated with the chain of subgroups

$$D_6 \to C_6 \to C_3 \to C_1.$$

It also shows that the symmetry breaking $D_6 \to C_3$ or $D_6 \to C_2$ cannot occur in a direct bifurcation.

Load versus displacement curves of this truss dome are shown in Fig. 6.8(b). The fundamental path has 3 bifurcation points A, B, and C. Point A is a simple bifurcation

[4]In Fig. 6.10, the bifurcation $C_n \dashrightarrow C_m$ at a double bifurcation point (m divides n and $n/m > 2$) exists only for a reciprocal system with a potential (Ikeda, Murota, and Fujii [33]; Ikeda and Murota [32]).

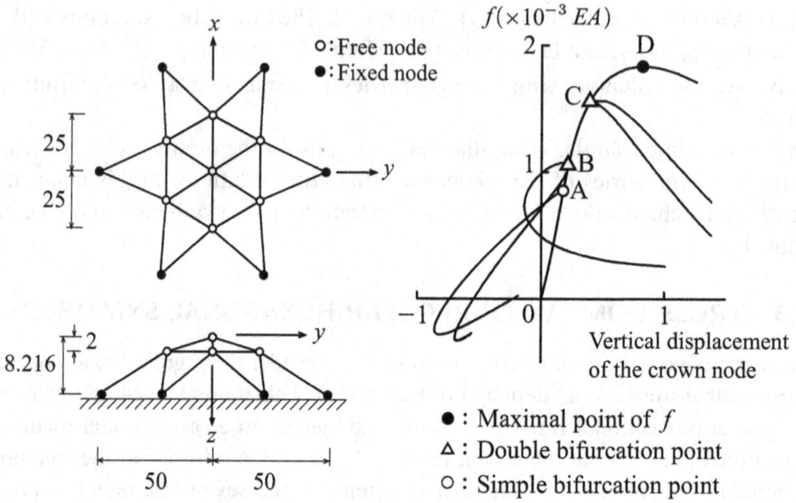

(a) Regular-hexagonal truss dome (b) Load versus displacement curves

Figure 6.8 A regular-hexagonal truss dome and its load versus displacement curves (Ikeda and Murota [32])

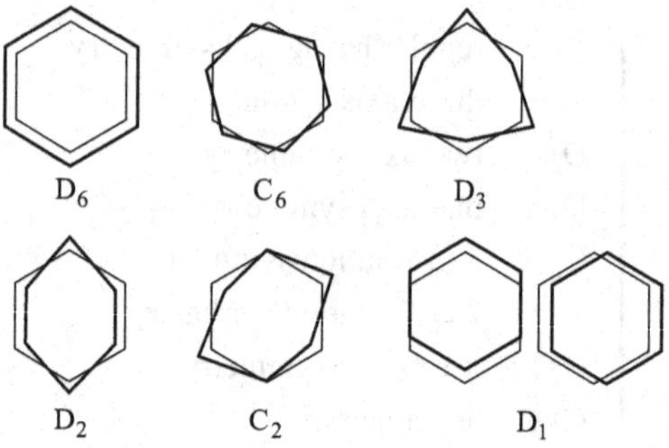

Figure 6.9 Deformation patterns of the regular-hexagonal free nodes of the hexagonal truss dome and the associated subgroups of D_6

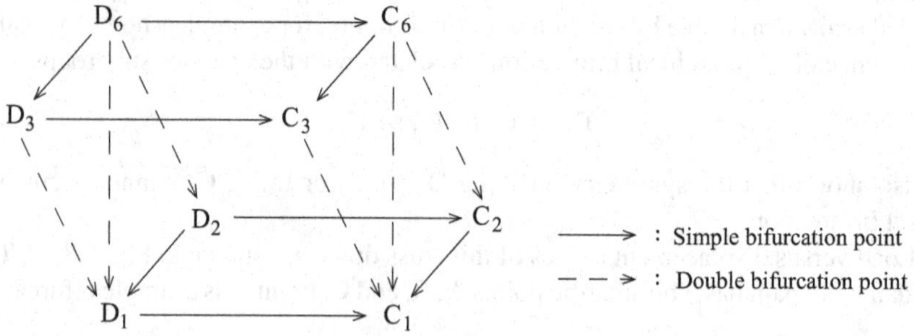

Figure 6.10 Rule of hierarchical bifurcation of a D_6-symmetric reciprocal system

point with a bifurcating path (two half branches). Points B and C are double bifurcation points with 3 and 6 bifurcating paths, respectively. These bifurcating paths are overlapping in this figure and look like two half branches for each point. Point D is the maximal point of f. The symmetry breaking at these bifurcation points is

$$\begin{cases} \mathrm{A}: \mathrm{D}_6 \to \mathrm{D}_3, \\ \mathrm{B}: \mathrm{D}_6 \to \mathrm{D}_2, \\ \mathrm{C}: \mathrm{D}_6 \to \mathrm{D}_1, \end{cases}$$

and is consistent with the bifurcation rule in Fig. 6.10.

6.5 STUDIES OF SYMMETRY AND BIFURCATION OF STRUCTURES

The mechanism of bifurcation of symmetric structures has been studied extensively in engineering and nonlinear mathematics. History of this study is reviewed in this section.

6.5.1 SYMMETRY IN STRUCTURAL MECHANICS

The underlying role of symmetry in structures undergoing bifurcation was gradually understood through the application of the asymptotic approach to individual structural systems. The symmetry systematically annihilates certain terms of the potential function to create coincidental and compound bifurcations, at which complex but interesting phenomena emerge. Studies of the mechanism of bifurcation of symmetric structures are reviewed below.

As the simplest symmetry, the reflection symmetry for two independent variables was imposed on the Taylor expansion of the potential function and was also on a twofold coincidental bifurcation point (Chilver [5]; Supple [63]; Mandadi and Huseyin [44]). Higher symmetry leads to complex bifurcation behavior. Complete spherical shells displayed rotationally-symmetric branching (Thompson [67]; Lange and Kriegsmann [42]). Diamond and other characteristic patterns were observed in cylindrical shells (Yamaki [78]; Yoshimura [79]; Esslinger and Geier [9]). Axisymmetric and regular-polygonal shells and domes have double bifurcation points, at which complex but interesting phenomena take place.

As a famous prototype model with symmetry, the Augusti model [2] was studied by the parametric sweep of a design parameter to generate secondary branching at near-coincidence and direct branching at strict-coincidence (Thompson and Hunt [71, 73]). At the coincidence, the model is endowed with three-axis symmetry, which is higher than two-axis symmetry at near-coincidence. Complex secondary bifurcations, often forming loops, were found in association with compound branching (Chilver [5]; Supple [63]; Thompson and Supple [74]), and the relation with symmetries was studied under the names of semi-symmetry (Thompson and Hunt [72]; Hunt [25, 26]) and hidden symmetry (Hunt [27]; Golubitsky, Marsden, and Shaeffer [14]; Hunt, Williams, and Cowell [28]).

Patterns appear ubiquitously for materials. The echelon mode was found in various materials: soils (Ikeda, Murota, and Nakano [34]; Ikeda and Murota [31]), rocks (Davis [8]), and metals (Poirier [54]; Bai and Dodd [3]). Periodic shear bands of materials forming an echelon mode were simulated (Petryk and Thermann [53]). The cross-checker pattern is found in metals (Voskamp and Hoolox [76]), and the zebra pattern is observed on the ocean floors (Nicolas [46]). A self-similar pattern model was introduced (Archambault et al. [1]). The instability of a shell surface was employed to produce intriguing patterns of plants, such as sunflower and snapdragon (Karam and Gibson [37, 38]; Green, Steele, and Rennich [19]; Green [18]; Steele [61]).

Honeycomb structures, which have regular-hexagonal and in-plane translational symmetries subjected to compression, were found to exhibit a series of characteristic deformation patterns during experiments.[5] A flower mode was observed experimentally (Papka and Kyriakides [51]) and simulated successfully by finite-element bifurcation and buckling analyses.[6] A gigantic honeycomb-like pattern[7] was found as a consequence of a secondary branching of an elasto-plastic honeycomb structure (Okumura, Ohno, and Noguchi [49]).

6.5.2 DEVELOPMENT OF CATASTROPHE THEORY

Critical points were classified by investigating the linear, quadratic, cubic, quartic, ... terms of the potential function (Thompson [66], Sewell [60], and Thompson and Hunt [71]). In particular, the normal forms of coincidental critical points were determined to show the diversity and complexity of these points, thereby overshadowing the systematics of the static perturbation method. The emergence of catastrophe theory[8] was quite timely and was quickly introduced into elastic stability theory to generalize the classification of critical points (Thompson and Hunt [72]).

The seven elementary catastrophes were correlated with structural problems. The fold catastrophe corresponds to a limit point or an asymmetric point of bifurcation, and the cusp catastrophe corresponds to a symmetric point of bifurcation (Thompson and Hunt [72]). Structural examples of the swallowtail and butterfly cuspoids (Hui and Hansen [23]), hyperbolic umbilic (Thompson and Hunt [72]; Thompson [69]), hyperbolic umbilic and elliptic umbilic (Thompson and Hunt [72]; Huseyin and Mandadi [29]; Hansen [21]), and parabolic umbilic catastrophes (Hui

[5]For the experiments of the honeycomb structures, see, e.g., Gibson and Ashby [12], Gibson et al. [13], Papka and Kyriakides [50], Triantafyllidis and Schraad [75], and Zhu and Mills [82].

[6]For the bifurcation and buckling analyses of the flower mode, see, e.g., Chung and Waas [6, 7], Guo and Gibson [20], Papka and Kyriakides [52], Ohno, Okumura, and Noguchi [47], Okumura, Ohno, and Noguchi [48], and Saiki, Ikeda, and Murota [56].

[7]Large circles are formed as an assemblage of a number of deformed hexagonal cells and such circles are regularly arranged in space to form this pattern.

[8]See, e.g., Chillingworth [4], Thom [65], Wassermann [80], Zeeman [81], Poston and Stewart [55], and Thompson [69].

and Hansen [24]) were found and investigated in detail. The set of umbilic catastrophes was classified (Thompson and Gaspar [70]) with reference to Zeeman's umbilic bracelet [81].

For double-cusp catastrophe, all cubic terms of the potential are annihilated by the symmetry in the combination of the two modes competing at the coincidental point (Thompson and Hunt [73]). Higher-order singularities, such as double-cusp catastrophe, falling beyond the seven elementary catastrophes, were found for structures (Poston and Stewart [55]).

6.5.3 MATHEMATICS ON SYMMETRY AND BIFURCATION

Mathematical treatment of symmetry in science can be found in introductory books (Stewart and Golubitsky [62]; Golubitsky and Stewart [16]). Group is an established means to describe (geometrical) symmetry, and its theoretical backgrounds are readily available (Serre [59]). The symmetry of molecules and crystals was studied in chemical crystallography by means of the point groups, which describe the spatial symmetry around a point (Kettle [39]; Ludwig and Falter [43]; Kim [40]).

Among these groups, particular attention was paid to the dihedral group D_n (for some integer n), which represents a regular-polygonal symmetry. The symmetry of the structures can indeed be labeled by groups as follows:

- The Augusti model: D_1-symmetry at near-coincidence and D_3-symmetry at strict-coincidence.
- Shells and domes of revolution and perfectly spherical shells: D_n-symmetry (n large) in the circumferential direction.

Bifurcation theory in nonlinear mathematics played a vital role in the classification and investigation of multiple bifurcation points of structures. Perfect and imperfect bifurcation behavior drew keen mathematical interest and blossomed into the concept of universal unfolding (Golubitsky and Schaeffer [15]) to describe general forms of imperfect systems at the presence of general initial imperfections.

Group-theoretic bifurcation theory was developed to describe the mechanism of pattern formation under symmetry. The loss of symmetry at the onset of bifurcation can be investigated theoretically by local bifurcation analysis using the Liapunov–Schmidt–Koiter reduction and the exploitation of the symmetry of the bifurcation equation (Sattinger [57], Golubitsky and Schaeffer [15], Golubitsky, Stewart, and Schaeffer [17], Ikeda and Murota [32]). The rule of such bifurcation can be determined by the symmetry of the system under consideration, and possible critical points and bifurcated solutions can be classified systematically.

The direct, secondary, tertiary, ... bifurcations of structures with axisymmetric and regular-polygonal symmetries, such as reticulated domes and cylindrical shells, were studied as a bifurcation problem of dihedral-group symmetry (Ikeda, Murota, and Fujii [33]; Ikeda and Murota [32]; Wohlever and Healey [77]; Healey [22]; Gatermann and Werner [11]). The emergence of the secondary branching, as was observed for the Augusti model, can be ascribed to the presence of an initial imperfection with partial symmetry (Ikeda and Murota [32]).

Hidden periodic symmetry was investigated by employing periodic boundary conditions. The bifurcation hierarchy of a rectangular plate was investigated by Ikeda and Nakazawa [35]. Bifurcation mechanism underlying echelon mode formation was made clear (Ikeda, Murota, and Nakano [34]; Ikeda et al. [30]; Murota, Ikeda, and Terada [45]). As a model for geometrical patterns of joints and folds, the bifurcation mechanism for pattern formation in three-dimensional uniform materials was studied (Tanaka, Saiki, and Ikeda [64]). Group-theoretic study of honeycomb patterns was conducted (Saiki, Ikeda, and Murota [56]).

6.6 PROBLEMS

Problem 6.1 Show the multiplication table for group D_4.

Problem 6.2 Obtain the rule of hierarchical bifurcation of a D_4-symmetric reciprocal system similar to the one in Fig. 6.10 for D_6.

REFERENCES

1. Archambault G, Rouleau A, Daigneault R, and Flamand R (1993), Progressive failure of rock masses by a self-similar anastomosing process of rupture at all scales and its scale effect on their shear strength, *Scale Effects in Rock Masses 93*, Lisbon, AP da Cunha (ed), Balkema, Rotterdam, 133–141.
2. Augusti G (1964), Stabilita di strutture elastische elementari in presenza di grandi spostamenri, *Atti. Accad. Sci. Fis. Mat.*, Napoli, Serie 3ª **4**(5).
3. Bai Y, and Dodd B (1992), *Adiabatic Shear Localization: Occurrence, Theories and Applications*, Pergamon, Oxford.
4. Chillingworth D (1975), The catastrophe of a buckling beam, *Dyna. Systems*, Warwick, 1974, A Manning (ed), Lecture Notes Math. **468**, Springer, Berlin.
5. Chilver AH (1967), Coupled modes of elastic buckling, *J. Mech. Phys. Solids* **15**(1), 15–28.
6. Chung J, and Waas AM (1999), Compressive response and failure of circular cell polycarbonate honeycombs under inplane uniaxial stresses, *ASME J. Eng. Mater. Tech.* **121**, 494–502.
7. Chung J, and Waas AM (2001), In-plane biaxial crush response of polycarbonate honeycombs, *J. Eng. Mech.* **127**, 180–193.
8. Davis GH (1984), *Structural Geology of Rocks & Regions*, Wiley, Singapore.
9. Esslinger M, and Geier B (1976), Calculated postbuckling loads as lower limits for the buckling loads of thin-walled circular cylinders, *Buckling of Struct., Proc. IUTAM Symp. on Buckling of Struct.*, Harvard University, Cambridge MA, B Budiansky (ed), Springer, Berlin, 274–290.
10. Fujii H, and Yamaguti M (1980), Structure of singularities and its numerical realization in nonlinear elasticity, *J. Math. Kyoto Univ.* **20**(3), 489–590.
11. Gatermann K, and Werner B (1994), Group theoretical mode interactions with different symmetries, *Int. J. Bifurcation Chaos* **4**(1), 177–191.
12. Gibson LJ, and Ashby MF (1997), *Cellular Solids: Structure and Properties,* 2nd ed., Cambridge University Press, Cambridge.
13. Gibson LJ, Ashby MF, Zhang J, and Triantafillou TC (1989), Failure surfaces for cellular materials under multiaxial loads: I. Modeling, *Int. J. Mech. Sci.* **31**, 635–663.

14. Golubitsky M, Marsden J, and Schaeffer D (1984), Bifurcation problems with hidden symmetries, *Partial Differential Equations and Dynamical Systems*, W Fitzgibbon (ed), Pitman, London, 181–210.
15. Golubitsky M, and Schaeffer DG (1985), *Singularities and Groups in Bifurcation Theory*, Vol. 1, Springer, New York.
16. Golubitsky M, and Stewart I (2002), *The Symmetry Perspective*, Birkhäuser-Verlag, Basel.
17. Golubitsky M, Stewart I, and Schaeffer DG (1988), *Singularities and Groups in Bifurcation Theory*, Vol. 2, Springer, New York.
18. Green PB (1999), Expression of pattern in plants: Combining molecular and calculus-based biophysical paradigms, *Am. J. Botany* **86**(8), 1059–1076.
19. Green PB, Steele CR, and Rennich SC (1996), Phyllotactic patterns: A biophysical mechanism for their origin, *Ann. Bot.* (London) **77**, 515–527.
20. Guo XE, and Gibson LJ (1999), Behavior of intact and damaged honeycomb: A finite element study, *Int. J. Mech. Sci.* **41**(1), 85–105.
21. Hansen JS (1977), Some two-mode buckling problems and their relations to catastrophe theory, *AIAA J.* **15**(1), 1638–1644.
22. Healey TJ (1988), A group theoretic approach to computational bifurcation problems with symmetry, *Comput. Methods Appl. Mech. Eng.* **67**(3), 257–295.
23. Hui D, and Hansen JS (1980), The swallowtail and butterfly cuspoids and their application in the initial post-buckling of single-mode structural systems, *Quart. Appl. Math.* **38**(1), 17–36.
24. Hui D, and Hansen JS (1980), Two-mode buckling of an elastically supported plate and its relation to catastrophe theory, *ASME J. Appl. Mech.* **47**(4), 607–612.
25. Hunt GW (1977), Imperfection-sensitivity of semi-symmetric branching, *Proc. Roy. Soc. London, Ser. A* **357**, 193–211.
26. Hunt GW (1981), An algorithm for the nonlinear analysis of compound bifurcation, *Phil. Trans. Roy. Soc. London, Ser. A* **300**, 443–471.
27. Hunt GW (1986), Hidden (a)symmetries of elastic and plastic bifurcation, *Appl. Mech. Rev.* **39**(8), 1165–1186.
28. Hunt GW, Williams KAJ, and Cowell RG (1986), Hidden symmetry concepts in the elastic buckling of axially-loaded cylinders, *Int. J. Solids Struct.* **22**(12), 1501–1515.
29. Huseyin K, and Mandadi V (1977), On the imperfection sensitivity of compound branching, *Ing.-Arch.* **46**, 213–222.
30. Ikeda K, Murakami S, Saiki I, Sano I, and Oguma N (2001), Image simulation of uniform materials subjected to recursive bifurcation, *Int. J. Eng. Sci.* **39**(17), 1963–1999.
31. Ikeda K, and Murota K (1997), Recursive bifurcation as sources of complexity in soil shearing behavior, *Soils & Foundations* **37**(3), 17–29.
32. Ikeda K, and Murota K (2019), *Imperfect Bifurcation in Structures and Materials: Engineering Use of Group-Theoretic Bifurcation Theory*, 3rd ed., Springer, New York.
33. Ikeda K, Murota K, and Fujii H (1991), Bifurcation hierarchy of symmetric structures, *Int. J. Solids Struct.* **27**(12), 1551–1573.
34. Ikeda K, Murota K, and Nakano M (1994), Echelon modes in uniform materials, *Int. J. Solids Struct.* **31**(19), 2709–2733.
35. Ikeda K, and Nakazawa M (1998), Bifurcation hierarchy of a rectangular plate, *Int. J. Solids Struct.* **35**(7-8), 593–617.
36. Ikeda K, Nishino F, Hartono W, and Torii K (1988), Bifurcation behavior of an axisymmetric elastic space truss, *Proc. Japan Soc. Civi Eng.* **392/I-9**, 231–234 (209s–212s).

37. Karam GN, and Gibson LJ (1995), Elastic buckling of cylindrical shells with elastic cores: I. Analysis, *Int. J. Solids Struct.* **32**(8-9), 1259–1283.
38. Karam GN, and Gibson LJ (1995), Elastic buckling of cylindrical shells with elastic cores: II. Experiments, *Int. J. Solids Struct.* **32**(8-9), 1285–1306.
39. Kettle SFA (1995), *Symmetry and Structure*, 2nd ed., Wiley, Chichester.
40. Kim SK (1999), *Group Theoretical Methods and Applications to Molecules and Crystals*, Cambridge University Press, Cambridge.
41. Koschmieder EL (1993), *Bénard Cells and Taylor Vortices*, Cambridge University Press, Cambridge.
42. Lange CG, and Kriegsmann GA (1981), The axisymmetric branching behavior of complete spherical shells, *Quart. Appl. Math.* **39**(2), 145–178.
43. Ludwig W, and Falter C (1996), *Symmetries in Physics: Group Theory Applied to Physical Problems*, 2nd ed., Springer, Berlin.
44. Mandadi V, and Huseyin K (1978), The effect of symmetry on the imperfection-sensitivity of coincident critical points, *Ing.-Arch.* **47**(1), 35–45.
45. Murota K, Ikeda K, and Terada K (1999), Bifurcation mechanism underlying echelon mode formation, *Comp. Methods Appl. Mech. Eng.* **170**(3–4), 423–448.
46. Nicolas A (1995), *The Mid-Oceanic Ridges: Mountains Below Sea Level*, Springer, Berlin.
47. Ohno N, Okumura D, and Noguchi H (2002), Microscopic symmetric bifurcation condition of cellular solids based on a homogenization theory of finite deformation, *J. Mech. Phys. Solids* **50**(5), 1125–1153.
48. Okumura D, Ohno N, and Noguchi H (2002), Post-buckling analysis of elastic honeycombs subject to in-plane biaxial compression, *Int. J. Solids Struct.* **39**(13-14), 3487–3503.
49. Okumura D, Ohno N, and Noguchi H (2004), Elastoplastic microscopic bifurcation and post-bifurcation behavior of periodic cellular solids, *J. Mech. Phys. Solids* **52**(3), 641–666.
50. Papka SD, and Kyriakides S (1994), In-plane compressive response and crushing of honeycomb, *J. Mech. Phys. Solids* **42**(10), 1499–1532.
51. Papka SD, and Kyriakides S (1999), Biaxial crushing of honeycombs—Part I: Experiments, *Int. J. Solids Struct.* **36**(29), 4367–4396.
52. Papka SD, and Kyriakides S (1999), Biaxial crushing of honeycombs—Part II: Analysis, *Int. J. Solids Struct.* **36**(29), 4397–4423.
53. Petryk H, and Thermann K (2002), Post-critical plastic deformation in incrementally nonlinear materials, *J. Mech. Phys. Solids* **50**(5), 925–954.
54. Poirier J-P (1985), *Creep of Crystals: High-Temperature Deformation Processes in Metals, Ceramics and Minerals*, Cambridge Earth Sci. Ser. **4**, Cambridge University Press, Cambridge.
55. Poston T, and Stewart I (1978), *Catastrophe Theory and Its Applications*, Survey and Ref. Works in Math., Pitman, London.
56. Saiki I, Ikeda K, and Murota K (2005), Flower patterns appearing on a honeycomb structure and their bifurcation mechanism, *Int. J. Bifurcation Chaos* **15**(2), 497–515.

57. Sattinger DH (1979), *Group Theoretic Methods in Bifurcation Theory*, Lecture Notes in Mathematics **762**, Springer, Berlin.
58. Sattinger DH (1980), Bifurcation and symmetry breaking in applied mathematics, *Bull. Amer. Math. Soc.* **3**(2), 779–819.
59. Serre J-P (1977), *Linear Representations of Finite Groups*, Graduate Texts in Math. **42**, Springer, New York.
60. Sewell MJ (1968), A general theory of equilibrium paths through critical points: Part I, *Proc. Roy. Soc. London, Ser. A* **306**, 201–223; Part II, *Proc. Roy. Soc. London, Ser. A* **306**, 225–238.
61. Steele CR (2000), Shell stability related to pattern formation in plants, *ASME J. Appl. Mech.* **67**(2), 237–247.
62. Stewart I, and Golubitsky M (1992), *Fearful Symmetry: Is God a Geometer?* Blackwell, Oxford.
63. Supple WJ (1967), Coupled branching configurations in the elastic buckling of symmetric structural system. *Int. J. Mech. Sci.* **9**(2), 97–112.
64. Tanaka R, Saiki I, and Ikeda K (2002), Group-theoretic bifurcation mechanism for pattern formation in three-dimensional uniform materials, *Int. J. Bifurcation Chaos* **12**(12), 2767–2797.
65. Thom R (1975), *Structural Stability and Morohogenesis* (English translation: DH Fowler), Benjamin, Reading MA.
66. Thompson JMT (1963), Basic principles in the general theory of elastic stability. *J. Mech. Phys. Solids* **11**(1), 13–20.
67. Thompson JMT (1964), The rotationally-symmetric branching behaviour of a complete spherical shell, *Proc. Kon. Ned. Akad. Wet., Ser. B* **67**, 295.
68. Thompson JMT (1975), Experiments in catastrophe, *Nature* **254**, 392–395.
69. Thompson JMT (1982), *Instabilities and Catastrophes in Science and Engineering*, Wiley, Chichester.
70. Thompson JMT, and Gaspar Z (1977), A buckling model for the set of umbilic catastrophes, *Math. Proc. Camb. Phil. Soc.* **82**, 497–507.
71. Thompson JMT, and Hunt GW (1973), *A General Theory of Elastic Stability*. Wiley, New York.
72. Thompson JMT, and Hunt GW (1975), Towards a unified bifurcation theory. *J. Appl. Math. Phys. (ZAMP)* **26**, 581–603.
73. Thompson JMT, and Hunt GW (1984), *Elastic Instability Phenomena*, Wiley, Chichester.
74. Thompson JMT, and Supple WJ (1973), Erosion of optimum designs by compound branching phenomena, *J. Mech. Phys. Solids* **21**(3), 135–144.
75. Triantafyllidis N, and Schraad MW (1998), Onset of failure in aluminum honeycombs under general in-plane loading, *J. Mech. Phys. Solids* **46**(6), 1089–1124.
76. Voskamp AP, and Hoolox GE (1998), Failsafe rating of ball bearing components: Effect of steel manufacturing processes on the quality of bearing steels, *ASTM STP* **987**, 102–112.
77. Wohlever JC, and Healey TJ (1995), A group theoretic approach to the global bifurcation analysis of an axially compressed cylindrical shell, *Comput. Methods Appl. Mech. Eng.* **122**(3-4), 315–349.
78. Yamaki N (1984), *Elastic Stability of Circular Cylindrical Shells*, Elsevier, Amsterdam.
79. Yoshimura Y (1951), On the mechanism of buckling of a cylindrical shell under axial compression, *Rep. Inst. Sci. Tech. University of Tokyo*, Tokyo, **5**(5) (in Japanese)

(English translation: Tech. Mem. **1390**, Nat. Advisory Committee on Aeron., Washington DC, 1955).

80. Wassermann G (1976), (r,s)-stable unfoldings and catastrophe theory, *Structural Stability, the Theory of Catastrophes, and Applications in the Sciences*, P Hilton (ed), Lecture Notes Math., Springer, Berlin, **525**, 253–262.
81. Zeeman EC (1976), The umbilic bracelet and the double-cusp catastrophe, *Structural Stability, the Theory of Catastrophes, and Applications in the Sciences*, P Hilton (ed), Lecture Notes Math., Springer Berlin, **525**, 328–366.
82. Zhu HX, and Mills NJ (2000), The in-plane non-linear compression of regular honeycombs, *Int. J. Solids Struct.* **37**(13), 1931–1949.

Part II

Buckling of Structures

We have studied a theoretical issue of buckling and bifurcation of a structure in Part I. In Part II, we march on to deal with an engineering issue of buckling analysis of a structure that obtains the buckling load and the buckling mode. We study in this part the buckling of beam-column members in Chapter 7 and the buckling of truss and frame structures in Chapters 8 and 9, respectively.

Bifurcation of a structure is a mathematical concept for an idealized perfect state of a structure. An actual structure inevitably has some deviations from the nominal values and, accordingly, does not undergo bifurcation in the strict mathematical sense. Such deviations are called initial imperfections. The buckling of a structural system with initial imperfections is studied in Chapter 10 as an advanced topic. Buckling and bifurcation theory in structural mechanics began with individual studies of particular structures, such as columns, beams, trusses, frames, and shells. The review of the history of this theory is presented in Chapter 11.

7 Member Buckling of Columns and Beams

7.1 SUMMARY

This chapter investigates the flexural buckling of the beam and column members. By solving the eigenvalue problem of the governing differential equation of a beam-column member, we obtain its buckling load and buckling mode. The influence of the boundary condition on the buckling load is investigated. While we have so far focused on the theoretical and mechanical aspects of bifurcation and buckling of structures, this chapter deals with various kinds of engineering aspects, such as cross-section design of members, initial deflection, and inelastic buckling.

Leonhard Euler carried out an analysis of a column subjected to an axial compression load and obtained the buckling load, called the *Euler buckling load*. Thereafter buckling of a beam-column member has been investigated extensively (e.g., Timoshenko and Gere [4]; Bažant and Cedolin [1]).

In this chapter, we consider the differential equation of columns and beams while we deal with a nonlinear equation describing a discrete system with finitely many unknown variables (or a finite-dimensional vector) in the rest of this book.

The differential equation of a beam-column member is introduced in Section 7.2. The solutions to this equation are obtained for different external loads in Sections 7.3–7.5. The buckling of a beam-column with an initial deflection is studied in Section 7.6 and inelastic buckling is investigated in Section 7.7. Basic facts about linear differential equations are given in Section 7.8.

Keywords: • Beam • Bifurcation • Buckling • Column • Euler buckling • Flexural buckling • Inelastic buckling • Linear differential equation

7.2 BEAM-COLUMN EQUATION

We derive the differential equation of a beam-column member[1] using the principle of stationary potential energy. As shown in Fig. 7.1, this member of a length L resides on an elastic foundation with a spring constant $k(x)$ $(0 \leq x \leq L)$ and is subjected to a distributed load $q(x)$ and an axial compression force $P > 0$.

We shall express the total potential energy U of this beam-column as a function in the vertical displacement $w(x)$ under the assumption that the axial deformation is relatively small.

[1] In this chapter, the terminology of a beam-column member expresses either a beam member or a column member, both of which have the same mechanical properties.

DOI: 10.1201/9781003112365-7

7.2.1 TOTAL POTENTIAL ENERGY

The total potential energy U of a beam-column member subjected to the axial force consists of several terms representing different kinds of energy.

There are two kinds of internal potential energy. The one is the energy due to bending of the beam-column given by

$$\int \frac{1}{2} EIw''(x)^2 \mathrm{d}x, \tag{7.1}$$

where w'' is the second-order derivative of w, E denotes the modulus of elasticity, and I the moment of inertia of a cross-sectional area of the member; both E and I are assumed to be constant along the x-axis throughout this chapter. The other internal energy is the energy stored in the elastic foundation, which is given by

$$\int \frac{1}{2} k(x) w(x)^2 \mathrm{d}x. \tag{7.2}$$

In addition to the two kinds of internal potential energy above, there are three kinds of external potential energy due to the axial force, the distributed load, and the shear forces and bending moments at the member ends.

In the evaluation of the external energy due to the axial force P, we need to know the amount of its axial shortening dL at each position x. This is expressed using the vertical displacement $w(x)$ and the axial displacement $u(x)$, as

$$dL = \left[\sqrt{(1+u')^2 + (w')^2} - 1\right] \mathrm{d}x,$$

where $(\cdot)'$ is the derivative with respect to x. The term u' is considered negligible in the present bending problem of a beam-column. Accordingly, the term u' is omitted, and the axial shortening dL above is approximated as

$$dL \approx \frac{1}{2}(w')^2 \mathrm{d}x. \tag{7.3}$$

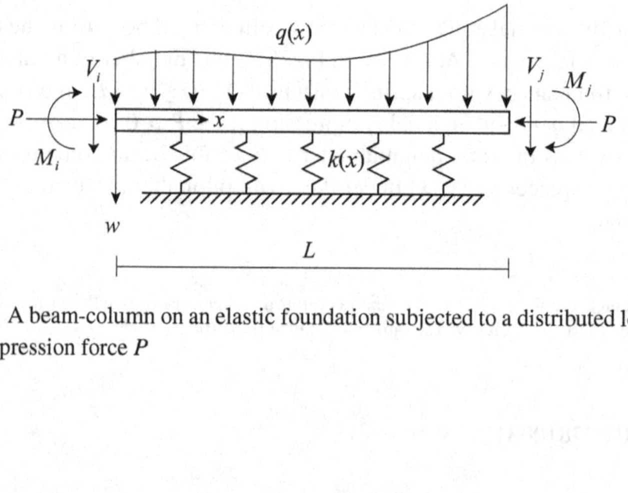

Figure 7.1 A beam-column on an elastic foundation subjected to a distributed load $q(x)$ and an axial compression force P

This approximation is valid when $w' \approx 0$. Accordingly, the external potential energy by the axial force P is given by

$$-P\int \mathrm{d}L = -P\int \frac{1}{2}w'(x)^2\mathrm{d}x. \tag{7.4}$$

Another external energy, due to the distributed load $q(x)$, is given by

$$-\int q(x)w(x)\mathrm{d}x. \tag{7.5}$$

The shear forces V_i and V_j and the bending moments M_i and M_j at the member ends i and j also make a contribution to the total potential energy, which is equal to

$$-\left[V_i w(0) + M_i w'(0) + V_j w(L) + M_j w'(L)\right]. \tag{7.6}$$

The total potential energy U of the beam-column is given as the sum of the five kinds of energy in (7.1), (7.2), and (7.4)–(7.6). That is,

$$\begin{aligned} U(w) = &\int_0^L \left[\frac{1}{2}EI(w'')^2 + \frac{1}{2}k(x)w^2 - \frac{1}{2}P(w')^2 - q(x)w\right]\mathrm{d}x \\ &- \left[V_i w(0) + M_i w'(0) + V_j w(L) + M_j w'(L)\right]. \end{aligned} \tag{7.7}$$

The total potential energy $U(w)$ is a function of a function $w = w(x)$. In mathematics, a function of a function is called a *functional*.

Remark 7.1 In the derivation of the total potential energy U in (7.7), we have focused on the flexural deformation and assumed that there is no shear deformation. This is called *Bernoulli–Euler's assumption*. The formulation of the buckling problem of a beam-column in this chapter employs this external energy (7.4) based on the approximation (7.3), which holds for a small displacement.[2] As demonstrated for a concrete example in Section 7.3.2, this formulation can give buckling loads and buckling modes of the beam-column, but cannot give information on post-bifurcation behavior. □

7.2.2 DERIVATION OF BEAM-COLUMN EQUATION

In this section, we derive the equation to be satisfied by the vertical displacement $w(x)$ by applying the variational principle to the functional $U(w)$ in (7.7).

We investigate the variation of the total potential energy $U(w)$ by introducing another function δw to perturb w to $w + \delta w$, where δw is an arbitrary smooth function such that the perturbed function $w + \delta w$ satisfies the given boundary condition. With the understanding that δw is small in magnitude, we denote the main part of $U(w + \delta w) - U(w)$ by δU, that is,

$$U(w + \delta w) - U(w) = \delta U + (\text{h.o.t.}),$$

[2]There are various variants of beam-column theory (Elishakoff, 2020 [3]).

where δU is linear in δw and (h.o.t.) denotes higher order terms such as $(\delta w)^2$, $((\delta w)')^2$, etc. Then the condition for the stationarity of $U(w)$ at w is given by $\delta U = 0$.

To derive a concrete form of the condition $\delta U = 0$, it is convenient to treat the integral part and the boundary-value part of U in (7.7) separately. By defining

$$A(w) = \int_0^L \left[\frac{1}{2}EI(w'')^2 + \frac{1}{2}k(x)w^2 - \frac{1}{2}P(w')^2 - q(x)w\right] \mathrm{d}x, \tag{7.8}$$

$$B(w) = -\left[V_i w(0) + M_i w'(0) + V_j w(L) + M_j w'(L)\right], \tag{7.9}$$

we have $U(w) = A(w) + B(w)$ by (7.7), and hence $\delta U = \delta A + \delta B$ (with the obvious meanings of δA and δB).

To calculate $\delta A \approx A(w+\delta w) - A(w)$ for A in (7.8), we note the relation

$$\int_0^L \left[\frac{1}{2}EI((w+\delta w)'')^2\right] \mathrm{d}x - \int_0^L \left[\frac{1}{2}EI(w'')^2\right] \mathrm{d}x = \int_0^L \left[EIw''(\delta w)''\right] \mathrm{d}x + \text{(h.o.t.)}$$

and similar relations for the other terms of the integrand in (7.8), to obtain

$$\begin{aligned} \delta A &= \int_0^L \left[EIw''(\delta w)'' + k(x)w\,\delta w - Pw'(\delta w)' - q(x)\delta w\right] \mathrm{d}x \\ &= \int_0^L \left[EIw'''' + Pw'' + k(x)w - q(x)\right] \delta w\, \mathrm{d}x \\ &\quad - \left[(EIw''' + Pw')\delta w\right]_0^L + \left[EIw''(\delta w)'\right]_0^L, \end{aligned} \tag{7.10}$$

where w', w'', w''', and w'''' are the first, the second, the third, and the fourth order derivatives of w, respectively; and the following identities (*integration by parts*) are used:

$$\begin{aligned} \int_0^L EIw''(\delta w)''\mathrm{d}x &= \left[EIw''(\delta w)'\right]_0^L - \int_0^L EIw'''(\delta w)'\mathrm{d}x \\ &= \left[EIw''(\delta w)'\right]_0^L - \left[EIw'''\delta w\right]_0^L + \int_0^L EIw''''\delta w\mathrm{d}x, \end{aligned}$$

$$\int_0^L Pw'(\delta w)'\mathrm{d}x = \left[Pw'\delta w\right]_0^L - \int_0^L Pw''\delta w\mathrm{d}x.$$

On the other hand, the definition of $B(w)$ in (7.9) immediately implies that

$$\delta B = -\left[V_i\delta w(0) + M_i(\delta w)'(0) + V_j\delta w(L) + M_j(\delta w)'(L)\right]. \tag{7.11}$$

With the use of (7.10) and (7.11) in $\delta U = \delta A + \delta B$, a concrete expression of the condition $\delta U = 0$ for the stationarity of U in (7.7) is given by

$$\begin{aligned} &\int_0^L \left[EIw'''' + Pw'' + k(x)w - q(x)\right] \delta w\, \mathrm{d}x \\ &+ \left[EIw'''(0) + Pw'(0) - V_i\right] \delta w(0) - \left[EIw''(0) + M_i\right] (\delta w)'(0) \\ &- \left[EIw'''(L) + Pw'(L) + V_j\right] \delta w(L) + \left[EIw''(L) - M_j\right] (\delta w)'(L) = 0. \end{aligned} \tag{7.12}$$

The equation (7.12) must be satisfied for any "admissible" perturbation δw such that the perturbed function $w+\delta w$ satisfies the given boundary condition. Therefore, the function w must satisfy the differential equation

$$EIw'''' + Pw'' + k(x)w = q(x). \tag{7.13}$$

This differential equation serves as the governing differential equation of the vertical displacement w of the beam-column. The boundary condition for this differential equation can be obtained as the condition that the equation

$$\begin{aligned} &\left[EIw'''(0) + Pw'(0) - V_i\right]\delta w(0) - \left[EIw''(0) + M_i\right](\delta w)'(0) \\ &- \left[EIw'''(L) + Pw'(L) + V_j\right]\delta w(L) + \left[EIw''(L) - M_j\right](\delta w)'(L) = 0 \end{aligned} \tag{7.14}$$

holds for any "admissible" δw and $(\delta w)'$ with respect to the given boundary condition.

For example, for a simply supported beam-column in Fig. 7.2, where both ends are fixed vertically but are free to rotate (tilt), we have $w(0) = w(L) = 0$ and $M_i = M_j = 0$ as a specified boundary condition. Since the perturbed function $w+\delta w$ must also satisfy this boundary condition, the condition $\delta w(0) = \delta w(L) = 0$ is imposed on "admissible" δw. Therefore, the equation (7.14) is simplified, in this case, to

$$-EIw''(0)(\delta w)'(0) + EIw''(L)(\delta w)'(L) = 0. \tag{7.15}$$

This, in turn, implies $w''(0) = w''(L) = 0$ since $(\delta w)'(0)$ and $(\delta w)'(L)$ are arbitrary for an "admissible" perturbation δw. Thus an additional boundary condition, i.e., $w''(0) = w''(L) = 0$, is derived from the variational principle. A boundary condition that can be derived from the variational principle is called a *natural boundary condition*. In contrast, the boundary condition $w(0) = w(L) = 0$ does not follow from (7.14) and must be specified explicitly. Such boundary condition is called an *essential boundary condition* in the variational method. To sum up, the simply supported beam-column in Fig. 7.2 is described by the governing differential equation (7.13) with the boundary conditions

$$w(0) = w(L) = 0, \qquad w''(0) = w''(L) = 0. \tag{7.16}$$

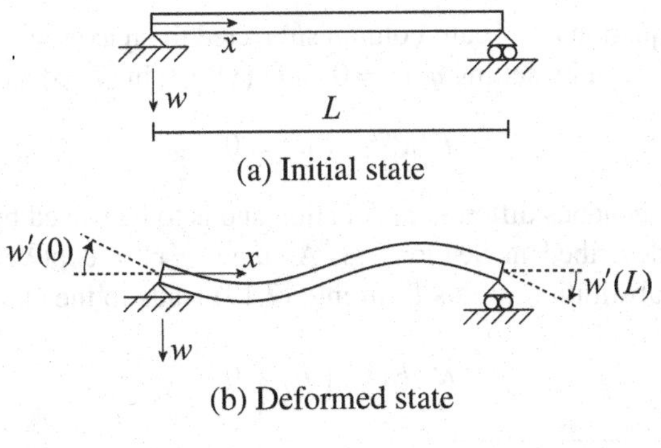

Figure 7.2 A simply supported beam-column

In Sections 7.3–7.5, we shall obtain solutions to the governing differential equation (7.13) of a beam-column for the cases (1) $q(x) = 0$, $k(x) = 0$, (2) $q(x) = 0$, $k(x) \neq 0$, (3) $q(x) \neq 0$, $k(x) \neq 0$ under various kinds of boundary conditions.

Remark 7.2 It is worth mentioning a standard procedure in *calculus of variations* (or *variational calculus*), which underlies the derivation of the beam-column equation above.[3] Suppose, for example, that we want to minimize a functional $U(w)$ in $w = w(x)$ represented in the form of

$$U(w) = \int_a^b G(x, w, w', \ldots, w^{(m)})\,\mathrm{d}x$$

under some specified boundary-value constraints on $w(a)$, $w(b)$, $w'(a)$, etc., where $w^{(m)}$ denotes the mth order derivative of w. The condition on w for the stationarity of $U(w)$ can be derived by considering $U(w+\delta w) - U(w)$ and calculating it to an expression of form

$$\int_a^b [\cdots]\delta w\,\mathrm{d}x + [\text{boundary-value terms}] + (\text{h.o.t.}) \tag{7.17}$$

by using integration by parts. From the vanishing of the integrand $[\cdots]$ in (7.17), we obtain the so-called *Euler–Lagrange equation*

$$\frac{\partial G}{\partial w} - \frac{\mathrm{d}}{\mathrm{d}x}\frac{\partial G}{\partial w'} + \cdots + (-1)^m \frac{\mathrm{d}^m}{\mathrm{d}x^m}\frac{\partial G}{\partial w^{(m)}} = 0, \tag{7.18}$$

which is an ordinary differential equation in $w(x)$ for which $U(w)$ is stationary. The governing equation (7.13) of a beam-column is a special case of this Euler–Lagrange equation. The vanishing of the boundary-value terms in (7.17) gives a boundary condition for this differential equation. If the specified condition on the boundary values is derived from this boundary condition, it is called a *natural boundary condition*; otherwise, it is referred to as an *essential boundary condition.* □

7.3 BEAM-COLUMN SUBJECTED TO AXIAL COMPRESSION

The governing equation of a beam-column subjected to an axial force only, depicted in Fig. 7.3, is obtained by setting $q(x) = 0$ and $k(x) = 0$ in (7.13) as

$$EIw'''' + Pw'' = 0. \tag{7.19}$$

This is a homogeneous differential equation and is to be solved by using the standard procedure described in Section 7.8. Assume $w(x) = \exp(\lambda x)$ for a complex number λ. The substitution of this form into (7.19) leads to the characteristic equation

$$\lambda^2(EI\lambda^2 + P) = 0,$$

[3] See, e.g., Courant and Hilbert [2] for details of the calculus of variations.

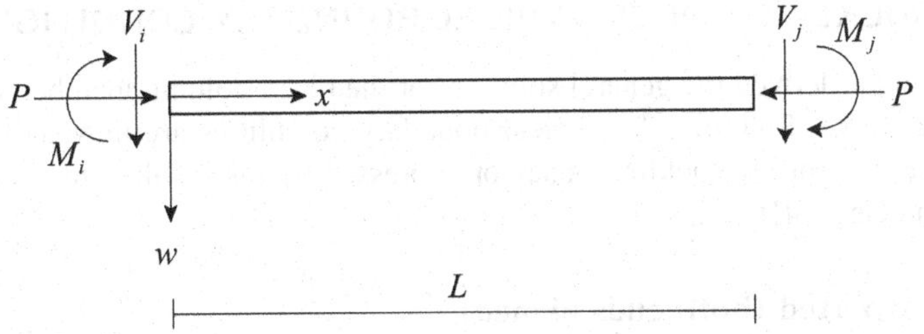

Figure 7.3 A beam-column subjected to an axial compression force P

and the solutions of this equation are $\lambda = 0$ (twice repeated) and $\lambda = \pm \mathrm{i}\mu$, where i is the imaginary unit and

$$\mu = \sqrt{\frac{P}{EI}}. \tag{7.20}$$

Accordingly, the general solution to the differential equation (7.19) is

$$w(x) = C_1 + C_2 x + C_3 \sin \mu x + C_4 \cos \mu x, \tag{7.21}$$

where $C_1, \ldots, C_4$ are constants.

7.3.1 PROCEDURE OF BUCKLING ANALYSIS

The substitution of $w(x)$ in (7.21) into the boundary condition, which varies with cases (see Section 7.3.2), leads to a system of linear simultaneous equations

$$A \begin{pmatrix} C_1 \\ C_2 \\ C_3 \\ C_4 \end{pmatrix} = \begin{pmatrix} 0 \\ 0 \\ 0 \\ 0 \end{pmatrix} \tag{7.22}$$

for the coefficients (C_1, C_2, C_3, C_4) in (7.21). Here $A = A(\mu)$ is a 4×4 matrix depending on μ in (7.20).

The buckling occurs when there is a nontrivial solution

$$(C_1, C_2, C_3, C_4) \neq (0, 0, 0, 0)^\top$$

to (7.22), which is the case if and only if the determinant of the matrix A in (7.22) vanishes (see (4.18) in Section 4.4). That is, the condition for the occurrence of buckling is given as

$$\det A(\mu) = 0. \tag{7.23}$$

For a solution $\mu = \mu_{\mathrm{c}}$ to (7.23), the buckling load is determined by (7.20) as

$$P_{\mathrm{c}} = EI\mu_{\mathrm{c}}^2. \tag{7.24}$$

As we see below, there are cases where the buckling condition (7.23) takes a simple form and the buckling load can be expressed explicitly. However, in general, the buckling load is to be obtained numerically.

7.3.2 BUCKLING UNDER TYPICAL BOUNDARY CONDITIONS

We have so far derived the general solution for the beam-column member subjected to an axial force. Solutions for typical boundary conditions are introduced below, with discussion on the buckling behavior expressed by these solutions (e.g., Timoshenko and Gere [4]).

Simply supported (both ends pinned)

As a special case of the general setting in Fig. 7.3 for a beam-column subjected to an axial force, we consider a simply supported beam-column (both ends pinned) in Fig. 7.4(a), and investigate its flexural buckling. The boundary conditions expressing the no-vertical-displacement and no-bending-moment conditions at the both ends are given by

$$w(0) = w(L) = 0, \qquad EIw''(0) = EIw''(L) = 0. \tag{7.25}$$

The substitution of the general solution in (7.21) to the boundary conditions in (7.25) leads to

$$C_1 = C_2 = C_4 = 0, \quad C_3 \sin \mu L = 0. \tag{7.26}$$

There are trivial and nontrivial solutions for (7.26). First, the trivial solution with $C_1 = C_2 = C_3 = C_4 = 0$ corresponds to the solution $w(x) = 0$ in (7.21).

The buckling occurs when (7.26) admits a nontrivial solution, which is the case when

$$\sin \mu L = \sin \sqrt{\frac{P}{EI}} L = 0. \tag{7.27}$$

This condition gives buckling loads

$$P_n = n^2 \frac{\pi^2 EI}{L^2}, \quad n = 1, 2, \ldots, \tag{7.28}$$

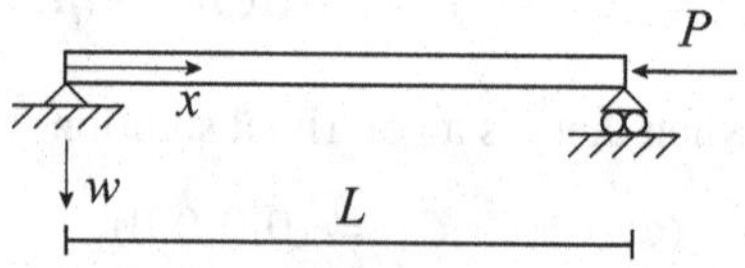

(a) Simply supported beam-column

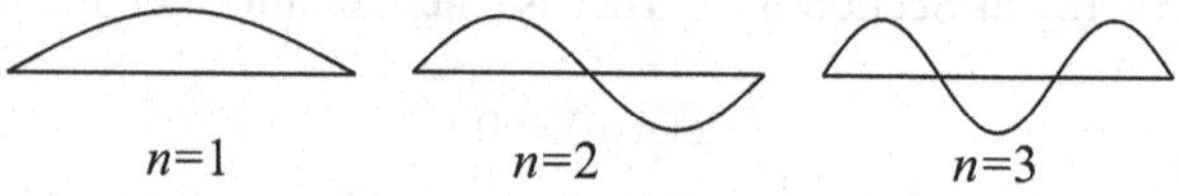

(b) Buckling modes w_n

Figure 7.4 The buckling of a simply supported beam-column

where P_n corresponds to $\mu_n = n\pi/L$ for $n = 1, 2, \ldots$. Accordingly, the buckling modes of the beam-column are given by (7.21) with $C_1 = C_2 = C_4 = 0$ as

$$w_n(x) = C_3 \sin n\pi\frac{x}{L}, \quad n = 1, 2, \ldots. \tag{7.29}$$

Figure 7.4(b) shows these buckling modes.

As the load P increases from zero, a buckling occurs when it reaches the minimum buckling load P_1. For later reference, we introduce notation

$$P_{\rm E} = \frac{\pi^2 EI}{L^2}, \tag{7.30}$$

for which we have $P_1 = P_{\rm E}$. The buckling mode for $n = 1$ is given by

$$w_1(x) = C_3 \sin \pi\frac{x}{L}.$$

The buckling load in (7.30) is linearly proportional to the flexural rigidity EI and is inversely proportional to the square of the length L. Accordingly, a slender beam-column with smaller I and longer L tends to undergo buckling for a smaller load.

Cantilever (one end fixed, the other end free)

As another typical boundary condition, we consider the cantilever (beam-column with one end fixed and the other end free) depicted in Fig. 7.5(a). We have the fixed end conditions

$$w(0) = 0, \qquad w'(0) = 0 \tag{7.31}$$

at $x = 0$ and the conditions of no shear force and no bending moment

$$EIw'''(L) + Pw'(L) = 0, \qquad EIw''(L) = 0 \tag{7.32}$$

at $x = L$.

Remark 7.3 It is noted that the conditions in (7.32) are natural boundary conditions that are derived from (7.14) since δw and $(\delta w)'$ are free and $V_j = 0$ and $M_j = 0$ at $x = L$. In contrast, the conditions in (7.31) are essential boundary conditions. □

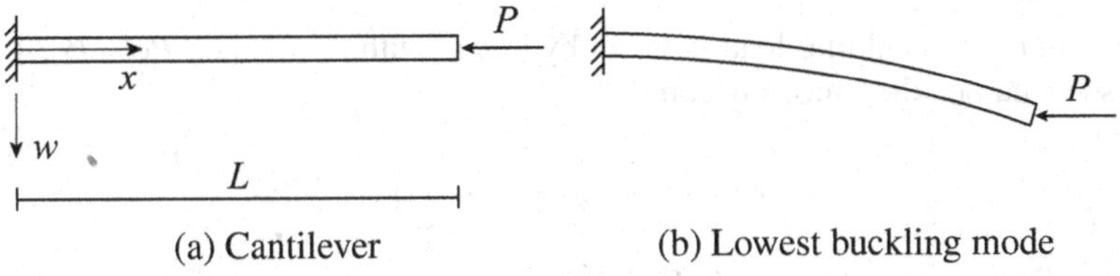

Figure 7.5 A cantilever and its lowest order buckling mode

The use of the expression (7.21) in the boundary conditions in (7.31) and (7.32) leads, in this order, to

$$C_1 + C_4 = 0, \quad C_2 + \mu C_3 = 0, \quad C_2 = 0, \quad C_3 \sin \mu L + C_4 \cos \mu L = 0. \tag{7.33}$$

Here $C_2 = 0$ is derived based on the following relation:

$$\begin{aligned} & EIw'''(L) + Pw'(L) \\ & = EI(-\mu^3 C_3 \cos \mu L - \mu^3 C_4 \sin \mu L) + P(C_2 + \mu C_3 \cos \mu L + \mu C_4 \sin \mu L) \\ & = PC_2, \end{aligned} \tag{7.34}$$

where $P = EI\mu^2$ is used (cf., (7.20)).

The conditions in (7.33) can be analyzed easily as explained below (see Remark 7.4 for a more systematic method). These conditions can be rewritten as

$$C_1 + C_4 = 0, \qquad C_2 = C_3 = 0, \qquad C_4 \cos \mu L = 0. \tag{7.35}$$

A nontrivial solution $(C_1, C_2, C_3, C_4) \neq (0,0,0,0)^\top$ exists when

$$\cos \mu L = 0 \tag{7.36}$$

is satisfied for some μ. Such $\mu = \mu_n$ is given by

$$\mu_n = \left(n - \frac{1}{2}\right)\frac{\pi}{L}, \quad n = 1, 2, \ldots.$$

Accordingly, the buckling loads are determined as

$$P_n = EI\mu_n^2 = \left(n - \frac{1}{2}\right)^2 \frac{\pi^2 EI}{L^2} = \left(n - \frac{1}{2}\right)^2 P_{\mathrm{E}}, \quad n = 1, 2, \ldots, \tag{7.37}$$

where $P_{\mathrm{E}} = \pi^2 EI/L^2$ in (7.30).

To obtain the buckling modes, we note that (7.35) with $\cos \mu L = 0$ reduces to the conditions $C_4 = -C_1$ and $C_2 = C_3 = 0$. Then (7.21) gives the buckling modes

$$w_n(x) = C_1 \left[1 - \cos \pi \left(n - \frac{1}{2}\right)\frac{x}{L}\right], \quad n = 1, 2, \ldots.$$

The minimum buckling load is given by (7.37) with $n = 1$, i.e., $P_1 = P_{\mathrm{E}}/4$, and the associated buckling mode becomes

$$w_1(x) = C_1 \left(1 - \cos \frac{\pi x}{2L}\right).$$

This buckling mode is depicted in Fig. 7.5(b).

Remark 7.4 A systematic method to derive the buckling condition in (7.36) is illustrated. The conditions in (7.33) are expressed in a matrix form as

$$\begin{pmatrix} 1 & 0 & 0 & 1 \\ 0 & 1 & \mu & 0 \\ 0 & 1 & 0 & 0 \\ 0 & 0 & \sin\mu L & \cos\mu L \end{pmatrix} \begin{pmatrix} C_1 \\ C_2 \\ C_3 \\ C_4 \end{pmatrix} = \begin{pmatrix} 0 \\ 0 \\ 0 \\ 0 \end{pmatrix}. \tag{7.38}$$

Then the buckling condition in (7.23) is given by the vanishing of the determinant

$$\begin{vmatrix} 1 & 0 & 0 & 1 \\ 0 & 1 & \mu & 0 \\ 0 & 1 & 0 & 0 \\ 0 & 0 & \sin\mu L & \cos\mu L \end{vmatrix} = \begin{vmatrix} 1 & \mu & 0 \\ 1 & 0 & 0 \\ 0 & \sin\mu L & \cos\mu L \end{vmatrix} = -\mu\cos\mu L.$$

Therefore, we obtain the condition $\cos\mu L = 0$ in (7.36). □

7.3.3 EFFECTIVE BUCKLING LENGTH

As we have seen in Section 7.3.2, the minimum buckling load P_1 varies with boundary conditions: $P_1 = P_E$ for the simply supported beam-column and $P_1 = P_E/4$ for the cantilever, where P_E is defined in (7.30). For the beam-column with both ends fixed at the left of Fig. 7.6(a) and the one at the left of (b) with one end fixed and the other end pinned, the minimum buckling loads are given, respectively, by

$$P_1 = \begin{cases} 4P_E, & \text{both ends fixed,} \\ 2.0457P_E, & \text{one end fixed and the other end pinned} \end{cases} \tag{7.39}$$

(see Problem 7.2).

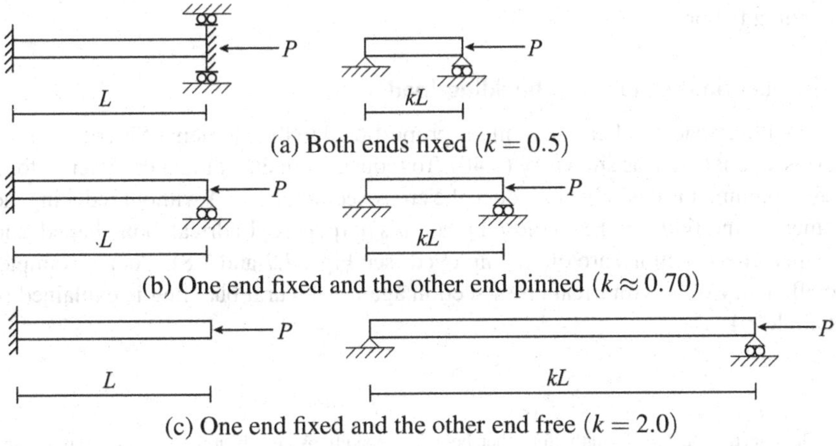

Figure 7.6 Beam-columns with various boundary conditions and their effective buckling length kL

In the argument of the buckling strength of beam-columns with different boundary conditions, it is conventional to express its minimum buckling load P_1 relative to the buckling load P_E defined in (7.30). For this purpose, we introduce a virtual length kL of a beam-column with a non-dimensional parameter k to express P_1 as

$$P_1 = \frac{\pi^2 EI}{(kL)^2} = \frac{1}{k^2} P_E. \tag{7.40}$$

In this expression, kL is called the *effective buckling length* and k is called the *coefficient of effective buckling length.* Note that k depends only on the boundary conditions and are independent of member properties.

Among the three beam-columns in Fig. 7.6, the beam-column in Fig. 7.6(a) with both ends fixed has the smallest k-value of $1/\sqrt{4} = 0.5$ by (7.39), the one in (b) with one end fixed and the other end pinned has $k = 1/\sqrt{2.0457} \approx 0.7$ by (7.39), and the one in (c) with one end fixed and the other end free has the largest k-value of $\sqrt{4} = 2.0$ by (7.37) with $n = 1$. The associated minimum buckling load decreases in this order. Thus, as the boundary condition becomes less restrictive, the value of k becomes larger, and the minimum buckling load decreases.[4]

A beam-column with coefficient k and length L has the same minimum buckling load as the simply supported beam-column with the length kL. For example, the beam-columns with various boundary conditions and the same length L depicted at the left of Figs. 7.6(a)–(c) have the same minimum buckling loads as the simply supported beam-columns with various lengths depicted at the right.

7.3.4 CROSS-SECTIONAL SHAPE AND BUCKLING STRESS

The predominant influence of member length L and the boundary conditions on the buckling load of a beam-column member has been demonstrated in Section 7.3.3. We march on to investigate the influence of cross-sectional shape as another important governing factor.

Cross-sectional shape and buckling load

The buckling load of a beam-column is proportional to the moment of inertia I of the cross-sectional area, as shown by (7.40). To reduce requisite amount of material for a beam-column, it is desirable to reduce the cross-sectional area A without reducing the moment of inertia I (cf., Example 7.1). For this purpose, cylindrical, box-shaped, and I-shaped cross-sections are often employed (see Figs. 7.7 and 7.8). As an example, the efficiency of a cylindrical cross-section against flexural buckling is explained in Example 7.1.

---

[4]The effective buckling lengths for other boundary conditions are given, e.g., in Problem 7.3 and Timoshenko and Gere [4].

Example 7.1 We compare a circular cross section and a cylindrical cross section with the same area $A = 9\pi\ \text{cm}^2$ in Fig. 7.7. The moment of inertia of these cross-sections are[5]

$$\begin{cases} \text{circular cross section:} & I_a = \dfrac{\pi}{4}3^4 = \dfrac{81\pi}{4}\ \text{cm}^4, \\ \text{cylindrical cross section:} & I_b = \dfrac{\pi}{4}(5^4 - 4^4) = \dfrac{369\pi}{4}\ \text{cm}^4. \end{cases}$$

Thus the cylindrical cross-section has a larger moment of inertia. The ratio of the buckling loads P_a and P_b for these two sections is given by the ratio of the moments of inertia as

$$\frac{P_b}{P_a} = \frac{\pi^2 EI_b}{(kL)^2} \div \frac{\pi^2 EI_a}{(kL)^2} = \frac{I_b}{I_a} \approx 4.56. \tag{7.41}$$

The cylindrical cross-section has a larger buckling load P_b. Although a cylindrical cross-section with a thinner plate is more efficient against flexural buckling, excessively thin plate thickness often entails local buckling causing bumps and dents on the cross-section. To avoid such local buckling, the minimum thickness is specified in the practice of member design. □

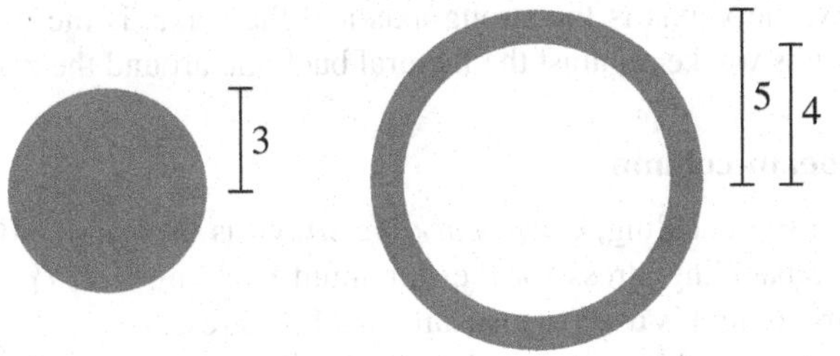

(a) Circular (b) Cylindrical

Figure 7.7 Two kinds of cross-sections with the same cross-sectional area (length unit in cm)

For box-shaped and I-shaped cross-sections (Fig 7.8), the moments of inertia I_y around the y-axis and I_z around the z-axis take different values. Accordingly, the flexural buckling loads around the y-axis and around the z-axis are different. The axis with a larger moment of inertia is called the *strong axis*, and the other axis is called the *weak axis*.

Example 7.2 The box-shaped cross-section shown in Fig 7.8(a) has two kinds of buckling loads: the buckling load P_y for the bending around the y-axis and the buckling load P_z for the bending around the z-axis. The moments of inertia are

$$I_y = \frac{8 \cdot 10^3}{12} - \frac{6 \cdot 8^3}{12} = \frac{1232}{3}\ \text{cm}^4, \qquad I_z = \frac{10 \cdot 8^3}{12} - \frac{8 \cdot 6^3}{12} = \frac{848}{3}\ \text{cm}^4$$

---

[5]The moment of inertia of a circular cross-section with a radius R is given by $\pi R^4/4$.

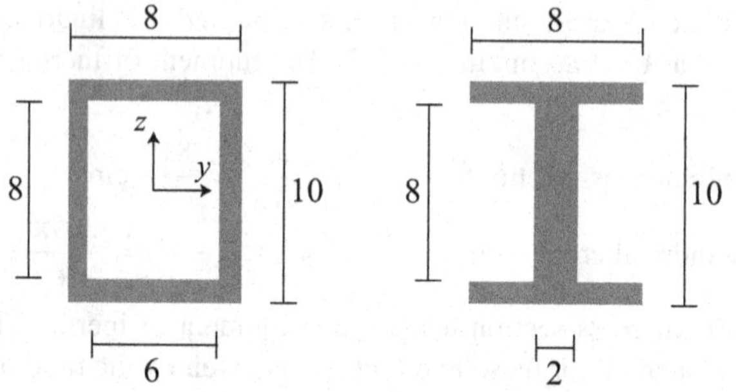

(a) Box-shaped (b) I-shaped

Figure 7.8 Box-shaped cross-section and I-shaped cross-section (length unit in cm)

and the buckling loads are

$$P_y = \frac{\pi^2 EI_y}{(kL)^2} = \frac{1232}{3}\frac{\pi^2 E}{(kL)^2}, \qquad P_z = \frac{\pi^2 EI_z}{(kL)^2} = \frac{848}{3}\frac{\pi^2 E}{(kL)^2}.$$

Accordingly, the y-axis is the strong axis and the z-axis is the weak axis, and this cross-section is weaker against the flexural buckling around the z-axis. □

Stress in beam-column

The stress at the buckling, called *buckling stress*, is used in the design of a beam-column. The buckling stress for the minimum buckling load P_{E} for a simply supported beam-column is used to measure buckling stress.

This buckling load is converted into the buckling stress as

$$\sigma_{\mathrm{E}} = \frac{\pi^2 EI}{L^2 A} = \frac{\pi^2 E}{\lambda^2}. \tag{7.42}$$

Here

$$\lambda = \frac{L}{r}, \qquad r = \sqrt{\frac{I}{A}}, \tag{7.43}$$

and λ is called the *slenderness ratio*, expressing how slender the beam-column is, and r is called the *radius of gyration*, characterizing the cross-sectional size. The σ_{E} versus λ curve expressed by (7.42) is plotted in Fig. 7.9. This curve is called the *Euler curve* and is used in the description of elastic buckling of a beam-column.

For other boundary conditions, the buckling stress in (7.42) is rewritten as

$$\sigma_{\mathrm{c}} = \frac{\sigma_{\mathrm{E}}}{k^2} = \frac{\pi^2 E}{k^2 \lambda^2} \tag{7.44}$$

using the coefficient k of effective buckling length.

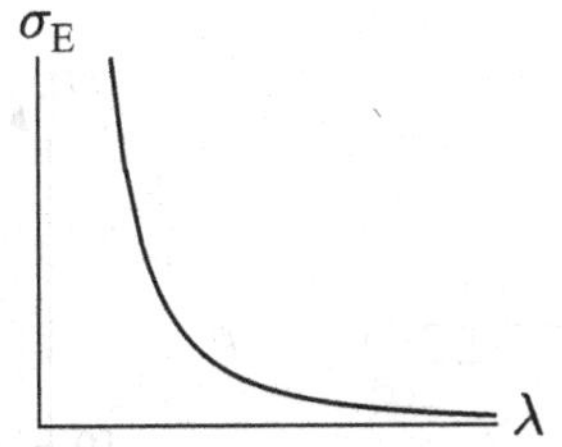

Figure 7.9 σ_E versus λ curve (Euler curve)

Example 7.3 The radius of gyration r for the circular cross-section and the cylindrical cross-section in Fig. 7.7 is

$$r = \sqrt{\frac{I}{A}} = \begin{cases} \sqrt{\frac{81\pi}{4}\frac{1}{9\pi}} = 1.5 \text{ cm}, & \text{circular cross-section}, \\ \sqrt{\frac{369\pi}{4}\frac{1}{9\pi}} = \frac{\sqrt{41}}{2} \approx 3.2 \text{ cm}, & \text{cylindrical cross-section}. \end{cases} \tag{7.45}$$

The cylindrical cross-section has a much larger radius of gyration r than the circular cross-section. Also, from (7.43), the cylindrical cross-section has a much smaller slenderness ratio λ than the circular cross-section for the same length L. Thus the cylindrical cross-section is stronger against buckling. □

7.4 BEAM-COLUMN ON ELASTIC FOUNDATION

As a special case of the general setting of the beam-column member in Fig. 7.1 (Section 7.2), we consider a *beam-column on elastic foundation* subjected to an axial compression force P depicted in Fig. 7.10(a). The governing differential equation of this beam-column is obtained by setting $q(x) = 0$ in (7.13) and $k(x) = k$ (= constant) as

$$EIw'''' + Pw'' + kw = 0. \tag{7.46}$$

We consider the boundary condition of simple support in (7.25), i.e.,

$$w(0) = 0, \quad w''(0) = 0; \qquad w(L) = 0, \quad w''(L) = 0. \tag{7.47}$$

The substitution of $w(x) = \exp(\lambda x)$ into (7.46) leads to a characteristic equation

$$EI\lambda^4 + P\lambda^2 + k = 0. \tag{7.48}$$

The solutions to this equation are

$$\lambda = \mathrm{i}\,\omega_-, \quad -\mathrm{i}\,\omega_-, \quad \mathrm{i}\,\omega_+, \quad -\mathrm{i}\,\omega_+ \tag{7.49}$$

with

$$\omega_- = \sqrt{\frac{P}{2EI} - \sqrt{\left(\frac{P}{2EI}\right)^2 - \frac{k}{EI}}}, \qquad \omega_+ = \sqrt{\frac{P}{2EI} + \sqrt{\left(\frac{P}{2EI}\right)^2 - \frac{k}{EI}}}, \tag{7.50}$$

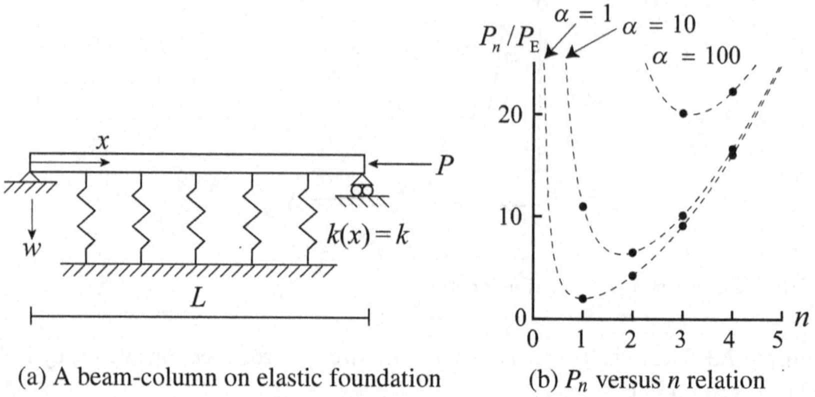

Figure 7.10 A beam-column on elastic foundation and its buckling property

where i is the imaginary unit. We hereafter assume

$$\left(\frac{P}{2EI}\right)^2 - \frac{k}{EI} > 0. \tag{7.51}$$

Then the general solution to the governing equation in (7.46) is expressed as

$$w(x) = C_1 \sin \omega_- x + C_2 \cos \omega_- x + C_3 \sin \omega_+ x + C_4 \cos \omega_+ x. \tag{7.52}$$

The substitution of (7.52) into the four boundary conditions in (7.47) leads, respectively, to

$$\begin{aligned}
&C_2 + C_4 = 0,\\
&-\omega_-^2 C_2 - \omega_+^2 C_4 = 0,\\
&(\sin \omega_- L)C_1 + (\cos \omega_- L)C_2 + (\sin \omega_+ L)C_3 + (\cos \omega_+ L)C_4 = 0,\\
&-\omega_-^2 (\sin \omega_- L)C_1 - \omega_-^2 (\cos \omega_- L)C_2 - \omega_+^2 (\sin \omega_+ L)C_3 - \omega_+^2 (\cos \omega_+ L)C_4 = 0.
\end{aligned} \tag{7.53}$$

The first two equations imply $C_2 = C_4 = 0$ since $\omega_+ > \omega_-$ by (7.50) and (7.51). Then we have[6]

$$\begin{pmatrix} \sin \omega_- L & \sin \omega_+ L \\ -\omega_-^2 \sin \omega_- L & -\omega_+^2 \sin \omega_+ L \end{pmatrix} \begin{pmatrix} C_1 \\ C_3 \end{pmatrix} = \begin{pmatrix} 0 \\ 0 \end{pmatrix}.$$

The buckling condition becomes

$$\begin{vmatrix} \sin \omega_- L & \sin \omega_+ L \\ -\omega_-^2 \sin \omega_- L & -\omega_+^2 \sin \omega_+ L \end{vmatrix} = (\omega_-^2 - \omega_+^2) \sin \omega_- L \, \sin \omega_+ L = 0. \tag{7.54}$$

[6]See Problem 7.4(1) for a more systematic derivation of the buckling condition.

That is, we have

$$\sin \omega_{-}L \, \sin \omega_{+}L = 0, \tag{7.55}$$

which gives

$$\omega_{-} = \frac{n\pi}{L} \text{ or } \omega_{+} = \frac{n\pi}{L}, \quad n = 1, 2, \ldots. \tag{7.56}$$

Then, from (7.50), the buckling load is evaluated to (see Problem 7.4(2))

$$\frac{P_n}{P_{\mathrm{E}}} = \frac{1}{P_{\mathrm{E}}}\left[EI\left(\frac{n\pi}{L}\right)^2 + k\left(\frac{L}{n\pi}\right)^2\right] = n^2 + \frac{\alpha}{n^2}, \quad n = 1, 2, \ldots, \tag{7.57}$$

where $P_{\mathrm{E}} = \pi^2 EI/L^2$ and $\alpha = kL^4/(\pi^4 EI)$. The associated buckling modes are

$$w_n(x) = C \sin n\pi \frac{x}{L}, \quad n = 1, 2, \ldots.$$

The P_n versus n relation is plotted for various values of α in Fig. 7.10(b) (note that n takes only integer values). As α increases as $\alpha = 1$, 10, 100, the value of n giving the minimum buckling load increases as $n = 1$, 2, 3. Thus the minimum buckling load is not associated with the lowest mode ($n = 1$) when the member is slender with large α. In contrast, the lowest mode always gives the minimum buckling load in the beam-column subjected to axial compression in Section 7.3.

7.5 BEAM-COLUMN SUBJECTED TO AXIAL FORCE AND DISTRIBUTED LOAD

As a special case of the general setting of the beam-column member in Fig. 7.1 (Section 7.2), we consider a simply supported beam-column subjected to compression by an axial force P and bending by a constant uniform load.

The differential equation of this beam-column is obtained by setting $k(x) = 0$ and $q(x) = q$ (constant) in (7.13) as

$$EIw'''' + Pw'' = q, \tag{7.58}$$

which is an inhomogeneous differential equation. This differential equation has a particular solution $qx^2/(2P)$ and its general solution is given as the sum of this particular solution with the solution in (7.21) as

$$w(x) = C_1 + C_2 x + C_3 \sin \mu x + C_4 \cos \mu x + \frac{qx^2}{2P}, \tag{7.59}$$

where $\mu = \sqrt{P/(EI)}$.

For the boundary conditions (7.25) of simple support, the solution to (7.58) becomes

$$w(x) = \frac{q}{2P}x(x - L) + \frac{q}{\mu^2 P}\left(\tan \frac{\mu L}{2} \sin \mu x + \cos \mu x - 1\right) \tag{7.60}$$

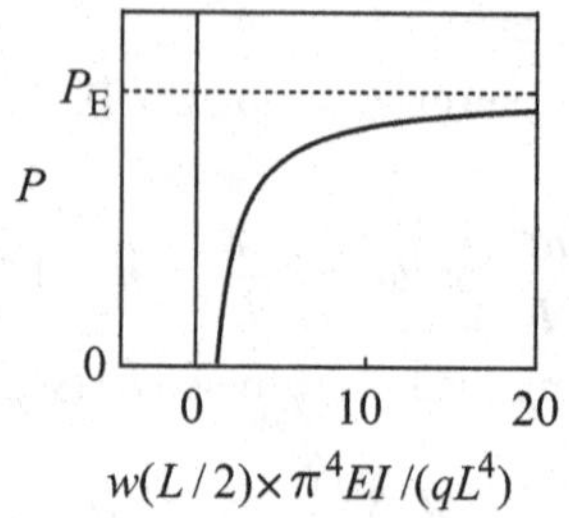

Figure 7.11 A load versus displacement curve of a beam-column subjected to an axial force and a uniformly distributed load

(Problem 7.5(1)). The vertical displacement at the midpoint ($x = L/2$) is evaluated to

$$w\left(\frac{L}{2}\right) = \frac{qL^4}{\pi^4 EI}\left(\frac{P}{P_{\rm E}}\right)^{-2}\left(\frac{1}{\cos\left(\frac{\pi}{2}\sqrt{\frac{P}{P_{\rm E}}}\right)} - 1 - \frac{\pi^2}{8}\frac{P}{P_{\rm E}}\right) \tag{7.61}$$

(Problem 7.5(2)), using $P_{\rm E} = \pi^2 EI/L^2$ in (7.30).

The axial load P is plotted against the displacement $w(L/2)$ in Fig. 7.11. The horizontal line at $P = P_{\rm E}$ is an asymptote. As the load P increases from zero, the displacement monotonically increases, whereas it diverges to infinity as the load P approaches $P_{\rm E}$ asymptotically. Accordingly, $P_{\rm E}$ is the upper bound of the buckling load of this beam-column.

7.6 INITIAL DEFLECTION

We investigate the influence of an *initial deflection* $\hat{w}(x)$ for a simply supported beam-column in Section 7.3 with $q(x) = 0$ and $k(x) = 0$. We assume that the axial force P stays in the range

$$0 < P < P_{\rm E} = \frac{\pi^2 EI}{L^2}. \tag{7.62}$$

At the initial state $w(x) = \hat{w}(x)$, there is no bending moment. At a displaced state represented by $w(x) \neq \hat{w}(x)$, there emerges bending moment in proportion to the additional displacement $w(x) - \hat{w}(x)$. Accordingly, the governing differential equation (7.13) becomes

$$EI(w - \hat{w})'''' + Pw'' = 0. \tag{7.63}$$

For an expanded form of the initial deflection

$$\hat{w}(x) = \sum_{i=1}^{n} \varepsilon_i \sin\frac{i\pi}{L}x \tag{7.64}$$

with some integer n, the governing equation (7.63) becomes

$$w'''' + \mu^2 w'' = \sum_{i=1}^{n} \varepsilon_i \left(\frac{i\pi}{L}\right)^4 \sin\frac{i\pi}{L}x \tag{7.65}$$

with the use of $\mu = \sqrt{P/(EI)}$ in (7.20). The general solution to this inhomogeneous differential equation is (Problem 7.6)

$$w(x) = C_1 + C_2 x + C_3 \sin\mu x + C_4 \cos\mu x + \sum_{i=1}^{n} \frac{\varepsilon_i}{1 - \dfrac{1}{i^2}\dfrac{P}{P_E}} \sin\frac{i\pi}{L}x. \tag{7.66}$$

The substitution of this expression of $w(x)$ into the boundary condition (7.25) of simple support leads to $C_1 = C_2 = C_3 = C_4 = 0$ since $\sin\mu L \neq 0$ by (7.62). Thus the solution satisfying the boundary condition is

$$w(x) = \sum_{i=1}^{n} \frac{\varepsilon_i}{1 - \dfrac{1}{i^2}\dfrac{P}{P_E}} \sin\frac{i\pi}{L}x. \tag{7.67}$$

As a specific initial deflection, we consider the component $\hat{w}(x) = \varepsilon_1 \sin(\pi x/L)$ for the first buckling mode associated with $\varepsilon_1 \neq 0$ and $\varepsilon_2 = \varepsilon_3 = \cdots = \varepsilon_n = 0$. Then (7.67) becomes

$$w(x) = \frac{\varepsilon_1}{1 - P/P_E} \sin\frac{\pi}{L}x.$$

The load versus displacement curve expressed by this relation is plotted in Fig. 7.12. The dashed (horizontal) line at $P = P_E$ denotes the curve for the perfect system without initial deflection ($\varepsilon_1 = 0$), and the solid lines denote the curves for imperfect systems with initial deflections. As the load P approaches $P = P_E$, the displacement $w(L/2)$ diverges to infinity. As the coefficient ε_1 expressing the magnitude of the initial deflection increases, the distance between the curves of the perfect system and an imperfect system enlarges.

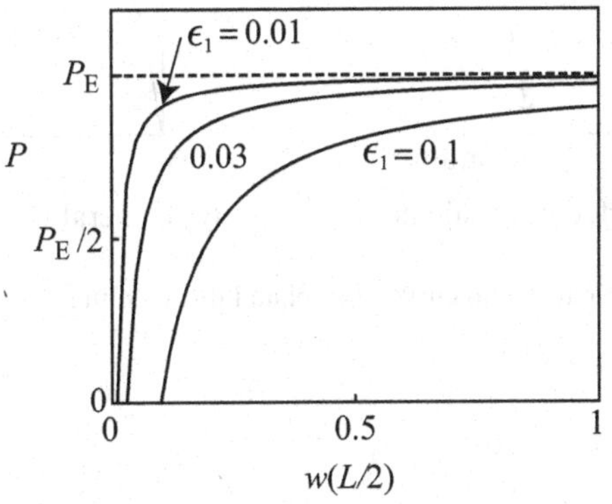

Figure 7.12 A load versus displacement curve of a beam-column with an initial deflection $\hat{w} = \varepsilon_1 \sin(\pi x/L)$

7.7 INELASTIC BUCKLING

We have so far dealt with a beam-column of a linear elastic body, for which the tangent of the stress versus strain curve is constant and is equal to the modulus of elasticity E. This is an idealization that is accurate when the stress is small.

Yet, when the stress reaches some limit value, the tangent modulus, which is the instantaneous slope of the stress versus strain curve, tends to decrease. Moreover, the slope becomes different when the strain increases (loading) and when it decreases (unloading); the slope is larger during unloading. The material displaying this kind of behavior is called an *inelastic body*. The buckling of a member of an inelastic body is called an *inelastic buckling*.

A typical stress versus strain curve of steel subjected to tension is illustrated in Fig. 7.13(a). In this figure, σ_Y is the *yield stress*, ε_Y is the associated strain, and σ_B is the *fracture stress*; the arrowed line represents the stress versus strain curve when the strain decreases. The elastic-perfectly plastic body is categorized as a special case of the inelastic body. An elastic-perfectly plastic body in Fig. 7.13(b) and a general inelastic body in (c) are typical modelings of the material property of steel.

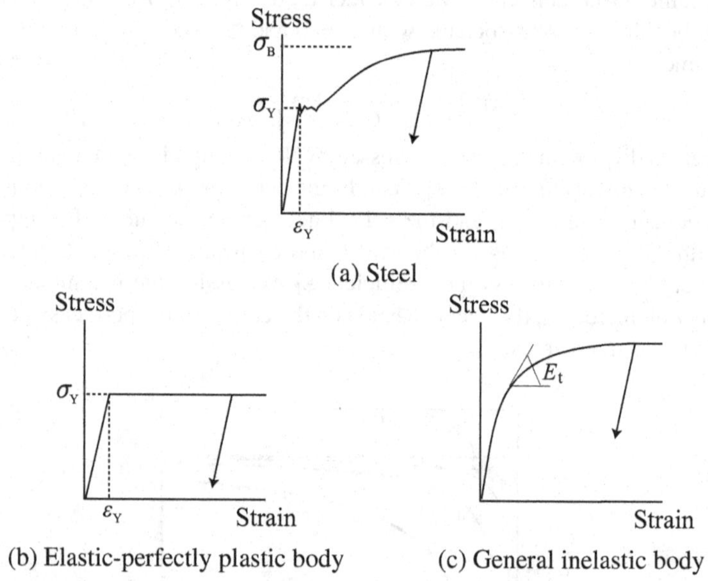

Figure 7.13 Stress versus strain curve of steel and its modeling

7.7.1 ELASTIC-PERFECTLY PLASTIC BODY

In an elastic-perfectly plastic body, the stress versus strain relation when the strain increases monotonically is defined by

$$\begin{cases} \sigma = E\varepsilon, & \text{elastic zone } (0 < \varepsilon < \varepsilon_Y), \\ \sigma = \sigma_Y, & \text{perfectly plastic zone } (\varepsilon \geq \varepsilon_Y), \end{cases} \tag{7.68}$$

where $\sigma_Y = E\varepsilon_Y$. This body displays a linear *elastic response* until the stress σ reaches the *yield stress* σ_Y, and perfectly *plastic response* thereafter ($\sigma = \sigma_Y$ and the strain increases).

The buckling stress σ_c of a beam-column member of an elastic-perfectly plastic body is given by the smaller value of the yield stress σ_Y and the elastic buckling stress $\sigma_E/k^2 = \pi^2 E/(k\lambda)^2$ in (7.44), that is,

$$\sigma_c = \min\left(\sigma_Y, \frac{\pi^2 E}{(k\lambda)^2}\right). \tag{7.69}$$

(Recall that k is the coefficient of effective buckling length and λ is the slenderness ratio introduced in Section 7.3.3.) The buckling curve of this beam-column is shown by the solid curve in Fig. 7.14(a).

The formula (7.69) of the buckling curve is normalized by the yield stress σ_Y as

$$\frac{\sigma_c}{\sigma_Y} = \min\left(1, \frac{1}{\bar{\lambda}^2}\right), \tag{7.70}$$

where

$$\bar{\lambda} = \frac{k\lambda}{\pi}\sqrt{\frac{\sigma_Y}{E}} = \frac{k\lambda}{\pi}\sqrt{\varepsilon_Y} \tag{7.71}$$

is called the *standard slenderness ratio*. The failure of a beam-column of an elastic-perfectly plastic body is described using $\bar{\lambda}$ as

$$\begin{cases} \bar{\lambda} > 1 & \Rightarrow \quad \text{elastic buckling}, \\ \bar{\lambda} < 1 & \Rightarrow \quad \text{yielding}. \end{cases}$$

A member is called a *long column* when $\bar{\lambda} > 1$, which is governed by elastic buckling, and is a *short column* when $\bar{\lambda} < 1$, which is governed by yielding.

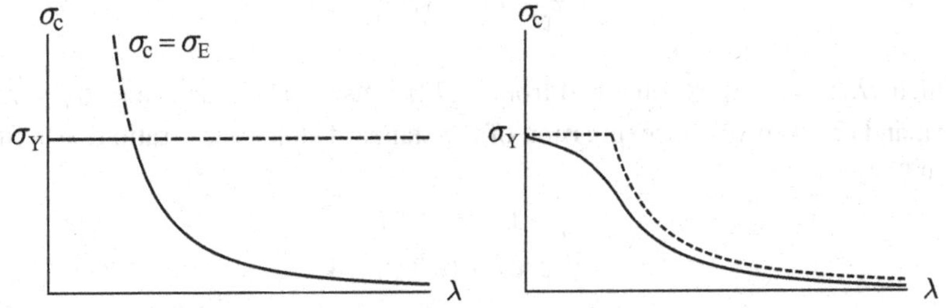

(a) Elastic-perfectly plastic body (b) General inelastic body

Figure 7.14 Buckling curves of a beam-column of an inelastic body

7.7.2 GENERAL INELASTIC BODY

In a general inelastic body, the stress σ is expressed as a function $\sigma = \sigma(\varepsilon)$ of the strain ε. The *tangent modulus* $E_t = d\sigma/d\varepsilon$, which is the slope of the stress versus strain curve, takes different values when the stress increases and when it decreases. In *tangent modulus theory*, the buckling load is obtained by replacing the modulus of elasticity E in (7.44) by the value of the tangent modulus at the buckling.[7]

By replacing the modulus of elasticity E in (7.44) by the tangent modulus E_t, we obtain the *tangent modulus buckling stress*

$$\sigma_c = \frac{\pi^2 E_t}{k^2\lambda^2} = \frac{E_t}{E}\frac{\sigma_E}{k^2}. \tag{7.72}$$

Note that $E = E_t|_{\varepsilon=0}$ is the tangent modulus at the initial state. The buckling stress curve expressed by (7.72) is shown in Fig. 7.14(b). Since $E_t/E \leq 1$ holds, the tangent modulus buckling stress is generally smaller than elastic buckling stress.

To illustrate the buckling behavior of a general inelastic body, we consider a beam-column with the material whose stress versus strain curve, for example, is given by

$$\begin{cases} \sigma = \sigma(\varepsilon) = \sigma_Y\left[\dfrac{\varepsilon}{\varepsilon_Y} - \dfrac{1}{4}\left(\dfrac{\varepsilon}{\varepsilon_Y}\right)^2\right], & 0 \leq \dfrac{\varepsilon}{\varepsilon_Y} \leq 2, \\ \sigma = \sigma_Y, & \dfrac{\varepsilon}{\varepsilon_Y} > 2, \end{cases} \tag{7.73}$$

where $\sigma_Y = E\varepsilon_Y$. This curve is plotted by the solid curve in Fig. 7.15(a). (For reference, the stress versus strain curve $\sigma = E\varepsilon$ of a linear elastic body is shown by the dashed line.) The tangent modulus is obtained by differentiating the stress versus strain relation (7.73) with respect to the strain ε as

$$E_t = \frac{d\sigma}{d\varepsilon} = E\left(1 - \frac{1}{2}\frac{\varepsilon}{\varepsilon_Y}\right), \qquad 0 \leq \frac{\varepsilon}{\varepsilon_Y} \leq 2. \tag{7.74}$$

The standard slenderness ratio $\bar{\lambda}$ in (7.71) is evaluated to

$$\bar{\lambda} = \frac{k\lambda}{\pi}\sqrt{\varepsilon_Y} = \sqrt{\frac{E_t\varepsilon_Y}{\sigma_c}},$$

in which $k\lambda/\pi = \sqrt{E_t/\sigma_c}$ obtained from (7.72) is used. Then the use of $\sigma_Y = E\varepsilon_Y$, (7.73), and (7.74) gives the expression of the standard slenderness ratio $\bar{\lambda}$ as a function in $\varepsilon/\varepsilon_Y$ as

$$\bar{\lambda} = \sqrt{\frac{1 - (\varepsilon/\varepsilon_Y)/2}{\varepsilon/\varepsilon_Y - (\varepsilon/\varepsilon_Y)^2/4}}.$$

[7] In *reduced modulus theory*, the reduced modulus considering the zone of decreasing stress in the cross-section is employed (Bažant and Cedolin [1]).

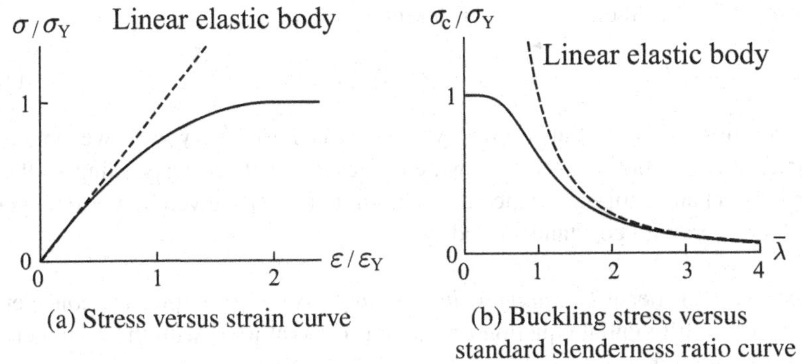

Figure 7.15 The stress versus strain curve in (7.73) and the buckling curve (σ_c versus $\bar{\lambda}$ curve)

Note that the buckling stress σ_c is also a function in $\varepsilon/\varepsilon_Y$ by (7.73). The buckling curve (σ_c versus $\bar{\lambda}$ curve) parametrized by $\varepsilon/\varepsilon_Y$ is shown by the solid curve in Fig. 7.15(b). This buckling curve generally gives a smaller buckling stress than the buckling curve of a linear elastic body.

7.8 APPENDIX: LINEAR ORDINARY DIFFERENTIAL EQUATIONS

We briefly introduce basic facts about linear differential equations and the boundary value problem of differential equations used in this chapter.

First, we consider the *homogeneous differential equation*

$$a_n y^{(n)} + a_{n-1} y^{(n-1)} + \cdots + a_1 y' + a_0 y = 0, \tag{7.75}$$

where $a_i = a_i(x)$ $(i = 0, 1, \ldots, n)$ are functions in x. The solution $y = y(x)$ to this differential equation is given by

$$y = \sum_{i=1}^{n} c_i y_i \tag{7.76}$$

as a linear combination of linearly independent n solutions $y_i = y_i(x)$ $(i = 1, \ldots, n)$ satisfying (7.75). Here c_i $(i = 1, \ldots, n)$ are constants.

Next, we consider the *inhomogeneous differential equation*

$$a_n y^{(n)} + a_{n-1} y^{(n-1)} + \cdots + a_1 y' + a_0 y = f(x). \tag{7.77}$$

If a solution $y_p = y_p(x)$ satisfying this equation is found, any solution $y = y(x)$ to this equation can be expressed as the sum of (7.76) and this particular solution as

$$y = \sum_{i=1}^{n} c_i y_i + y_p. \tag{7.78}$$

Example 7.4 The inhomogeneous differential equation

$$y'' - y = 2 - x^2 \tag{7.79}$$

has a solution $y_p = x^2$. Substituting $y = \exp(\lambda x)$ into $y'' - y = 0$, we obtain the characteristic equation $\lambda^2 - 1 = 0$, which gives $\lambda = \pm 1$, corresponding to the solutions $\exp(x)$ and $\exp(-x)$. Hence the solution to (7.79) is given by $y = c_1 \exp(x) + c_2 \exp(-x) + x^2$ with constants c_1 and c_2. □

Last, we consider a *boundary value problem*. As an example, we consider the homogeneous differential equation on the unit interval $[0, 1]$ with the homogeneous boundary condition

$$y'' + fy = 0 \qquad y(0) = y(1) = 0,$$

where $f > 0$ is a parameter. The solution to this equation is given as

$$y(x) = c_1 \sin\left(\sqrt{f}x\right) + c_2 \cos\left(\sqrt{f}x\right).$$

The use of this solution in the boundary condition leads to

$$c_1 \sin\sqrt{f} = 0, \quad c_2 = 0.$$

If $\sin\sqrt{f} \neq 0$, we have $c_1 = 0$, which gives the trivial solution $y = 0$. Therefore, a nontrivial solution to this differential equation exists only when $\sin\sqrt{f} = 0$, for which the parameter f must take special values of

$$\sqrt{f} = i\pi, \quad i = 1, 2, \ldots.$$

The associated solutions are $y(x) = c_1 \sin(i\pi x)$, where $i = 1, 2, \ldots$.

7.9 PROBLEMS

Problem 7.1 Solve the following differential equations. We assume $f > 0$ in (2).

(1) $y'' + 3y' + 2y = 2x^2 + 6x + 2$.

(2) $y'' + 4fy = 0, \qquad y(0) = 0,\ y'(1) = 0$.

Problem 7.2 Obtain the minimum buckling load of a beam-column for the following boundary conditions: (1) both ends fixed (Fig. 7.6(a)) and (2) one end fixed and the other end pinned (Fig. 7.6(b)).

Problem 7.3 Consider a beam-column with one end fixed and the other end elastically supported (Fig. 7.16). (1) Derive the buckling condition. (2) Obtain the buckling conditions for the two cases where spring constant $k \to 0$ and $k \to \infty$, and investigate the correspondence to other boundary conditions.

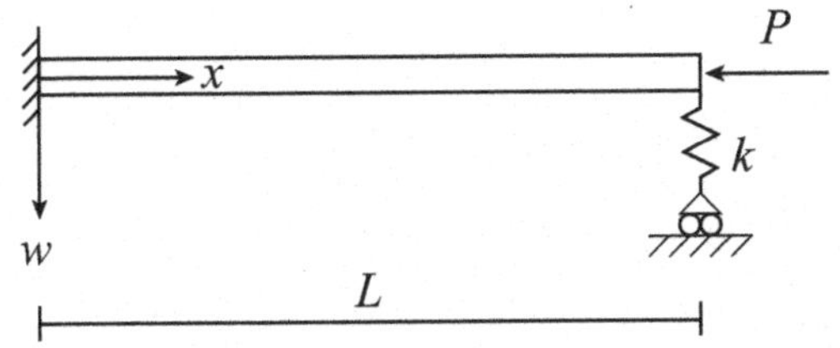

Figure 7.16 A beam-column with one end fixed and the other end elastically supported

Problem 7.4 (1) Derive (7.55) using the buckling condition in (7.23). (2) Derive (7.57).

Problem 7.5 (1) Derive (7.60). (2) Derive (7.61).

Problem 7.6 Derive (7.66).

Problem 7.7 For the I-beam cross-section in Fig. 7.8(b), obtain the moment of inertia around the y-axis and that around z-axis, and determine which is the weak axis.

Problem 7.8 We consider a beam-column member with the cylindrical cross-section in Fig. 7.7(b) with the length $L = 200$ cm, the yield stress $\sigma_Y = 3.0 \times 10^3$kgf/cm^2, and Young's modulus $E = 2.0 \times 10^6$kgf/cm^2. (1) Consider the material property of an elastic-perfectly plastic body in (7.68) and three types of boundary conditions in Fig. 7.6. For each boundary condition, obtain the standard slenderness ratio and investigate whether the failure of the member is governed by elastic buckling or yielding with reference to this ratio. (2) Consider another material property in (7.73) and obtain the buckling stress for the boundary condition of simple support.

Problem 7.9 For beam-columns with the circular cross-section and the cylindrical cross-section (Fig. 7.7) of an elastic-perfectly plastic body, we denote by L_a and L_b the maximum member lengths undergoing yielding. Then obtain the ratio L_a/L_b.

REFERENCES

1. Bažant ZP, and Cedolin L (1991), *Stability of Structures*, Oxford University Press, New York.
2. Courant R, and Hilbert D (1953), *Methods of Mathematical Physics*, Vol. 1, Interscience Publishers, New York.
3. Elishakoff, I (2020), *Handbook on Timoshenko–Ehrenfest Beam and Uflyand–Mindlin Plate Theories*, World Scientific, Singapore.
4. Timoshenko SP, and Gere JM (1961), *Theory of Elastic Stability*, 2nd ed., McGraw-Hill, New York.

8 Structural Buckling I: Truss

8.1 SUMMARY

Following the study of member buckling of beam-columns in Chapter 7, we deal with the structural buckling of truss structures. We present three kinds of analysis methods: finite displacement analysis, small-displacement analysis, and linear buckling analysis. These methods are presented consistently with the three-dimensional finite element analysis. In addition, flexural buckling of members in truss structures is investigated, based on Chapter 7. The member stiffness matrix of a truss member presented in this chapter forms a foundation of the derivation of the member stiffness matrix of a frame member in Chapter 9. Chapter 3 of path tracing and Chapter 4 of branch switching are prerequisites for the finite displacement bifurcation analysis of truss structures.

Table 8.1
Comparison of three kinds of analysis methods for truss structures

	Finite displacement	Small displacement	Linear buckling
Nonlinearity	Strong	Absent	Weak
Geometric stiffness	Nonlinear	Absent	Linear
Buckling load/mode	Obtainable (more accurate)	Unobtainable	Obtainable (less accurate)
Post-buckling behavior	Obtainable	Unobtainable	Unobtainable

We present finite displacement equilibrium equations for each member and for the whole structure in Section 8.2 as a foundation of this chapter. The other two analysis methods are presented in Sections 8.3 and 8.4. The simplicity of a method of analysis is inversely proportional to the amount and accuracy of information to be obtained (see Table 8.1 for the comparison of these three methods). The equilibrium equation of the small displacement analysis is a linear simultaneous equation, which is easy to handle but fails to express the buckling behavior. The linear buckling analysis[1] can give the buckling load and mode but cannot offer information about post-buckling behavior. The finite displacement method can elucidate post-bifurcation behavior at

[1]The linear buckling analysis of frame structures is presented in Chapter 9 and also in the literature (e.g., Iyengar [3] and Bažant and Cedolin [1]).

DOI: 10.1201/9781003112365-8

the cost of a complicated analysis procedure (see, e.g., Crisfield [2] for more advanced issues of this method). Buckling of truss members is studied in Section 8.5.

Keywords: • Bifurcation • Buckling • Finite displacement analysis • Geometric stiffness matrix • Linear buckling analysis • Member stiffness equation • Small displacement analysis • Stiffness matrix • Truss structure

8.2 FINITE DISPLACEMENT ANALYSIS

As the first of the three kinds of analysis methods for a truss structure in Sections 8.2–8.4, we formulate the finite displacement equilibrium equations for each member and for the whole structure. We introduce a finite displacement analysis method for directly solving the nonlinear finite-displacement equilibrium equation of a truss structure. The truss member is assumed not to buckle and has linearly elastic material property.

We consider a truss structure with n nodes (x_i, y_i, z_i) $(i = 1, \ldots, n)$ and M members indexed by $m = 1, \ldots, M$. The *nodal displacement* of the ith node is denoted by

$$\boldsymbol{u}_i = \begin{pmatrix} u_i \\ v_i \\ w_i \end{pmatrix} \tag{8.1}$$

and the *member end forces* of the mth member, assumed to connect the ith and jth nodes, are denoted by

$$\boldsymbol{p}_i^{(m)} = \begin{pmatrix} X_i^{(m)} \\ Y_i^{(m)} \\ Z_i^{(m)} \end{pmatrix}, \qquad \boldsymbol{p}_j^{(m)} = \begin{pmatrix} X_j^{(m)} \\ Y_j^{(m)} \\ Z_j^{(m)} \end{pmatrix}. \tag{8.2}$$

8.2.1 MEMBER STIFFNESS EQUATION

First, we deal with a single *truss member*, say the mth member connecting the ith and jth nodes, as depicted in Fig. 8.1. The nodes i and j are initially located at (x_i, y_i, z_i) and (x_j, y_j, z_j). With reference to the variables in (8.1) and (8.2), we define variables $\boldsymbol{u}^{(m)}$ and $\boldsymbol{p}^{(m)}$ for the mth member as

$$\boldsymbol{u}^{(m)} = \begin{pmatrix} \boldsymbol{u}_i \\ \boldsymbol{u}_j \end{pmatrix} = \begin{pmatrix} u_i^{(m)} \\ v_i^{(m)} \\ w_i^{(m)} \\ u_j^{(m)} \\ v_j^{(m)} \\ w_j^{(m)} \end{pmatrix}, \qquad \boldsymbol{p}^{(m)} = \begin{pmatrix} \boldsymbol{p}_i^{(m)} \\ \boldsymbol{p}_j^{(m)} \end{pmatrix} = \begin{pmatrix} X_i^{(m)} \\ Y_i^{(m)} \\ Z_i^{(m)} \\ X_j^{(m)} \\ Y_j^{(m)} \\ Z_j^{(m)} \end{pmatrix}, \tag{8.3}$$

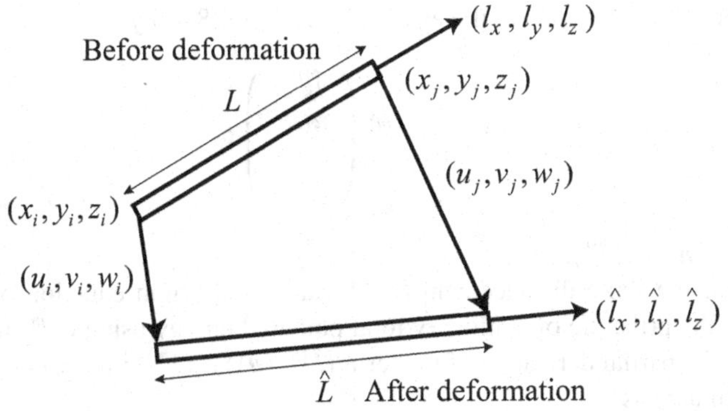

Figure 8.1 A truss member

where $\boldsymbol{u}^{(m)}$ represents member end displacements and $\boldsymbol{p}^{(m)}$ stands for member end forces.

The *axial force* $N^{(m)} = N^{(m)}(\boldsymbol{u}^{(m)})$ that emerges in the member is expressed by *Hooke's law* as

$$N^{(m)} = EA^{(m)} \frac{\hat{L}^{(m)} - L^{(m)}}{L^{(m)}}, \tag{8.4}$$

where E is the modulus of elasticity, which is common to all the members, $A^{(m)}$ is the cross-sectional area, $L^{(m)}$ and $\hat{L}^{(m)} = \hat{L}^{(m)}(\boldsymbol{u}^{(m)})$ are initial and deformed lengths defined, respectively, as

$$L^{(m)} = [(x_j - x_i)^2 + (y_j - y_i)^2 + (z_j - z_i)^2]^{1/2}, \tag{8.5}$$

$$\hat{L}^{(m)} = [(u_j - u_i + x_j - x_i)^2 + (v_j - v_i + y_j - y_i)^2 + (w_j - w_i + z_j - z_i)^2]^{1/2}. \tag{8.6}$$

Note that a tensile axial force is defined to be positive in (8.4).

If we regard the member end force $\boldsymbol{p}^{(m)}$ as an external force, the total potential energy stored in the truss member is expressed as

$$\begin{aligned} U^{(m)}(\boldsymbol{u}^{(m)}, \boldsymbol{p}^{(m)}) &= \int (\text{Axial force})\,\mathrm{d}(\text{Axial displacement}) \\ &\quad - \sum (\text{Member end force}) \times (\text{Member end displacement}) \\ &= \frac{EA^{(m)}}{2L^{(m)}} \left(\hat{L}^{(m)}(\boldsymbol{u}^{(m)}) - L^{(m)}\right)^2 - (\boldsymbol{p}^{(m)})^\top \boldsymbol{u}^{(m)}. \end{aligned} \tag{8.7}$$

For the derivation of the governing equation, we need to define the *direction cosine* $(\hat{l}_x^{(m)}, \hat{l}_y^{(m)}, \hat{l}_z^{(m)})$ of the deformed member from i to j by

$$\begin{pmatrix} \hat{l}_x^{(m)} \\ \hat{l}_y^{(m)} \\ \hat{l}_z^{(m)} \end{pmatrix} = \frac{1}{\hat{L}^{(m)}} \begin{pmatrix} u_j - u_i + x_j - x_i \\ v_j - v_i + y_j - y_i \\ w_j - w_i + z_j - z_i \end{pmatrix}, \tag{8.8}$$

and a vector expression $\boldsymbol{n}^{(m)}$ of the axial force $N^{(m)}$ in (8.4) by

$$\boldsymbol{n}^{(m)} = N^{(m)} \begin{pmatrix} \hat{l}_x^{(m)} \\ \hat{l}_y^{(m)} \\ \hat{l}_z^{(m)} \end{pmatrix}, \tag{8.9}$$

where $\boldsymbol{n}^{(m)} = \boldsymbol{n}^{(m)}(\boldsymbol{u}^{(m)})$.

We derive the finite displacement (nonlinear) equilibrium equation of the truss member by the principle of stationary total potential energy using $U^{(m)}$ in (8.7). To begin with, the partial derivative of the term $(\hat{L}^{(m)}(\boldsymbol{u})) - L^{(m)})^2$ in (8.7) with respect to u_i is calculated as

$$\frac{\partial}{\partial u_i}\left(\hat{L}^{(m)} - L^{(m)}\right)^2 = 2\left(\hat{L}^{(m)} - L^{(m)}\right)\left(-\frac{u_j - u_i + x_j - x_i}{\hat{L}^{(m)}}\right) = -2\left(\hat{L}^{(m)} - L^{(m)}\right)\hat{l}_x^{(m)}$$

by (8.6) and (8.8). Then, with (8.4), we have

$$\frac{\partial}{\partial u_i}\frac{EA^{(m)}}{2L^{(m)}}\left(\hat{L}^{(m)} - L^{(m)}\right)^2 = -N^{(m)}\hat{l}_x^{(m)}.$$

By similar calculations, we obtain[2]

$$\left(\frac{\partial}{\partial \boldsymbol{u}_i}\frac{EA^{(m)}}{2L^{(m)}}\left(\hat{L}^{(m)} - L^{(m)}\right)^2\right)^\top = -\boldsymbol{n}^{(m)},$$

$$\left(\frac{\partial}{\partial \boldsymbol{u}_j}\frac{EA^{(m)}}{2L^{(m)}}\left(\hat{L}^{(m)} - L^{(m)}\right)^2\right)^\top = \boldsymbol{n}^{(m)}.$$

Therefore, the partial derivative of $U^{(m)}$ in (8.7) with respect to $\boldsymbol{u}^{(m)}$, which we denote by $\boldsymbol{F}^{(m)} = \boldsymbol{F}^{(m)}(\boldsymbol{u}^{(m)}, \boldsymbol{p}^{(m)})$, is given as

$$\boldsymbol{F}^{(m)} = \begin{pmatrix} -\boldsymbol{n}^{(m)} \\ \boldsymbol{n}^{(m)} \end{pmatrix} - \begin{pmatrix} \boldsymbol{p}_i^{(m)} \\ \boldsymbol{p}_j^{(m)} \end{pmatrix} = \boldsymbol{N}^{(m)}(\boldsymbol{u}^{(m)}) - \boldsymbol{p}^{(m)} \tag{8.10}$$

with $\boldsymbol{N}^{(m)} = \boldsymbol{N}^{(m)}(\boldsymbol{u}^{(m)})$ defined as

$$\boldsymbol{N}^{(m)} = \begin{pmatrix} -\boldsymbol{n}^{(m)} \\ \boldsymbol{n}^{(m)} \end{pmatrix}. \tag{8.11}$$

Then the equilibrium equation is given by

$$\boldsymbol{F}^{(m)}(\boldsymbol{u}^{(m)}, \boldsymbol{p}^{(m)}) \equiv \boldsymbol{N}^{(m)}(\boldsymbol{u}^{(m)}) - \boldsymbol{p}^{(m)} = \boldsymbol{0}. \tag{8.12}$$

---

[2]For any function φ, the gradient vector $\partial\varphi/\partial\boldsymbol{u}_i$ is defined to be a row vector, i.e., $\partial\varphi/\partial\boldsymbol{u}_i = (\partial\varphi/\partial u_i, \partial\varphi/\partial v_i, \partial\varphi/\partial w_i)$.

Since the axial force and the direction cosine are nonlinear functions in displacements, the variables $\boldsymbol{n}^{(m)}$, $\boldsymbol{N}^{(m)}$, and $\boldsymbol{F}^{(m)}$ are also nonlinear functions in $\boldsymbol{u}^{(m)}$.

The tangent stiffness matrix (Jacobian matrix) $K^{(m)} = K^{(m)}(\boldsymbol{u}^{(m)})$ of a truss member is obtained as the derivative of $\boldsymbol{F}^{(m)}$ with respect to $\boldsymbol{u}^{(m)}$, which takes the form of

$$K^{(m)} = \begin{pmatrix} k^{(m)} & -k^{(m)} \\ -k^{(m)} & k^{(m)} \end{pmatrix} \tag{8.13}$$

with a symmetric matrix $k^{(m)}$ defined as

$$k^{(m)} = -\frac{\partial \boldsymbol{n}^{(m)}}{\partial \boldsymbol{u}_i} = \frac{\partial \boldsymbol{n}^{(m)}}{\partial \boldsymbol{u}_j}. \tag{8.14}$$

For later use (cf., Sections 8.3 and 8.4), we introduce the following two alternative expressions of $k^{(m)}$:

$$k^{(m)} = \frac{EA^{(m)}}{\hat{L}^{(m)}} h^{(m)} + \frac{N^{(m)}}{\hat{L}^{(m)}} I_3 \tag{8.15}$$

$$= \frac{EA^{(m)}}{L^{(m)}} h^{(m)} + \frac{N^{(m)}}{\hat{L}^{(m)}} \left(I_3 - h^{(m)} \right), \tag{8.16}$$

where

$$I_3 = \begin{pmatrix} 1 & 0 & 0 \\ 0 & 1 & 0 \\ 0 & 0 & 1 \end{pmatrix}, \qquad h^{(m)} = \begin{pmatrix} (\hat{l}_x^{(m)})^2 & \hat{l}_x^{(m)} \hat{l}_y^{(m)} & \hat{l}_x^{(m)} \hat{l}_z^{(m)} \\ \hat{l}_x^{(m)} \hat{l}_y^{(m)} & (\hat{l}_y^{(m)})^2 & \hat{l}_y^{(m)} \hat{l}_z^{(m)} \\ \hat{l}_x^{(m)} \hat{l}_z^{(m)} & \hat{l}_y^{(m)} \hat{l}_z^{(m)} & (\hat{l}_z^{(m)})^2 \end{pmatrix}.$$

The derivation of (8.15) and (8.16) is addressed in Problem 8.3. As can be seen from (8.15), a negative (compressive) axial force $N^{(m)}$ reduces the diagonal elements of $k^{(m)}$ and often leads to bifurcation and buckling.

8.2.2 STRUCTURAL EQUILIBRIUM EQUATION

We derive the finite displacement equilibrium equation of the whole truss structure with n nodes (x_i, y_i, z_i) $(i = 1, \ldots, n)$ and M members indexed by $m = 1, \ldots, M$. This equation and the tangent stiffness matrix are derived through an assembly of the equilibrium equations of the members. We define a nodal displacement vector $\boldsymbol{u}$ and a nodal external force vector $\boldsymbol{f}$ as

$$\boldsymbol{u} = \begin{pmatrix} \boldsymbol{u}_1 \\ \vdots \\ \boldsymbol{u}_n \end{pmatrix}, \qquad \boldsymbol{f} = \begin{pmatrix} \boldsymbol{f}_1 \\ \vdots \\ \boldsymbol{f}_n. \end{pmatrix},$$

where $\boldsymbol{u}_i$ is the displacement vector of the ith node and $\boldsymbol{f}_i$ is the external force vector at the ith node. Note that each $\boldsymbol{u}^{(m)}$ is a subvector of $\boldsymbol{u}$. Furthermore, we introduce vectors $\tilde{\boldsymbol{N}}^{(m)}$ and $\tilde{\boldsymbol{p}}^{(m)}$ for the whole structure by embedding the vectors $\boldsymbol{N}^{(m)}$ in (8.11) and $\boldsymbol{p}^{(m)}$ in (8.3) as

$$\tilde{\boldsymbol{N}}^{(m)} = \begin{matrix} \\ i \\ \\ j \\ \\ \end{matrix}\begin{pmatrix} \mathbf{0} \\ -\boldsymbol{n}^{(m)} \\ \mathbf{0} \\ \boldsymbol{n}^{(m)} \\ \mathbf{0} \end{pmatrix}, \qquad \tilde{\boldsymbol{p}}^{(m)} = \begin{matrix} \\ i \\ \\ j \\ \\ \end{matrix}\begin{pmatrix} \mathbf{0} \\ \boldsymbol{p}_i^{(m)} \\ \mathbf{0} \\ \boldsymbol{p}_j^{(m)} \\ \mathbf{0} \end{pmatrix}, \tag{8.17}$$

which have nonzero components only for the ith and jth nodes.

The potential U of this truss structure is obtained as a *superposition* of $U^{(m)}$ in (8.7) for all members as

$$U(\boldsymbol{u}) = \sum_{m=1}^{M} U^{(m)}(\boldsymbol{u}^{(m)}, \boldsymbol{p}^{(m)}). \tag{8.18}$$

By the equilibrium of external forces and internal forces, we have

$$\boldsymbol{f} = \sum_{m=1}^{M} \tilde{\boldsymbol{p}}^{(m)}. \tag{8.19}$$

The equilibrium equation is obtained by differentiating the potential U in (8.18) with respect to $\boldsymbol{u}$ and using (8.10) and (8.19) as[3]

$$\boldsymbol{F}(\boldsymbol{u},\boldsymbol{f}) = \left(\frac{\partial U}{\partial \boldsymbol{u}}\right)^{\top} = \sum_{m=1}^{M} \left(\tilde{\boldsymbol{N}}^{(m)}(\boldsymbol{u}^{(m)}) - \tilde{\boldsymbol{p}}^{(m)}\right) = \sum_{m=1}^{M} \tilde{\boldsymbol{N}}^{(m)}(\boldsymbol{u}^{(m)}) - \boldsymbol{f}. \tag{8.20}$$

Then the equilibrium equation is given by

$$\boldsymbol{F}(\boldsymbol{u},\boldsymbol{f}) \equiv \sum_{m=1}^{M} \tilde{\boldsymbol{N}}^{(m)}(\boldsymbol{u}^{(m)}) - \boldsymbol{f} = \mathbf{0}. \tag{8.21}$$

This equation is to be solved together with the displacement boundary conditions.

We consider the most common boundary conditions in which some nodes are free and others are fixed, and external forces are applied to some or all of the free nodes and reaction forces emerge at the fixed nodes. Let I denote the set of free nodes, where $I \subset \{1,2,\ldots,n\}$; the set of the fixed nodes is the complement of I, to be denoted by $\bar{I}$. Then the boundary conditions are expressed as

$$\boldsymbol{f}_i \text{ is specified for } i \in I, \tag{8.22}$$

$$\boldsymbol{u}_i = \mathbf{0} \text{ is specified for } i \in \bar{I}. \tag{8.23}$$

[3]Recall that $\partial U/\partial \boldsymbol{u}$ is a row vector (cf., (2.21)).

It should be clear that $\boldsymbol{u}_i$ for free nodes $(i \in I)$ and $\boldsymbol{f}_i$ for fixed nodes $(i \in \bar{I})$ are determined from (8.21) using (8.22) and (8.23). For a fixed node i, the reaction force $\boldsymbol{f}_i$ is given by

$$\boldsymbol{f}_i = \sum_{m \in M^+(i)} \boldsymbol{n}^{(m)} - \sum_{m \in M^-(i)} \boldsymbol{n}^{(m)}, \tag{8.24}$$

where $M^+(i)$ and $M^-(i)$ denote the sets of members directing to node i and from node i, respectively.

Since the function $\boldsymbol{F}$ in (8.20) is obtained by an assembly of $\boldsymbol{F}^{(m)}$ in (8.10), the Jacobian matrix of $\boldsymbol{F}$ with respect to $\boldsymbol{u}$ can be obtained by an assembly of the Jacobian matrices of $\boldsymbol{F}^{(m)}$. That is, the tangent stiffness matrix $K^\circ = K^\circ(\boldsymbol{u})$ of the whole structure is represented as

$$K^\circ = \sum_{m=1}^{M} \tilde{K}^{(m)}, \tag{8.25}$$

where $\tilde{K}^{(m)}$ is the whole structure counterpart of the tangent stiffness matrix $K^{(m)}$ in (8.13) and has nonzero blocks only for the ith and the jth nodes.[4]

In accordance with the classification into free and fixed nodes, the tangent stiffness matrix K° is regarded as a 2×2 block matrix, with rows (and columns) partitioned into I and $\bar{I}$. Assuming $I = \{1, \ldots, |I|\}$ for simplicity of presentation, the matrix K° is expressed as

$$K^\circ = \begin{pmatrix} K & K^\triangle \\ K^\triangledown & K^\bullet \end{pmatrix}, \tag{8.26}$$

where K is an $|I| \times |I|$ matrix corresponding to the free nodes in I and $K^\bullet$ is an $|\bar{I}| \times |\bar{I}|$ matrix corresponding to the fixed nodes in $\bar{I}$.

Remark 8.1 A *proportional loading*, which has a constant *load pattern vector* $\boldsymbol{f}_0$ and varying magnitude f, is usually used in finite displacement analysis. The finite-displacement equilibrium equation (8.21) reduces to

$$\boldsymbol{F}(\boldsymbol{u}, f) = \sum_{m=1}^{M} \tilde{\boldsymbol{N}}^{(m)}(\boldsymbol{u}^{(m)}) - f \cdot \boldsymbol{f}_0 = \boldsymbol{0} \tag{8.27}$$

in this case. □

8.2.3 STRUCTURAL EXAMPLE

A finite displacement analysis is conducted on the three-bar truss tent depicted in Fig. 8.2(a).[5] All members have the same cross-sectional rigidity EA. The initial locations of the nodes are given by

$$(x_1, y_1, z_1) = (1, 0, 3), \qquad (x_2, y_2, z_2) = (-1/2, -\sqrt{3}/2, 3),$$
$$(x_3, y_3, z_3) = (-1/2, \sqrt{3}/2, 3), \qquad (x_4, y_4, z_4) = (0, 0, 0).$$

[4] In this chapter, we use $\tilde{\cdot}$ to indicate embedding of a member characteristic into the whole structure.

[5] We employ a simple structure in this chapter, for which the equilibrium path can be obtained easily. See Chapter 3 of path tracing and Chapter 4 of branch switching for large-scale analysis of truss structures.

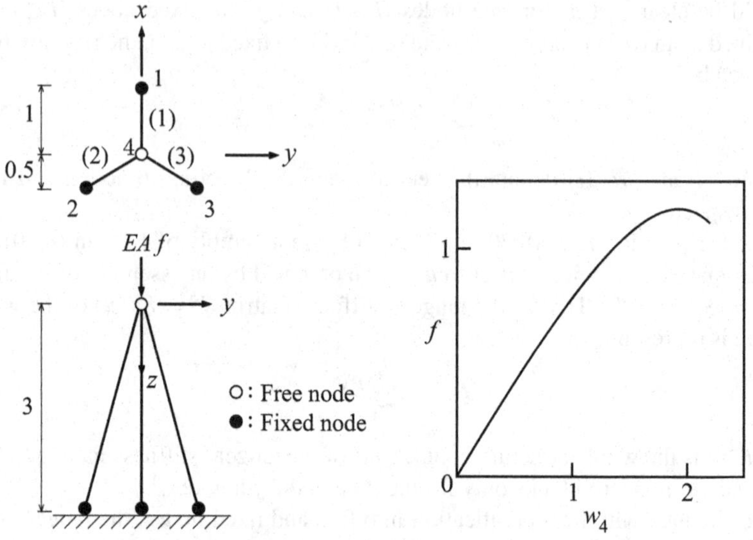

(a) Three-bar truss tent (b) Load versus displacement curve

Figure 8.2 A three-bar truss tent and its load versus displacement curve

The crown node 4 is located initially at the origin $(0,0,0)$ and its displacement is expressed by vector (u_4, v_4, w_4). Since the crown node 4 is free and the other three are fixed, we have $I = \{4\}$ and $\bar{I} = \{1,2,3\}$ in (8.22) and (8.23). The equilibrium equation of this truss tent is given by (8.21) as

$$\boldsymbol{F}(\boldsymbol{u},\boldsymbol{f}) = \begin{pmatrix} -\boldsymbol{n}^{(1)} \\ -\boldsymbol{n}^{(2)} \\ -\boldsymbol{n}^{(3)} \\ \boldsymbol{n}^{(1)} + \boldsymbol{n}^{(2)} + \boldsymbol{n}^{(3)} \end{pmatrix} - \begin{pmatrix} \boldsymbol{f}_1 \\ \boldsymbol{f}_2 \\ \boldsymbol{f}_3 \\ \boldsymbol{f}_4 \end{pmatrix} = \boldsymbol{0}, \tag{8.28}$$

where the members are directed from nodes 1, 2, and 3 to node 4. The tangent stiffness matrix K° in (8.25) is given as

$$K^\circ = \left(\begin{array}{ccc|c} k^{(1)} & & & -k^{(1)} \\ & k^{(2)} & & -k^{(2)} \\ & & k^{(3)} & -k^{(3)} \\ \hline -k^{(1)} & -k^{(2)} & -k^{(3)} & k^{(1)} + k^{(2)} + k^{(3)} \end{array}\right).$$

For $I = \{4\}$ and $\bar{I} = \{1,2,3\}$, we have

$$K = k^{(1)} + k^{(2)} + k^{(3)}, \qquad K^\bullet = \begin{pmatrix} k^{(1)} & & \\ & k^{(2)} & \\ & & k^{(3)} \end{pmatrix}$$

in the notation of (8.26).

By the boundary conditions $\boldsymbol{u}_1 = \boldsymbol{u}_2 = \boldsymbol{u}_3 = \boldsymbol{0}$ for the fixed nodes 1, 2 and 3 and a given external force $\boldsymbol{f}_4 = (0, 0, EAf)^\top$, (8.28) can be decomposed into the equilibrium equation

$$\boldsymbol{n}^{(1)} + \boldsymbol{n}^{(2)} + \boldsymbol{n}^{(3)} = \boldsymbol{f}_4 \tag{8.29}$$

for the free node 4 and the equations for fixed end forces

$$\boldsymbol{f}_1 = -\boldsymbol{n}^{(1)}, \quad \boldsymbol{f}_2 = -\boldsymbol{n}^{(2)}, \quad \boldsymbol{f}_3 = -\boldsymbol{n}^{(3)}.$$

The concrete form of the equilibrium equation (8.29) is obtained, using (8.8) and (8.9), as

$$\sum_{m=1}^{3} EA^{(m)} \left(\frac{1}{L^{(m)}} - \frac{1}{\hat{L}^{(m)}} \right) \begin{pmatrix} u_4 - x_m \\ v_4 - y_m \\ w_4 - z_m \end{pmatrix} = EAf \begin{pmatrix} 0 \\ 0 \\ 1 \end{pmatrix}, \tag{8.30}$$

where

$$A^{(1)} = A^{(2)} = A^{(3)} = A, \qquad L^{(1)} = L^{(2)} = L^{(3)} = L = \sqrt{10},$$
$$\hat{L}^{(m)} = \sqrt{(u_4 - x_m)^2 + (v_4 - y_m)^2 + (w_4 - z_m)^2}, \quad m = 1, 2, 3.$$

The solution (u_4, v_4, w_4, f) to (8.30) is obtained as $u_4 = v_4 = 0$ and

$$f = \sum_{m=1}^{3} \left(\frac{1}{L} - \frac{1}{\hat{L}^{(m)}} \right) (w_4 - z_m) = 3 \left(\frac{1}{\sqrt{10}} - \frac{1}{\sqrt{1 + (w_4 - 3)^2}} \right) (w_4 - 3).$$

Figure 8.2(b) plots the load versus displacement curve expressed by this $f - w_4$ relation.

8.3 SMALL DISPLACEMENT ANALYSIS

The small-displacement equilibrium equation is a linear simultaneous equation expressing the equilibrium of the system undergoing deformation. This equilibrium equation is easy to deal with and gives unique displacements for given external forces. This equation gives the first-order approximation to the mechanical response of a truss structure and cannot express buckling and bifurcation. In this section, we introduce the formulation of the small displacement analysis of a truss structure and an application to a structural system.

8.3.1 MEMBER STIFFNESS EQUATION

We derive the small-displacement equilibrium equation of a truss member, which is also called the stiffness equation. For this purpose, we adopt a linear approximation of the finite-displacement nonlinear equilibrium equation (8.10):

$$\boldsymbol{F}^{(m)} = \boldsymbol{N}^{(m)}(\boldsymbol{u}^{(m)}) - \boldsymbol{p}^{(m)}. \tag{8.31}$$

To be more precise, we approximate this equation in two ways:

- The direction cosine $(\hat{l}_x^{(m)}, \hat{l}_y^{(m)}, \hat{l}_z^{(m)})$ in (8.8) for the member after deformation is approximated by the direction cosine

$$l_x^{(m)} = \frac{x_j - x_i}{L^{(m)}}, \qquad l_y^{(m)} = \frac{y_j - y_i}{L^{(m)}}, \qquad l_z^{(m)} = \frac{z_j - z_i}{L^{(m)}} \tag{8.32}$$

at the initial state.

- Approximate the member axial force $N^{(m)}$ in (8.4) as

$$\begin{aligned} N^{(m)} &= \frac{EA^{(m)}}{L^{(m)}} (\hat{L}^{(m)} - L^{(m)}) \\ &\approx \frac{EA^{(m)}}{L^{(m)}} \{l_x^{(m)}(u_j - u_i) + l_y^{(m)}(v_j - v_i) + l_z^{(m)}(w_j - w_i)\} \\ &= \frac{EA^{(m)}}{L^{(m)}} (-l_x^{(m)}, -l_y^{(m)}, -l_z^{(m)}, l_x^{(m)}, l_y^{(m)}, l_z^{(m)})\, \boldsymbol{u}^{(m)}. \end{aligned} \tag{8.33}$$

See Problem 8.4 for the derivation of this relation.

Using these two approximations, the axial force vector $\boldsymbol{n}^{(m)}$ in (8.9) is approximated as

$$\begin{aligned} \boldsymbol{n}^{(m)} &= N^{(m)} (\hat{l}_x^{(m)}, \hat{l}_y^{(m)}, \hat{l}_z^{(m)})^\top \\ &\approx \frac{EA^{(m)}}{L^{(m)}} \begin{pmatrix} l_x^{(m)} \\ l_y^{(m)} \\ l_z^{(m)} \end{pmatrix} (-l_x^{(m)}, -l_y^{(m)}, -l_z^{(m)}, l_x^{(m)}, l_y^{(m)}, l_z^{(m)})\, \boldsymbol{u}^{(m)}. \end{aligned}$$

That is,

$$\boldsymbol{n}^{(m)} \approx \begin{pmatrix} -k_{\rm E}^{(m)} & k_{\rm E}^{(m)} \end{pmatrix} \boldsymbol{u}^{(m)} \tag{8.34}$$

with $k_{\rm E}^{(m)}$ defined as

$$k_{\rm E}^{(m)} = \frac{EA^{(m)}}{L^{(m)}} \begin{pmatrix} (l_x^{(m)})^2 & l_x^{(m)} l_y^{(m)} & l_x^{(m)} l_z^{(m)} \\ l_x^{(m)} l_y^{(m)} & (l_y^{(m)})^2 & l_y^{(m)} l_z^{(m)} \\ l_x^{(m)} l_z^{(m)} & l_y^{(m)} l_z^{(m)} & (l_z^{(m)})^2 \end{pmatrix}. \tag{8.35}$$

The vector $\boldsymbol{N}^{(m)}$ in (8.11) is approximated as

$$\boldsymbol{N}^{(m)} = \begin{pmatrix} -\boldsymbol{n}^{(m)} \\ \boldsymbol{n}^{(m)} \end{pmatrix} \approx K_{\rm E}^{(m)} \boldsymbol{u}^{(m)}, \tag{8.36}$$

where $K_{\rm E}^{(m)}$ is the *member stiffness matrix* whose concrete form is given by

$$K_{\rm E}^{(m)} = \begin{pmatrix} k_{\rm E}^{(m)} & -k_{\rm E}^{(m)} \\ -k_{\rm E}^{(m)} & k_{\rm E}^{(m)} \end{pmatrix}. \tag{8.37}$$

The use of the approximation in (8.36) in the finite displacement equilibrium equation in (8.31) leads to the *small-displacement equilibrium equation*

$$\boldsymbol{F}^{(m)} \approx K_{\mathrm{E}}^{(m)} \boldsymbol{u}^{(m)} - \boldsymbol{p}^{(m)} = \boldsymbol{0}, \tag{8.38}$$

which is a linear equation in $\boldsymbol{u}^{(m)}$.

8.3.2 STRUCTURAL STIFFNESS EQUATION

We continue to consider a truss structure with n nodes (x_i, y_i, z_i) $(i = 1, \dots, n)$ and M members indexed by $m = 1, \dots, M$. Similarly to Section 8.2.2 for the finite displacement analysis, we can derive the structural stiffness equation of the whole truss structure from an assembly of the equation (8.38) for each member. The resulting equation may be represented as

$$\boldsymbol{F} \approx K_{\mathrm{E}}^{\circ} \boldsymbol{u} - \boldsymbol{f} = \boldsymbol{0}, \tag{8.39}$$

where $\boldsymbol{f}$ is given by (8.19) and K_{E}° is a $3n \times 3n$ *stiffness matrix*, which is given by

$$K_{\mathrm{E}}^{\circ} = \sum_{m=1}^{M} \tilde{K}_{\mathrm{E}}^{(m)}. \tag{8.40}$$

Here $\tilde{K}_{\mathrm{E}}^{(m)}$ denotes the $3n \times 3n$ matrix obtained by embedding the 6×6 member stiffness matrix $K_{\mathrm{E}}^{(m)}$ in (8.37) compatibly with the whole structure; this matrix has nonzero blocks only for those related to ith and jth nodes at the ends of the mth member.

The small-displacement equilibrium equation (8.39) is a linear equation with respect to $\boldsymbol{u}$. Accordingly, for given external forces and boundary conditions, the nodal displacements of the truss structure can be determined uniquely from $K_{\mathrm{E}}^{\circ} \boldsymbol{u} - \boldsymbol{f} = \boldsymbol{0}$ in (8.39).

8.3.3 STRUCTURAL EXAMPLE

A small displacement analysis is conducted on the two-bar truss arch in Fig. 8.3(a). The two truss members have the same cross-sectional rigidity EA. Note that the finite displacement analysis of this truss arch is available in Section 3.4.

The direction cosines of the two truss members before deformation are given by (8.32) as

$$l_x^{(1)} = \frac{1}{\sqrt{2}}, \quad l_y^{(1)} = -\frac{1}{\sqrt{2}}, \quad l_z^{(1)} = 0; \qquad l_x^{(2)} = l_y^{(2)} = -\frac{1}{\sqrt{2}}, \quad l_z^{(2)} = 0.$$

The member stiffness matrices $k_{\mathrm{E}}^{(1)}$ and $k_{\mathrm{E}}^{(2)}$ are given by (8.35) as

$$k_{\mathrm{E}}^{(1)} = \frac{EA}{2\sqrt{2}} \begin{pmatrix} 1 & -1 \\ -1 & 1 \end{pmatrix}, \quad k_{\mathrm{E}}^{(2)} = \frac{EA}{2\sqrt{2}} \begin{pmatrix} 1 & 1 \\ 1 & 1 \end{pmatrix}, \tag{8.41}$$

where the z-directional components are omitted since this truss arch is a two-dimensional structure in the xy-plane. The stiffness matrix $K_{\rm E}^{\circ}$ in (8.40) of this truss arch is given as a superposition of the member stiffness matrices as

$$K_{\rm E}^{\circ} = \tilde{K}_{\rm E}^{(1)} + \tilde{K}_{\rm E}^{(2)} = \begin{pmatrix} k_{\rm E}^{(1)} & O & -k_{\rm E}^{(1)} \\ O & O & O \\ -k_{\rm E}^{(1)} & O & k_{\rm E}^{(1)} \end{pmatrix} + \begin{pmatrix} O & O & O \\ O & k_{\rm E}^{(2)} & -k_{\rm E}^{(2)} \\ O & -k_{\rm E}^{(2)} & k_{\rm E}^{(2)} \end{pmatrix}.$$

Accordingly, the equilibrium equation (8.39) for this truss arch is

$$\begin{pmatrix} k_{\rm E}^{(1)} & O & -k_{\rm E}^{(1)} \\ O & k_{\rm E}^{(2)} & -k_{\rm E}^{(2)} \\ -k_{\rm E}^{(1)} & -k_{\rm E}^{(2)} & k_{\rm E}^{(1)} + k_{\rm E}^{(2)} \end{pmatrix} \begin{pmatrix} \boldsymbol{u}_1 \\ \boldsymbol{u}_2 \\ \boldsymbol{u}_3 \end{pmatrix} - \begin{pmatrix} \boldsymbol{f}_1 \\ \boldsymbol{f}_2 \\ \boldsymbol{f}_3 \end{pmatrix} = \boldsymbol{0}. \tag{8.42}$$

Since node 3 is free and nodes 1 and 2 are fixed, we have $I = \{3\}$ and $\bar{I} = \{1,2\}$ in the notation of Section 8.2.2. The submatrix K in (8.26) is given as $K = k_{\rm E}^{(1)} + k_{\rm E}^{(2)}$.

First, we obtain the unknown displacement vector $\boldsymbol{u}_3$ at the free node 3. Using the boundary conditions $\boldsymbol{u}_1 = \boldsymbol{u}_2 = \boldsymbol{0}$ in the third equation of (8.42), we obtain the equation

$$(k_{\rm E}^{(1)} + k_{\rm E}^{(2)})\,\boldsymbol{u}_3 - \boldsymbol{f}_3 = \boldsymbol{0}. \tag{8.43}$$

With the use of $k_{\rm E}^{(1)}$ and $k_{\rm E}^{(2)}$ in (8.41) and the external force $\boldsymbol{f}_3 = (0, EAf)^\top$, the equation (8.43) becomes

$$\frac{EA}{\sqrt{2}} \begin{pmatrix} 1 & 0 \\ 0 & 1 \end{pmatrix} \begin{pmatrix} u_3 \\ v_3 \end{pmatrix} - EAf \begin{pmatrix} 0 \\ 1 \end{pmatrix} = \begin{pmatrix} 0 \\ 0 \end{pmatrix}, \tag{8.44}$$

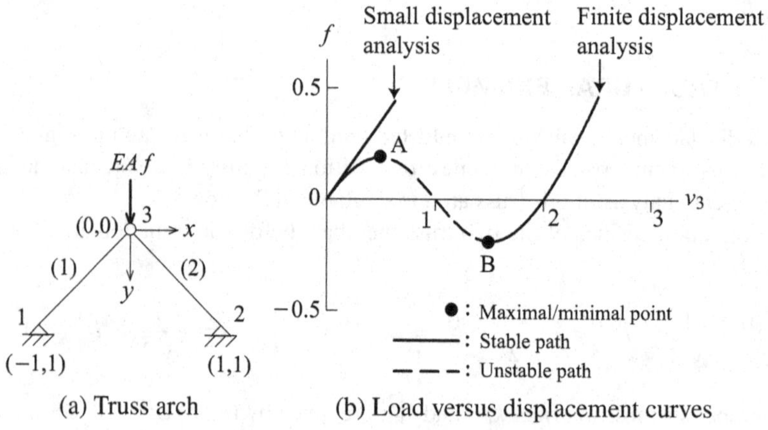

Figure 8.3 A two-bar truss arch and its load versus displacement curves

from which we obtain the displacement of the free node 3 as $\boldsymbol{u}_3 = (u_3, v_3)^\top = (0, \sqrt{2}f)^\top$. Next, the reaction forces at the fixed nodes 1 and 2 are obtained from the first and second equations of (8.42) with $\boldsymbol{u}_1 = \boldsymbol{u}_2 = \boldsymbol{0}$ as

$$\boldsymbol{f}_1 = -k_{\mathrm{E}}^{(1)} \boldsymbol{u}_3 = \frac{EAf}{2} \begin{pmatrix} 1 \\ -1 \end{pmatrix}, \qquad \boldsymbol{f}_2 = -k_{\mathrm{E}}^{(2)} \boldsymbol{u}_3 = \frac{EAf}{2} \begin{pmatrix} -1 \\ -1 \end{pmatrix}. \tag{8.45}$$

Figure 8.3(b) compares the load versus displacement relation $f = v_3/\sqrt{2}$ obtained by the small displacement analysis with the load versus displacement curve by the finite displacement analysis (see Fig. 3.4(b) in Section 3.4). The former represents the tangent of the curve obtained by the finite displacement analysis. The linearized relation $f = v_3/\sqrt{2}$ has a large error for a large displacement v_3, and cannot represent the maximal point A nor the loss of stability thereafter captured by the finite displacement analysis.

8.4 LINEAR BUCKLING ANALYSIS

We introduce the method of *linear buckling analysis* (linearized finite displacement analysis) to obtain buckling loads and buckling modes. Nonlinear terms of the tangent stiffness matrix are linearized to arrive at an eigenvalue problem, which gives buckling loads and buckling modes. However, this method cannot afford information about post-bifurcation behavior.

8.4.1 FORMULATION

In the linear buckling analysis, we introduce the following approximations in the expression of $k^{(m)}$ in (8.16):

- By the formula (8.33), the axial force $N^{(m)}$ is approximated to

$$N_0^{(m)} = \frac{EA^{(m)}}{L^{(m)}} (-l_x^{(m)}, -l_y^{(m)}, -l_z^{(m)}, l_x^{(m)}, l_y^{(m)}, l_z^{(m)})\, \boldsymbol{u}^{(m)}, \tag{8.46}$$

  where $\boldsymbol{u}^{(m)}$ is to be obtained by the small displacement analysis.
- The member length $\hat{L}^{(m)}$ after deformation is approximated by the initial member length $L^{(m)}$.
- The direction cosine $(\hat{l}_x^{(m)}, \hat{l}_y^{(m)}, \hat{l}_z^{(m)})$ in (8.8) after deformation is approximated by the direction cosine $(l_x^{(m)}, l_y^{(m)}, l_z^{(m)})$ in (8.32) at the initial state.

Then the *member tangent stiffness matrix* $K^{(m)}$ in (8.13) with the submatrix $k^{(m)}$ represented as (8.16) is approximated as

$$K^{(m)} \approx K_{\mathrm{E}}^{(m)} + K_{\mathrm{G}}^{(m)}. \tag{8.47}$$

Here $K_{\mathrm{E}}^{(m)}$ is the member stiffness matrix of the small displacement analysis defined in (8.37), that is,

$$K_{\mathrm{E}}^{(m)} = \begin{pmatrix} k_{\mathrm{E}}^{(m)} & -k_{\mathrm{E}}^{(m)} \\ -k_{\mathrm{E}}^{(m)} & k_{\mathrm{E}}^{(m)} \end{pmatrix}$$

with $k_{\mathrm{E}}^{(m)}$ in (8.35), and $K_{\mathrm{G}}^{(m)}$ is the *member geometric stiffness matrix*, being defined as

$$K_{\mathrm{G}}^{(m)} = \begin{pmatrix} k_{\mathrm{G}}^{(m)} & -k_{\mathrm{G}}^{(m)} \\ -k_{\mathrm{G}}^{(m)} & k_{\mathrm{G}}^{(m)} \end{pmatrix} \tag{8.48}$$

with

$$k_{\mathrm{G}}^{(m)} = \frac{N_0^{(m)}}{L^{(m)}} \left[\begin{pmatrix} 1 & 0 & 0 \\ 0 & 1 & 0 \\ 0 & 0 & 1 \end{pmatrix} - \begin{pmatrix} (l_x^{(m)})^2 & l_x^{(m)} l_y^{(m)} & l_x^{(m)} l_z^{(m)} \\ l_x^{(m)} l_y^{(m)} & (l_y^{(m)})^2 & l_y^{(m)} l_z^{(m)} \\ l_x^{(m)} l_z^{(m)} & l_y^{(m)} l_z^{(m)} & (l_z^{(m)})^2 \end{pmatrix} \right]. \tag{8.49}$$

It is noted that $k_{\mathrm{E}}^{(m)}$ and $k_{\mathrm{G}}^{(m)}$ are approximations, respectively, to the first and second terms of (8.16).

The linearized form of the tangent stiffness matrix K° in (8.25) for the whole structure is obtained as follows. Recall

$$K_{\mathrm{E}}^{\circ} = \sum_{m=1}^{M} \tilde{K}_{\mathrm{E}}^{(m)}$$

in (8.40) for the superposition of the member stiffness matrices. Similarly, we obtain the geometric stiffness matrix

$$K_{\mathrm{G}}^{\circ} = \sum_{m=1}^{M} \tilde{K}_{\mathrm{G}}^{(m)},$$

where $\tilde{K}_{\mathrm{G}}^{(m)}$ is the $3n \times 3n$ matrix for the whole structure obtained by embedding the 6×6 member geometric stiffness matrix $K_{\mathrm{G}}^{(m)}$. Then we can approximate the tangent stiffness matrix K° in (8.25), using (8.47), as

$$K^{\circ} = \sum_{m=1}^{M} \tilde{K}^{(m)} \approx \sum_{m=1}^{M} (\tilde{K}_{\mathrm{E}}^{(m)} + \tilde{K}_{\mathrm{G}}^{(m)}) = K_{\mathrm{E}}^{\circ} + K_{\mathrm{G}}^{\circ}.$$

That is,

$$K^{\circ} \approx K_{\mathrm{E}}^{\circ} + K_{\mathrm{G}}^{\circ}. \tag{8.50}$$

After considering the boundary conditions in (8.22) and (8.23), the relation (8.50) for $3n \times 3n$ matrices are condensed into the relation

$$K \approx K_{\mathrm{E}} + K_{\mathrm{G}} \tag{8.51}$$

for $3n' \times 3n'$ matrices, where n' is the number of free nodes and K is a submatrix of K° as in (8.26). By the eigenanalysis of the matrix $K_{\mathrm{E}} + K_{\mathrm{G}}$, we can approximate buckling loads and buckling modes.

Remark 8.2 If we choose the direction cosine $(l_x^{(m)}, l_y^{(m)}, l_z^{(m)}) = (1,0,0)$, the member stiffness matrix $K_{\rm E}^{(m)}$ in (8.37) with $k_{\rm E}^{(m)}$ in (8.35) reduces to the well-known *local stiffness matrix*

$$\bar{K}_{\rm E}^{(m)} = \frac{EA^{(m)}}{L^{(m)}} \begin{pmatrix} 1 & 0 & 0 & -1 & 0 & 0 \\ 0 & 0 & 0 & 0 & 0 & 0 \\ 0 & 0 & 0 & 0 & 0 & 0 \\ -1 & 0 & 0 & 1 & 0 & 0 \\ 0 & 0 & 0 & 0 & 0 & 0 \\ 0 & 0 & 0 & 0 & 0 & 0 \end{pmatrix}, \tag{8.52}$$

and the member geometric stiffness matrix $K_{\rm G}^{(m)}$ in (8.48) with $k_{\rm G}^{(m)}$ in (8.49) reduces to

$$\bar{K}_{\rm G}^{(m)} = \frac{N_0^{(m)}}{L^{(m)}} \begin{pmatrix} 0 & 0 & 0 & 0 & 0 & 0 \\ 0 & 1 & 0 & 0 & -1 & 0 \\ 0 & 0 & 1 & 0 & 0 & -1 \\ 0 & 0 & 0 & 0 & 0 & 0 \\ 0 & -1 & 0 & 0 & 1 & 0 \\ 0 & 0 & -1 & 0 & 0 & 1 \end{pmatrix},$$

which is called the *local geometric stiffness matrix*. □

8.4.2 STRUCTURAL EXAMPLE

We conduct the linear buckling analysis for the truss arch in Fig. 8.4 with the height h. Note that the truss arch in Fig. 8.3(a) in Section 8.3.3 corresponds to a special case of $h = 1$, and some of the results of the small displacement analysis of this truss arch in Section 8.3.3 are utilized below.

We set

$$L = \sqrt{1+h^2}, \qquad c = \frac{1}{L}, \qquad s = \frac{h}{L}.$$

Note that $c^2 + s^2 = 1$. Then the direction cosines of the two truss members before

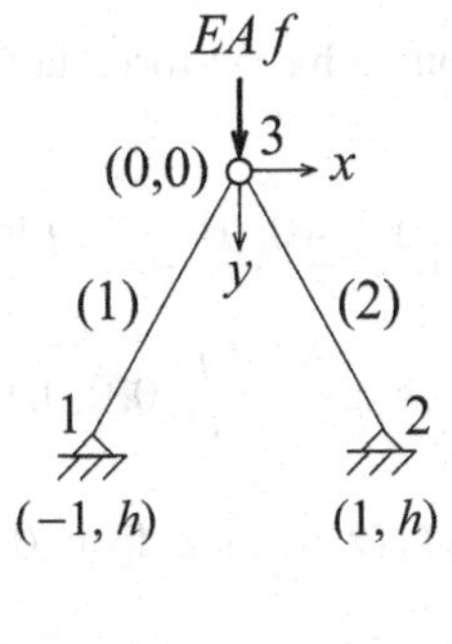

Figure 8.4 A two-bar truss arch

deformation are given by (8.32) as

$$l_x^{(1)} = c, \quad l_y^{(1)} = -s, \quad l_z^{(1)} = 0; \qquad l_x^{(2)} = -c, \quad l_y^{(2)} = -s, \quad l_z^{(2)} = 0,$$

where $(i, j) = (1,3)$ for member 1 and $(i, j) = (2,3)$ for member 2. We hereafter omit the z-directional components since this truss arch is a two-dimensional structure in the xy-plane. The member stiffness matrices are given by (8.35) as

$$k_{\mathrm{E}}^{(1)} = \frac{EA}{L}\begin{pmatrix} c^2 & -sc \\ -sc & s^2 \end{pmatrix}, \qquad k_{\mathrm{E}}^{(2)} = \frac{EA}{L}\begin{pmatrix} c^2 & sc \\ sc & s^2 \end{pmatrix}, \tag{8.53}$$

and the member geometric stiffness matrices are given by (8.49) as

$$k_{\mathrm{G}}^{(1)} = \frac{N_0^{(1)}}{L}\begin{pmatrix} s^2 & sc \\ sc & c^2 \end{pmatrix}, \qquad k_{\mathrm{G}}^{(2)} = \frac{N_0^{(2)}}{L}\begin{pmatrix} s^2 & -sc \\ -sc & c^2 \end{pmatrix}. \tag{8.54}$$

First, the small displacement analysis is conducted to obtain the axial forces $N_0^{(m)}$ that appear in the geometric stiffness matrix. After using the boundary conditions

$$\boldsymbol{u}_1 = \boldsymbol{u}_2 = \boldsymbol{0}, \qquad \boldsymbol{f}_3 = EAf\begin{pmatrix} 0 \\ 1 \end{pmatrix},$$

the equilibrium equation of this truss arch is given by

$$K_{\mathrm{E}}\boldsymbol{u}_3 = EAf\begin{pmatrix} 0 \\ 1 \end{pmatrix} \tag{8.55}$$

with

$$K_{\mathrm{E}} = k_{\mathrm{E}}^{(1)} + k_{\mathrm{E}}^{(2)} = \frac{2EA}{L}\begin{pmatrix} c^2 & 0 \\ 0 & s^2 \end{pmatrix} \tag{8.56}$$

(cf., (8.43) and (8.53)). The equation (8.55) is solved for $\boldsymbol{u}_3$ as

$$\boldsymbol{u}_3 = \frac{L}{2EA}\begin{pmatrix} c^2 & 0 \\ 0 & s^2 \end{pmatrix}^{-1} EAf\begin{pmatrix} 0 \\ 1 \end{pmatrix} = \frac{Lf}{2s^2}\begin{pmatrix} 0 \\ 1 \end{pmatrix}.$$

The axial force $N_0^{(1)}$ is computed by the formula (8.46), which is adapted to this two-dimensional truss arch as

$$\begin{aligned} N_0^{(1)} &= \frac{EA}{L}(-l_x^{(1)}, -l_y^{(1)}, l_x^{(1)}, l_y^{(1)})\begin{pmatrix} \boldsymbol{u}_1 \\ \boldsymbol{u}_3 \end{pmatrix} \\ &= \frac{EA}{L}(-c, s, c, -s)\frac{Lf}{2s^2}(0,0,0,1)^{\top} = -\frac{EAf}{2s}. \end{aligned}$$

Similarly, we have $N_0^{(2)} = -EAf/(2s)$. Then, from (8.54), we have

$$K_{\mathrm{G}} = k_{\mathrm{G}}^{(1)} + k_{\mathrm{G}}^{(2)} = -\frac{EAf}{Ls}\begin{pmatrix} s^2 & 0 \\ 0 & c^2 \end{pmatrix}. \tag{8.57}$$

The condition for buckling is

$$\det(K_{\rm E}+K_{\rm G})=0.$$

Using (8.56) and (8.57), this condition is evaluated to

$$\det\left\{\frac{2EA}{L}\begin{pmatrix}c^2 & 0\\ 0 & s^2\end{pmatrix}-\frac{EAf}{Ls}\begin{pmatrix}s^2 & 0\\ 0 & c^2\end{pmatrix}\right\}=0.$$

Hence buckling loads are obtained as

$$f_x=\frac{2c^2}{s}=\frac{2}{hL},\qquad f_y=\frac{2s^3}{c^2}=\frac{2h^3}{L}. \tag{8.58}$$

The buckling mode for f_x is $(1,0)$, expressing the bifurcation, and that for f_y is $(0,1)$, expressing snap-through buckling. The smallest buckling load is given by

$$\begin{cases} f_x & \text{for } h>1,\\ f_x \text{ and } f_y & \text{for } h=1,\\ f_y & \text{for } h<1.\end{cases}$$

According to the nonlinear analysis (Section 2.7), which gives more accurate quantitative information, the threshold value between bifurcation buckling and snap-through buckling is equal to $h=\sqrt{7}\approx 2.646$, which is significantly larger than $h=1$ obtained above.

8.5 BUCKLING OF TRUSS MEMBERS

The buckling of a structural system comprising a set of members is classified into the bucking of the whole structure and the buckling of a member. We have so far dealt with the structural buckling of truss structures.

In this section, we deal with the member buckling of a truss member. A truss member, which is supposed to transmit an axial force only, is in fact a simply supported beam-column member and, accordingly, has the possibility of undergoing the flexural (Euler) buckling (Section 7.3). The flexural buckling of a truss member within a structure is investigated here using a pair of structural examples.

Example 8.1 We refer to the propped cantilever depicted in Fig. 8.5(a), comprising a column supported by a linearly elastic spring. The column has the flexural rigidity EI and the spring constant is k. We hereby conduct a buckling analysis under the assumption that the value of the spring constant k is very large. This structural system has two kinds of buckling: (1) the buckling as a structural system (Fig. 8.5(b)) and (2) the flexural buckling of the column member (Fig. 8.5(c)).

The buckling load as a structural system is $f_{\rm c}=1$ (cf., (1.23) in Section 1.5). The flexural buckling load is given as $kLf_{\rm c}=\pi^2EI/L^2$ (i.e., $f_{\rm c}=\pi^2EI/(kL^3)$) by the formula (7.30) for a simply supported column (Section 7.3.2).[6] The minimum

[6]For simplicity, the boundary condition of simple support is assumed since k is assumed to be large.

buckling load of this structural system is given by the smaller value of $f_c = 1$ for the structural buckling and $f_c = \pi^2 EI/(kL^3)$ for the member buckling. □

Example 8.2 We investigate the member buckling of the truss structure subjected to two vertical loads $P > 0$ (Fig. 8.6(a)). All members have the same axial and flexural rigidities of EA and EI. The small-displacement analysis (cf, Section 8.3) is conducted to arrive at axial forces of members depicted in Fig 8.6(b), where compressive axial force is defined to be negative. Members 2, 5, 6 are under compression and have the possibility to undergo flexural buckling.

The members 5 and 6 with the length L are both subjected to the same compressive load $-P$. Accordingly, by the formula (7.30) for the flexural buckling, we see that these members undergo buckling at the load:

$$P = P_c^{(5)} = P_c^{(6)} = \frac{\pi^2 EI}{L^2}.$$

The member 2 has a length of $\sqrt{2}L$ and is subjected to the compressive load of

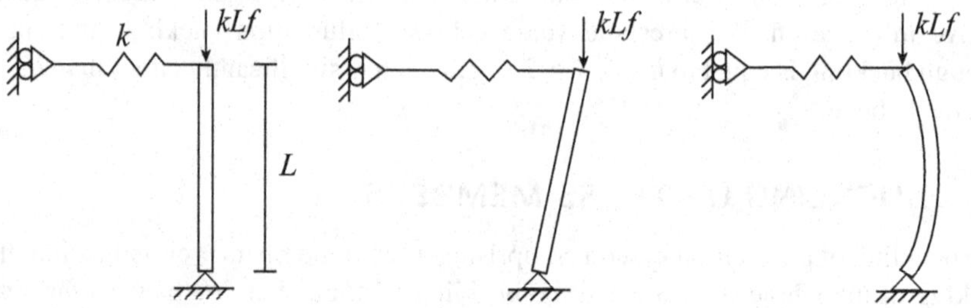

(a) Initial or pre-buckling state (b) Buckling of a structural system (c) Buckling of a column member

Figure 8.5 Structural buckling and member buckling of a propped cantilever

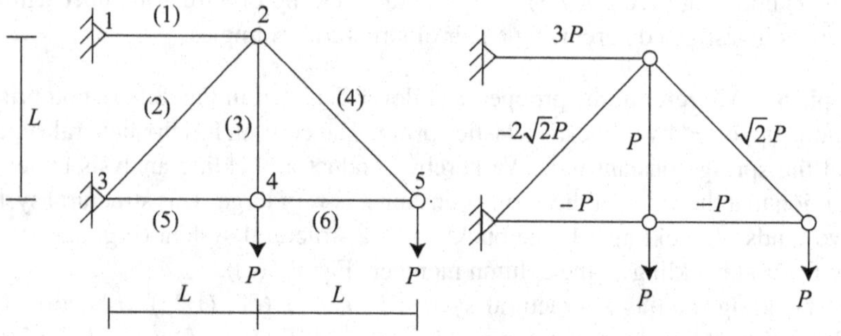

(a) Geometry (b) Axial force distribution (compressive force is negative)

Figure 8.6 A truss structure and the distribution of axial forces

$-2\sqrt{2}P$. The load $P = P_c^{(2)}$ for the buckling of this member is determined from

$$2\sqrt{2}P_c^{(2)} = \frac{\pi^2 EI}{(\sqrt{2}L)^2},$$

that is,

$$P_c^{(2)} = \frac{\sqrt{2}}{8}\frac{\pi^2 EI}{L^2}.$$

When the load P is increased from zero, member 2, with the smallest buckling load, buckles first. Member flexural buckling does not take place if the condition

$$P < P_c^{(2)} = \frac{\sqrt{2}}{8}\frac{\pi^2 EI}{L^2}$$

is satisfied. □

8.6 PROBLEMS

Problem 8.1 Conduct the small displacement analysis for the two-dimensional cross truss structure depicted in Fig. 8.7. All members have the same length L and the cross-sectional rigidity EA.

Problem 8.2 Consider the truss tent structure in Fig. 8.8. (1) Express the stiffness matrix K_E° using $k_E^{(m)}$ $(m = 1,\ldots,7)$. (2) Express the equilibrium equation for the small displacement analysis using $k_E^{(m)}$ $(m = 1,\ldots,7)$.

Problem 8.3 Derive (8.15) and (8.16).

Problem 8.4 Derive (8.33).

Problem 8.5 Conduct the linear buckling analysis of the two-bar truss arch shown in Fig. 8.3(a).

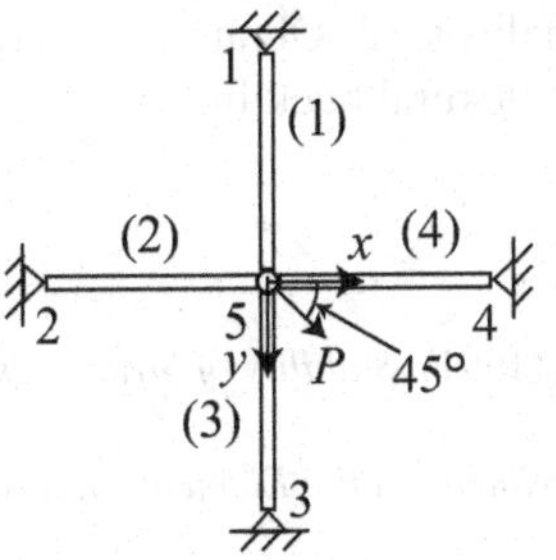

Figure 8.7 A two-dimensional cross truss structure

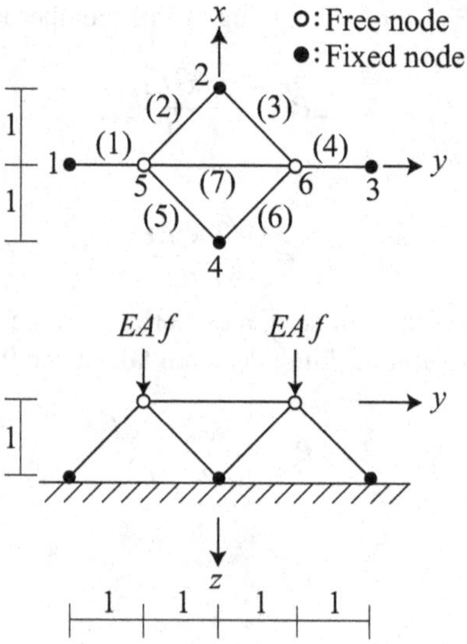

Figure 8.8 A truss tent structure

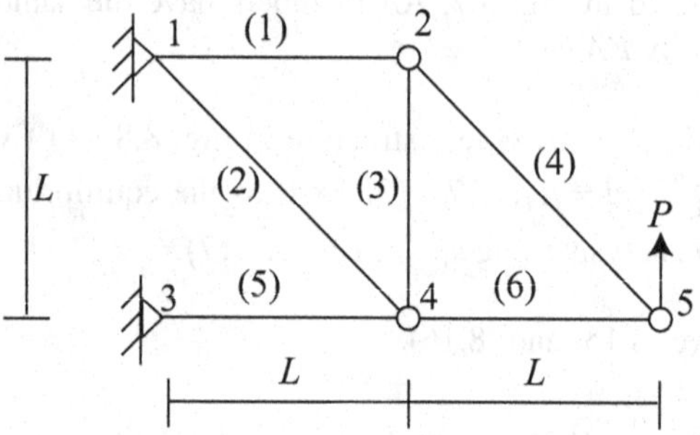

Figure 8.9 A truss structure

Problem 8.6 Consider the truss structure in Fig 8.9. All members have the same axial and flexural rigidities EA and EI. (1) Obtain the axial forces in the truss members by the small displacement analysis. (2) Obtain the condition on the load P (> 0) for which no member undergoes flexural buckling.

REFERENCES

1. Bažant ZP, and Cedolin L (1991), *Stability of Structures*, Oxford University Press, New York.
2. Crisfield MA (1991), *Non-linear Finite Element Analysis of Solids and Structures*, Vol. 1, Wiley, Chichester.
3. Iyengar NGR (1988), *Structural Stability of Columns and Plates*, Ellis Horwood Series in Civil Engineering, Chichester.

9 Structural Buckling II: Frame

9.1 SUMMARY

Chapter 7 has dealt with the buckling of a beam-column member and Chapter 8 with the buckling of a truss structure. This chapter studies the buckling of a two-dimensional frame structure comprising beam-column members subjected to axial forces and bending moments. We focus on the method of linear buckling analysis introduced in Chapter 8 and present a formulation consistent with the finite element method. Results and terminology of the buckling of a beam-column member in Chapter 7 are employed from time to time.

The methods of buckling analysis for a frame are classified into linear buckling analysis and finite displacement analysis. In this chapter, we focus on the linear buckling method (Iyengar [2]; Bažant and Cedolin [1]), for which the buckling condition is formulated for the tangent stiffness matrix of the whole structure and buckling loads and buckling modes are obtained from this condition. Though not giving information about post-buckling behavior, this method is useful in approximating buckling loads and buckling modes. The finite displacement analysis is to be consulted for post-buckling behavior.

We use a cubic polynomial representation of the deflection to obtain a member stiffness equation (a discretized equilibrium equation) of a beam-column member, and to obtain the stiffness equation of a whole structure by assembling member stiffness equations (Section 9.2). Linear buckling analysis is illustrated in Section 9.3 for a simple structural example undergoing only flexural deformation. However, it is to be noted that members in a frame structure undergo flexural deformation as beam-columns and axial deformation as trusses. We, accordingly, combine the small-displacement equilibrium equation of a truss member (Section 8.3) and that of a beam-column to obtain member and structural stiffness equations (Sections 9.4 and 9.6). The structural stiffness equation obtained in this manner is used in the linear buckling analysis to obtain buckling loads and buckling modes.

For a frame structure only with vertical and horizontal members (Fig. 9.1(a)), a linear buckling analysis with a simple formulation is applicable. The assemblage of member stiffness equations can be conducted in a straightforward manner and then the resulting buckling condition gives the buckling loads and buckling modes. This simple linear buckling analysis is illustrated for an L-shaped frame in Section 9.5.

For a more general frame structure with tilted members (Fig. 9.1(b)), the linear buckling analysis is refined by using the transformation from the local coordinate system for each member to the global coordinate system for a whole structure (Section 9.6).

Keywords: • Beam • Bifurcation • Buckling • Column • Frame • Geometric stiffness matrix • Linear buckling analysis • Stiffness matrix

DOI: 10.1201/9781003112365-9

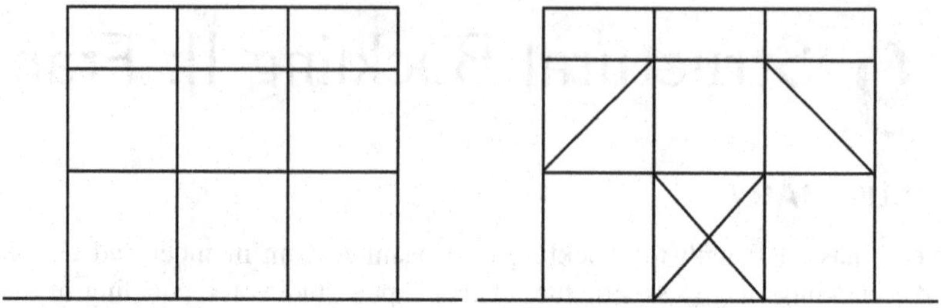

(a) Without tilted member (b) With tilted members

Figure 9.1 Frame structures

9.2 STIFFNESS EQUATIONS OF BEAM-COLUMN

The member stiffness equation of a beam-column member is derived and is assembled for a structure to obtain a structural stiffness equation. Only the flexural deformation is considered in the formulation in this section, while an axial deformation is to be implemented in Sections 9.4 and 9.6.

9.2.1 MEMBER STIFFNESS EQUATION

We consider the beam-column member in Fig. 9.2, which is assumed to connect the nodes i and j. We express the total potential energy of this beam-column as a function in the vertical displacement $w(s)$ under the assumption that the axial deformation is relatively small as explained in Section 7.2.1. Here s denotes the coordinate along the member axis.

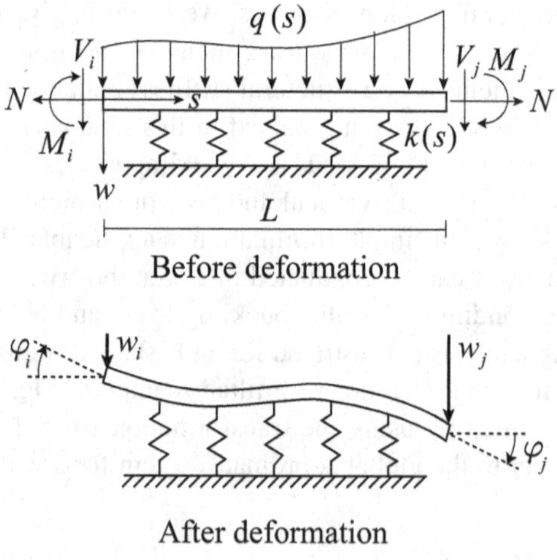

Figure 9.2 A beam-column member

We revive the *total potential energy* of the beam-column that comprises the following five kinds of energy in Section 7.2.1:

- Internal energy $\int \frac{1}{2}EIw''(s)^2 \mathrm{d}s$ by the bending of the beam-column, where E denotes the modulus of elasticity and I expresses the moment of inertia of a cross-sectional area.
- Internal energy $\int \frac{1}{2}k(s)w(s)^2 \mathrm{d}s$ stored in the elastic foundation with the spring constant $k(s)$.
- External potential energy $\int[-\frac{1}{2}Pw'(s)^2]\mathrm{d}s$ by the axial force P (compression is positive), where $P = -N$ in Fig. 9.2.
- External potential energy $\int[-q(s)w(s)]\mathrm{d}s$ by the distributed load $q(s)$.
- External potential energy

$$-\left[V_i w(0) + M_i w'(0) + V_j w(L) + M_j w'(L)\right]$$

by the shear forces V_i and V_j and the bending moments M_i and M_j at the member ends i and j.

The total potential energy of the beam-column is given as the sum of these five terms as

$$U(w) = \int_0^L \left[\frac{1}{2}EI(w'')^2 + \frac{1}{2}k(s)w^2 + \frac{1}{2}N(w')^2 - q(s)w\right] \mathrm{d}s$$
$$- \left[V_i w(0) + M_i w'(0) + V_j w(L) + M_j w'(L)\right]. \tag{9.1}$$

We introduce several notations with the superscript (m) to distinguish different members. The *member end displacement vector* $\boldsymbol{u}^{(m)}$ and the *member end force vector* $\boldsymbol{p}^{(m)}$ are defined, respectively, by

$$\boldsymbol{u}^{(m)} = (w_i^{(m)}, \varphi_i^{(m)}, w_j^{(m)}, \varphi_j^{(m)})^\top, \qquad \boldsymbol{p}^{(m)} = (V_i^{(m)}, M_i^{(m)}, V_j^{(m)}, M_j^{(m)})^\top \tag{9.2}$$

with

$$\begin{pmatrix} w_i^{(m)} \\ \varphi_i^{(m)} \\ w_j^{(m)} \\ \varphi_j^{(m)} \end{pmatrix} = \begin{pmatrix} w(0) \\ \dfrac{\mathrm{d}w}{\mathrm{d}s}(0) \\ w(L) \\ \dfrac{\mathrm{d}w}{\mathrm{d}s}(L) \end{pmatrix}, \tag{9.3}$$

where $L = L^{(m)}$. With these notations, the total potential energy in (9.1) is rewritten as

$$U^{(m)}(w) = \int_0^L \left[\frac{1}{2}EI^{(m)}(w'')^2 + \frac{1}{2}k^{(m)}(s)w^2 + \frac{1}{2}N^{(m)}(w')^2 - q^{(m)}(s)w\right] \mathrm{d}s$$
$$- (\boldsymbol{u}^{(m)})^\top \boldsymbol{p}^{(m)}. \tag{9.4}$$

We approximate the vertical deflection $w(s)$ of the member by a cubic function in s. That is, we assume

$$w(s) = c_0 + c_1 s + c_2 s^2 + c_3 s^3 = (1, s, s^2, s^3)\boldsymbol{c}, \qquad 0 \le s \le L, \tag{9.5}$$

where $\boldsymbol{c} = (c_0, c_1, c_2, c_3)^\top$. Then the member end displacements in (9.3) are expressed as

$$\begin{pmatrix} w_i{}^{(m)} \\ \varphi_i^{(m)} \\ w_j^{(m)} \\ \varphi_j^{(m)} \end{pmatrix} = \begin{pmatrix} 1 & 0 & 0 & 0 \\ 0 & 1 & 0 & 0 \\ 1 & L & L^2 & L^3 \\ 0 & 1 & 2L & 3L^2 \end{pmatrix} \begin{pmatrix} c_0 \\ c_1 \\ c_2 \\ c_3 \end{pmatrix}. \tag{9.6}$$

By solving (9.6) for $\boldsymbol{c} = (c_0, c_1, c_2, c_3)^\top$ and substituting this $\boldsymbol{c}$ into (9.5), we can express $w(s)$ in terms of the member end displacements $\boldsymbol{u}^{(m)} = (w_i{}^{(m)}, \varphi_i^{(m)}, w_j^{(m)}, \varphi_j^{(m)})^\top$ as

$$w(s) = \boldsymbol{H}(s)^\top \boldsymbol{u}^{(m)}, \tag{9.7}$$

where

$$\boldsymbol{H} = \boldsymbol{H}(s) = \left(1 - 3\frac{s^2}{L^2} + 2\frac{s^3}{L^3},\ s - 2\frac{s^2}{L} + \frac{s^3}{L^2},\ 3\frac{s^2}{L^2} - 2\frac{s^3}{L^3},\ -\frac{s^2}{L} + \frac{s^3}{L^2}\right)^\top \tag{9.8}$$

is called the *Hermite type shape function* (see Problem 9.1).

Using (9.8), we introduce three 4×4 matrices and a 4-dimensional vector:

- The *stiffness matrix*,

$$K_{\mathrm{E}}^{(m)} = \int_0^L EI^{(m)} \frac{\mathrm{d}^2\boldsymbol{H}}{\mathrm{d}s^2} \frac{\mathrm{d}^2\boldsymbol{H}}{\mathrm{d}s^2}^\top \mathrm{d}s. \tag{9.9}$$

- The *geometric stiffness matrix*,

$$K_{\mathrm{G}}^{(m)} = \int_0^L N^{(m)} \frac{\mathrm{d}\boldsymbol{H}}{\mathrm{d}s} \frac{\mathrm{d}\boldsymbol{H}}{\mathrm{d}s}^\top \mathrm{d}s, \tag{9.10}$$

  expressing the influence of the axial force $N^{(m)}$ induced by the geometrical deformation of the member.

- The matrix

$$K_{\mathrm{S}}^{(m)} = \int_0^L k^{(m)} \boldsymbol{H}\boldsymbol{H}^\top \mathrm{d}s$$

  expressing the spring stiffness (S means the spring).

- The vector

$$\boldsymbol{q}^{(m)} = \int_0^L q^{(m)} \boldsymbol{H} \mathrm{d}s$$

  expressing the *equivalent nodal load* of the distributed load.

Then from (9.4), we obtain

$$U^{(m)} = U^{(m)}(\boldsymbol{u}^{(m)}) = \frac{1}{2}(\boldsymbol{u}^{(m)})^\top (K_{\mathrm{E}}^{(m)} + K_{\mathrm{G}}^{(m)} + K_{\mathrm{S}}^{(m)}) \boldsymbol{u}^{(m)} - (\boldsymbol{u}^{(m)})^\top (\boldsymbol{q}^{(m)} + \boldsymbol{p}^{(m)}). \tag{9.11}$$

We hereafter assume that the flexural rigidity $EI^{(m)}$ and the axial force $N^{(m)}$ are constant along the axial direction of the member. Then the concrete form of the stiffness matrix $K_{\mathrm{E}}^{(m)}$ is obtained from (9.9) as

$$K_{\mathrm{E}}^{(m)} = \frac{2EI^{(m)}}{L^2} \left(\begin{array}{cc|cc} 6/L & 3 & -6/L & 3 \\ 3 & 2L & -3 & L \\ \hline -6/L & -3 & 6/L & -3 \\ 3 & L & -3 & 2L \end{array} \right). \tag{9.12}$$

(see Problem 9.2), where $L = L^{(m)}$, and the concrete form of the geometric stiffness matrix $K_{\mathrm{G}}^{(m)}$ is obtained from (9.10) as

$$K_{\mathrm{G}}^{(m)} = \frac{N^{(m)}}{30} \left(\begin{array}{cc|cc} 36/L & 3 & -36/L & 3 \\ 3 & 4L & -3 & -L \\ \hline -36/L & -3 & 36/L & -3 \\ 3 & -L & -3 & 4L \end{array} \right). \tag{9.13}$$

Applying the principle of stationary total potential energy to $U^{(m)}$ in (9.11), we obtain the member stiffness (equilibrium) equation

$$(K_{\mathrm{E}}^{(m)} + K_{\mathrm{G}}^{(m)} + K_{\mathrm{S}}^{(m)})\,\boldsymbol{u}^{(m)} - \boldsymbol{q}^{(m)} - \boldsymbol{p}^{(m)} = \boldsymbol{0}. \tag{9.14}$$

In the subsequent structural analysis, we refer mainly to $K_{\mathrm{E}}^{(m)}$ and $K_{\mathrm{G}}^{(m)}$, as well as $\boldsymbol{p}^{(m)}$, while omitting $K_{\mathrm{S}}^{(m)}$ and $\boldsymbol{q}^{(m)}$ (corresponding to $k^{(m)} = 0$ and $q^{(m)} = 0$). That is, we consider the equilibrium equation of the form

$$(K_{\mathrm{E}}^{(m)} + K_{\mathrm{G}}^{(m)})\,\boldsymbol{u}^{(m)} = \boldsymbol{p}^{(m)}. \tag{9.15}$$

9.2.2 STRUCTURAL STIFFNESS MATRIX

We assemble the member matrices, $K_{\mathrm{E}}^{(m)}$ and $K_{\mathrm{G}}^{(m)}$ in (9.15), to fabricate the matrices K_{E}° and K_{G}° for the whole structure. We also assemble $\boldsymbol{p}^{(m)}$ to obtain the nodal force vector $\boldsymbol{p}^{\circ}$ for the whole structure. Then we can obtain the stiffness equation of the form

$$(K_{\mathrm{E}}^{\circ} + K_{\mathrm{G}}^{\circ})\,\boldsymbol{u}^{\circ} = \boldsymbol{p}^{\circ}, \tag{9.16}$$

where $\boldsymbol{u}^{\circ}$ is the displacement vector for the whole structure.

We assume that some components of $\boldsymbol{u}^{\circ}$ are fixed to zero and the remaining ones are free and that the external forces are given for these free nodes. We, accordingly, decompose the displacement vector $\boldsymbol{u}^{\circ}$ into the unknown variable vector $\boldsymbol{u}$ and the given vector $\boldsymbol{u}^{\bullet} = \boldsymbol{0}$, and the force vector $\boldsymbol{p}^{\circ}$ into the known force vector $\boldsymbol{p}$ and the unknown force vector $\boldsymbol{p}^{\bullet}$. That is,

$$\boldsymbol{u}^{\circ} = \begin{pmatrix} \boldsymbol{u} \\ \boldsymbol{u}^{\bullet} \end{pmatrix}, \qquad \boldsymbol{p}^{\circ} = \begin{pmatrix} \boldsymbol{p} \\ \boldsymbol{p}^{\bullet} \end{pmatrix}. \tag{9.17}$$

The vector $\boldsymbol{p}$ is to be specified by force boundary conditions and the vector $\boldsymbol{p}^{\circ}$ expresses reaction forces at the nodes of fixed displacements.

The matrices $K_{\rm E}^{\circ}$ and $K_{\rm G}^{\circ}$ are partitioned accordingly as

$$K_{\rm E}^{\circ} = \begin{pmatrix} K_{\rm E} & K_{\rm E}^{\triangle} \\ K_{\rm E}^{\triangledown} & K_{\rm E}^{\bullet} \end{pmatrix}, \qquad K_{\rm G}^{\circ} = \begin{pmatrix} K_{\rm G} & K_{\rm G}^{\triangle} \\ K_{\rm G}^{\triangledown} & K_{\rm G}^{\bullet} \end{pmatrix}, \tag{9.18}$$

and the stiffness equation in (9.16), with $\boldsymbol{u}^{\bullet} = \boldsymbol{0}$, can be rewritten as

$$\left[\begin{pmatrix} K_{\rm E} & K_{\rm E}^{\triangle} \\ K_{\rm E}^{\triangledown} & K_{\rm E}^{\bullet} \end{pmatrix} + \begin{pmatrix} K_{\rm G} & K_{\rm G}^{\triangle} \\ K_{\rm G}^{\triangledown} & K_{\rm G}^{\bullet} \end{pmatrix} \right] \begin{pmatrix} \boldsymbol{u} \\ \boldsymbol{0} \end{pmatrix} = \begin{pmatrix} \boldsymbol{p} \\ \boldsymbol{p}^{\bullet} \end{pmatrix}. \tag{9.19}$$

Thus we can construct a stiffness equation of smaller size:

$$(K_{\rm E} + K_{\rm G})\, \boldsymbol{u} = \boldsymbol{p}, \tag{9.20}$$

where $K_{\rm E}$ and $K_{\rm G}$ are the submatrices of $K_{\rm E}^{\circ}$ and $K_{\rm G}^{\circ}$, respectively, corresponding to the unknown variable vector $\boldsymbol{u}$ and the known force vector $\boldsymbol{p}$ as in (9.19).

We shall consider two kinds of analysis for the equation in (9.19).

- The small-displacement analysis ignoring the geometric stiffness matrix. By ignoring $K_{\rm G}^{(m)}$ in (9.15) for each member, we obtain

  $$K_{\rm E}\, \boldsymbol{u} = \boldsymbol{p} \tag{9.21}$$

  in place of (9.20). The reaction force vector $\boldsymbol{p}^{\bullet}$ is determined from (9.19) as

  $$\boldsymbol{p}^{\bullet} = K_{\rm E}^{\triangledown}\, \boldsymbol{u}. \tag{9.22}$$

  This analysis gives the information on deformation of a frame structure but cannot give information on buckling (Section 9.2.3).
- The linear buckling analysis considering the geometric stiffness matrix $K_{\rm G}^{(m)}$ in (9.15). With this method, we can obtain the buckling load from the buckling condition

  $$\det(K_{\rm E} + K_{\rm G}) = 0. \tag{9.23}$$

  The buckling mode, to be denoted by $\boldsymbol{\eta}$, is given as an eigenvector of $K_{\rm E} + K_{\rm G}$ for the zero eigenvalue, that is,

  $$(K_{\rm E} + K_{\rm G})\boldsymbol{\eta} = \boldsymbol{0}. \tag{9.24}$$

  This method is presented in Sections 9.3, 9.4, and 9.6.

Remark 9.1 When all members have a common length L, it is convenient, in both analytical study and numerical analysis, to employ the following normalized variables[1]

$$\hat{\boldsymbol{u}}^{(m)} = (w_i^{(m)}/L, \varphi_i^{(m)}, w_j^{(m)}/L, \varphi_j^{(m)})^{\top}, \quad \hat{\boldsymbol{p}}^{(m)} = (V_i^{(m)}, M_i^{(m)}/L, V_j^{(m)}, M_j^{(m)}/L)^{\top}, \tag{9.25}$$

---

[1]Throughout this chapter, normalized variables are denoted with (ˆ).

for which the equation (9.15) is replaced by

$$(\hat{K}_{\rm E}^{(m)} + \hat{K}_{\rm G}^{(m)})\,\hat{\boldsymbol{u}}^{(m)} = \hat{\boldsymbol{p}}^{(m)} \tag{9.26}$$

with simpler matrices

$$\hat{K}_{\rm E}^{(m)} = \frac{2EI^{(m)}}{L^2}\left(\begin{array}{cc|cc} 6 & 3 & -6 & 3 \\ 3 & 2 & -3 & 1 \\ \hline -6 & -3 & 6 & -3 \\ 3 & 1 & -3 & 2 \end{array}\right), \qquad \hat{K}_{\rm G}^{(m)} = \frac{N^{(m)}}{30}\left(\begin{array}{cc|cc} 36 & 3 & -36 & 3 \\ 3 & 4 & -3 & -1 \\ \hline -36 & -3 & 36 & -3 \\ 3 & -1 & -3 & 4 \end{array}\right). \tag{9.27}$$

We denote the normalized versions of $\boldsymbol{u}$, $\boldsymbol{p}$, $K_{\rm E}^{\circ}$, $K_{\rm G}^{\circ}$, etc. by $\hat{\boldsymbol{u}}$, $\hat{\boldsymbol{p}}$, $\hat{K}_{\rm E}^{\circ}$, $\hat{K}_{\rm G}^{\circ}$, etc. Then the normalized versions of (9.16), (9.20), (9.21), (9.22), (9.23), and (9.24) are given, respectively, by

$$(9.16):\quad (\hat{K}_{\rm E}^{\circ} + \hat{K}_{\rm G}^{\circ})\,\hat{\boldsymbol{u}}^{\circ} = \hat{\boldsymbol{p}}^{\circ}, \tag{9.28}$$

$$(9.20):\quad (\hat{K}_{\rm E} + \hat{K}_{\rm G})\,\hat{\boldsymbol{u}} = \hat{\boldsymbol{p}}, \tag{9.29}$$

$$(9.21):\quad \hat{K}_{\rm E}\,\hat{\boldsymbol{u}} = \hat{\boldsymbol{p}}, \tag{9.30}$$

$$(9.22):\quad \hat{\boldsymbol{p}}^{\bullet} = \hat{K}_{\rm E}^{\triangledown}\,\hat{\boldsymbol{u}}, \tag{9.31}$$

$$(9.23):\quad \det(\hat{K}_{\rm E} + \hat{K}_{\rm G}) = 0, \tag{9.32}$$

$$(9.24):\quad (\hat{K}_{\rm E} + \hat{K}_{\rm G})\,\hat{\boldsymbol{\eta}} = \boldsymbol{0}, \tag{9.33}$$

where

$$\hat{K}_{\rm E}^{\circ} = \begin{pmatrix} \hat{K}_{\rm E} & \hat{K}_{\rm E}^{\vartriangle} \\ \hat{K}_{\rm E}^{\triangledown} & \hat{K}_{\rm E}^{\bullet} \end{pmatrix}, \qquad \hat{K}_{\rm G}^{\circ} = \begin{pmatrix} \hat{K}_{\rm G} & \hat{K}_{\rm G}^{\vartriangle} \\ \hat{K}_{\rm G}^{\triangledown} & \hat{K}_{\rm G}^{\bullet} \end{pmatrix}. \tag{9.34}$$

These normalized variants are often used later. □

9.2.3 STRUCTURAL EXAMPLE OF SMALL-DISPLACEMENT ANALYSIS

A small displacement analysis is demonstrated for a structure not undergoing buckling. We consider the two-element beam structure depicted in Fig. 9.3. This beam

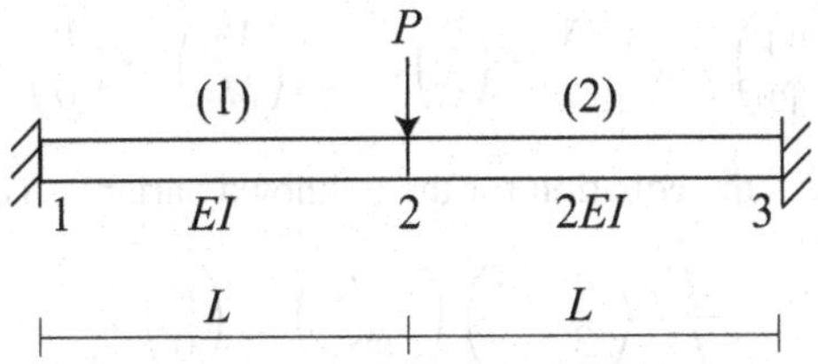

Figure 9.3 A two-element beam structure subjected to a vertical force P

structure is subjected to bending by a concentrated vertical force P and comprises two beam members 1 and 2 with the same member length L and different flexural rigidities EI and $2EI$, respectively. Since the axial force is absent in both members, the geometric stiffness is absent, and this beam structure does not buckle. We employ the formulation for the normalized variables in Remark 9.1, which is applicable when all members have a common length.

By omitting the matrix $\hat{K}_{\mathrm{G}}^{(m)}$ in (9.26), we obtain the (normalized) member stiffness equation

$$\hat{K}_{\mathrm{E}}^{(m)} \hat{\boldsymbol{u}}^{(m)} = \hat{\boldsymbol{p}}^{(m)}, \quad m = 1,2$$

with

$$\hat{K}_{\mathrm{E}}^{(1)} = \frac{2EI}{L^2} \left(\begin{array}{cc|cc} 6 & 3 & -6 & 3 \\ 3 & 2 & -3 & 1 \\ \hline -6 & -3 & 6 & -3 \\ 3 & 1 & -3 & 2 \end{array} \right), \quad \hat{K}_{\mathrm{E}}^{(2)} = 2\hat{K}_{\mathrm{E}}^{(1)} = \left(\begin{array}{cc|cc} 12 & 6 & -12 & 6 \\ 6 & 4 & -6 & 2 \\ \hline -12 & -6 & 12 & -6 \\ 6 & 2 & -6 & 4 \end{array} \right).$$

The stiffness matrix $\hat{K}_{\mathrm{E}}^{\circ}$ for the whole structure is obtained by superposing these member stiffness matrices as

$$\hat{K}_{\mathrm{E}}^{\circ} = \frac{2EI}{L^2} \left(\begin{array}{cc|cc|cc} 6 & 3 & -6 & 3 & 0 & 0 \\ 3 & 2 & -3 & 1 & 0 & 0 \\ \hline -6 & -3 & 6 & -3 & 0 & 0 \\ 3 & 1 & -3 & 2 & 0 & 0 \\ \hline 0 & 0 & 0 & 0 & 0 & 0 \\ 0 & 0 & 0 & 0 & 0 & 0 \end{array} \right) + \frac{2EI}{L^2} \left(\begin{array}{cc|cc|cc} 0 & 0 & 0 & 0 & 0 & 0 \\ 0 & 0 & 0 & 0 & 0 & 0 \\ \hline 0 & 0 & 12 & 6 & -12 & 6 \\ 0 & 0 & 6 & 4 & -6 & 2 \\ \hline 0 & 0 & -12 & -6 & 12 & -6 \\ 0 & 0 & 6 & 2 & -6 & 4 \end{array} \right).$$

Then we obtain the normalized stiffness equation $\hat{K}_{\mathrm{E}}^{\circ}\hat{\boldsymbol{u}}^{\circ} = \hat{\boldsymbol{p}}^{\circ}$ of the structural system as

$$\frac{2EI}{L^2} \left(\begin{array}{cc|cc|cc} 6 & 3 & -6 & 3 & 0 & 0 \\ 3 & 2 & -3 & 1 & 0 & 0 \\ \hline -6 & -3 & 18 & 3 & -12 & 6 \\ 3 & 1 & 3 & 6 & -6 & 2 \\ \hline 0 & 0 & -12 & -6 & 12 & -6 \\ 0 & 0 & 6 & 2 & -6 & 4 \end{array} \right) \left(\begin{array}{c} w_1/L \\ \varphi_1 \\ \hline w_2/L \\ \varphi_2 \\ \hline w_3/L \\ \varphi_3 \end{array} \right) = \left(\begin{array}{c} V_1 \\ M_1/L \\ \hline V_2 \\ M_2/L \\ \hline V_3 \\ M_3/L \end{array} \right). \tag{9.35}$$

Using the displacement boundary conditions and given external forces

$$\begin{pmatrix} w_1 \\ \varphi_1 \end{pmatrix} = \begin{pmatrix} w_3 \\ \varphi_3 \end{pmatrix} = \begin{pmatrix} 0 \\ 0 \end{pmatrix}, \quad \begin{pmatrix} V_2 \\ M_2 \end{pmatrix} = \begin{pmatrix} P \\ 0 \end{pmatrix}, \tag{9.36}$$

we obtain, from (9.35), the equation for the unknown variables $(w_2/L, \varphi_2)$ as

$$\frac{2EI}{L^2} \begin{pmatrix} 18 & 3 \\ 3 & 6 \end{pmatrix} \begin{pmatrix} w_2/L \\ \varphi_2 \end{pmatrix} = \begin{pmatrix} P \\ 0 \end{pmatrix}. \tag{9.37}$$

This corresponds to the normalized stiffness equation $\hat{K}_{\mathrm{E}}\, \hat{\boldsymbol{u}} = \hat{\boldsymbol{p}}$ in (9.30).

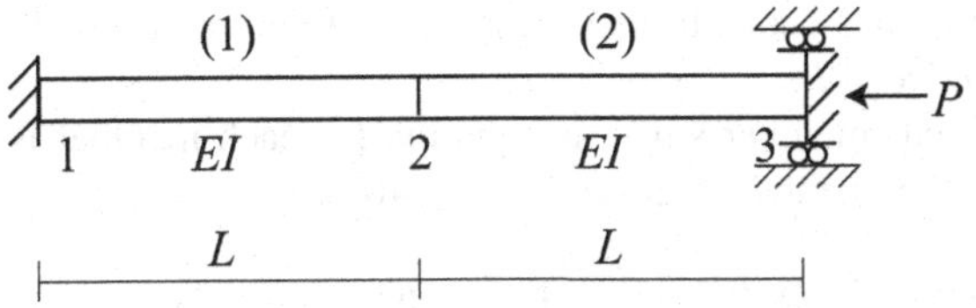

Figure 9.4 A two-element beam structure subjected to an axial force P

The equation (9.37) gives

$$\begin{pmatrix} w_2/L \\ \varphi_2 \end{pmatrix} = \frac{L^2}{2EI}\begin{pmatrix} 18 & 3 \\ 3 & 6 \end{pmatrix}^{-1}\begin{pmatrix} P \\ 0 \end{pmatrix} = \frac{PL^2}{66EI}\begin{pmatrix} 2 \\ -1 \end{pmatrix}.$$

Substituting these values and the boundary conditions (9.36) into the formula $\hat{\boldsymbol{p}}^{\bullet} = \hat{K}_{\mathrm{E}}^{\nabla}\,\hat{\boldsymbol{u}}$ in (9.31) for the reaction force vector, we obtain the reaction forces as

$$\begin{pmatrix} V_1 \\ M_1/L \\ \hline V_3 \\ M_3/L \end{pmatrix} = \frac{2EI}{L^2}\begin{pmatrix} -6 & 3 \\ -3 & 1 \\ \hline -12 & -6 \\ 6 & 2 \end{pmatrix} \times \frac{PL^2}{66EI}\begin{pmatrix} 2 \\ -1 \end{pmatrix} = \frac{P}{33}\begin{pmatrix} -15 \\ -7 \\ \hline -18 \\ 10 \end{pmatrix}. \tag{9.38}$$

As can been seen from this beam structural example, when the geometric stiffness is absent, the buckling does not occur, and the unknown variables can be determined uniquely for a given load P.

9.3 LINEAR BUCKLING ANALYSIS: INTRODUCTORY EXAMPLE

The method of linear buckling analysis is introduced using the beam structure with both ends fixed subjected to an axial force; the fixed end at the right has a roller to move horizontally (Fig. 9.4). We consider only flexural deformation and ignore axial deformation of beam members. The member stiffness equation (9.15) is assembled for the whole structural system to derive the stiffness matrix for the structure. On introducing the boundary condition, we can obtain the buckling condition, which gives buckling loads and buckling modes.

The beam structure has three nodes 1, 2, and 3 and two members 1 and 2 with the same length L. The displacement boundary conditions are given by

$$\begin{pmatrix} w_1 \\ \varphi_1 \end{pmatrix} = \begin{pmatrix} w_3 \\ \varphi_3 \end{pmatrix} = \begin{pmatrix} 0 \\ 0 \end{pmatrix}. \tag{9.39}$$

The external force P produces axial forces in both members as[2]

$$N^{(1)} = N^{(2)} = -P. \tag{9.40}$$

[2]For this beam structure, the member axial forces can be obtained without conducting the small displacement analysis.

Since this structure has geometric stiffness due to the presence of P, it can undergo buckling for some load P.

The (normalized) member stiffness equation for each member is obtained from (9.26) and using the member forces $N^{(m)}$ in (9.40) as

$$(\hat{K}_{\mathrm{E}}^{(m)} + \hat{K}_{\mathrm{G}}^{(m)})\,\hat{\boldsymbol{u}}^{(m)} = \mathbf{0}, \quad m = 1,2$$

with

$$\hat{K}_{\mathrm{E}}^{(1)} = \hat{K}_{\mathrm{E}}^{(2)} = \frac{2EI}{L^2}\left(\begin{array}{cc|cc} 6 & 3 & -6 & 3 \\ 3 & 2 & -3 & 1 \\ \hline -6 & -3 & 6 & -3 \\ 3 & 1 & -3 & 2 \end{array}\right),$$

$$\hat{K}_{\mathrm{G}}^{(1)} = \hat{K}_{\mathrm{G}}^{(2)} = \frac{-P}{30}\left(\begin{array}{cc|cc} 36 & 3 & -36 & 3 \\ 3 & 4 & -3 & -1 \\ \hline -36 & -3 & 36 & -3 \\ 3 & -1 & -3 & 4 \end{array}\right).$$

By assembling these member matrices, we obtain the matrices for the whole structure as

$$\hat{K}_{\mathrm{E}}^{\circ} = \frac{2EI}{L^2}\left(\begin{array}{cc|cc|cc} 6 & 3 & -6 & 3 & 0 & 0 \\ 3 & 2 & -3 & 1 & 0 & 0 \\ \hline -6 & -3 & 12 & 0 & -6 & 3 \\ 3 & 1 & 0 & 4 & -3 & 1 \\ \hline 0 & 0 & -6 & -3 & 6 & -3 \\ 0 & 0 & 3 & 1 & -3 & 2 \end{array}\right), \tag{9.41}$$

$$\hat{K}_{\mathrm{G}}^{\circ} = \frac{-P}{30}\left(\begin{array}{cc|cc|cc} 36 & 3 & -36 & 3 & 0 & 0 \\ 3 & 4 & -3 & 1 & 0 & 0 \\ \hline -36 & -3 & 72 & 0 & -36 & 3 \\ 3 & 1 & 0 & 8 & -3 & -1 \\ \hline 0 & 0 & -36 & -3 & 36 & -3 \\ 0 & 0 & 3 & -1 & -3 & 4 \end{array}\right). \tag{9.42}$$

Using the displacement boundary conditions $(w_1, \varphi_1) = (w_3, \varphi_3) = (0,0)$ in (9.39), we obtain the equilibrium equation for the free node 2 as

$$\left[\frac{2EI}{L^2}\begin{pmatrix} 12 & 0 \\ 0 & 4 \end{pmatrix} - \frac{P}{30}\begin{pmatrix} 72 & 0 \\ 0 & 8 \end{pmatrix}\right]\begin{pmatrix} w_2/L \\ \varphi_2 \end{pmatrix} = \begin{pmatrix} 0 \\ 0 \end{pmatrix},$$

which corresponds to the normalized stiffness equation $(\hat{K}_{\mathrm{E}} + \hat{K}_{\mathrm{G}})\hat{\boldsymbol{u}} = \hat{\boldsymbol{p}}$ in (9.29). The buckling condition $\det(\hat{K}_{\mathrm{E}} + \hat{K}_{\mathrm{G}}) = 0$ in (9.32) is given by

$$\det\left[\frac{2EI}{L^2}\begin{pmatrix} 12 & 0 \\ 0 & 4 \end{pmatrix} - \frac{P}{30}\begin{pmatrix} 72 & 0 \\ 0 & 8 \end{pmatrix}\right] = \left(24\frac{EI}{L^2} - \frac{12}{5}P\right)\left(8\frac{EI}{L^2} - \frac{4}{15}P\right) = 0.$$

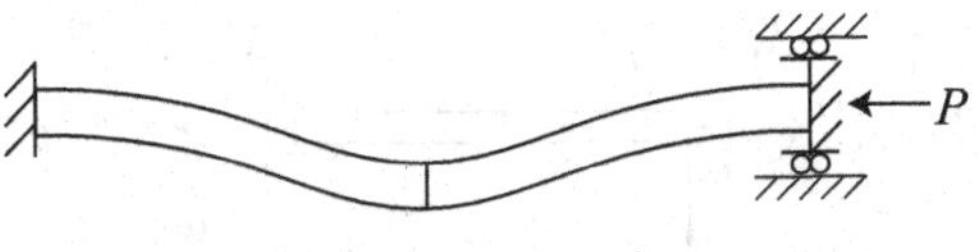

First buckling mode $\hat{\eta}_1$

Second buckling mode $\hat{\eta}_2$

Figure 9.5 Buckling modes of a beam with both ends fixed (the fixed end at the right has a roller to move horizontally)

Accordingly, the buckling loads P_1 and P_2 are

$$P_1 = 10\frac{EI}{L^2}, \qquad P_2 = 30\frac{EI}{L^2}. \tag{9.43}$$

We can determine the associated buckling modes $\hat{\boldsymbol{\eta}} = (\mathrm{d}w_2/L, \mathrm{d}\varphi_2)^\top$ from $(\hat{K}_\mathrm{E} + \hat{K}_\mathrm{G})\hat{\boldsymbol{\eta}} = \mathbf{0}$ in (9.33) as

$$\hat{\boldsymbol{\eta}}_1 = \begin{pmatrix} 1 \\ 0 \end{pmatrix}, \qquad \hat{\boldsymbol{\eta}}_2 = \begin{pmatrix} 0 \\ 1 \end{pmatrix}.$$

These buckling modes are depicted in Fig. 9.5.

Remark 9.2 The minimum buckling load P_1 obtained in this manner is compared with the theoretical buckling load discussed in Section 7.3.3. The theoretical minimum buckling load P_1^* of the beam-column with both ends fixed with the member length $2L$ is obtained as

$$P_1^* = 4P_\mathrm{E} = 4\pi^2\frac{EI}{(2L)^2} = \pi^2\frac{EI}{L^2} \approx 9.87\frac{EI}{L^2},$$

which follows from (7.39) using (7.30) for the member length $2L$. Accordingly, the minimum buckling load $P_1 = 10EI/L^2$ in (9.43) has an error of slightly over 1%, which is due to the approximation of the deflection $w(s)$ in each member by the cubic polynomial function in (9.5). □

9.4 FORMULATION IMPLEMENTING AXIAL DEFORMATION

Members in a frame structure undergo flexural deformation as beam-columns and axial deformation as trusses. While we have considered only the flexural deformation up to now, we present a formulation implementing axial deformation in this section to upgrade the accuracy of the buckling analysis.

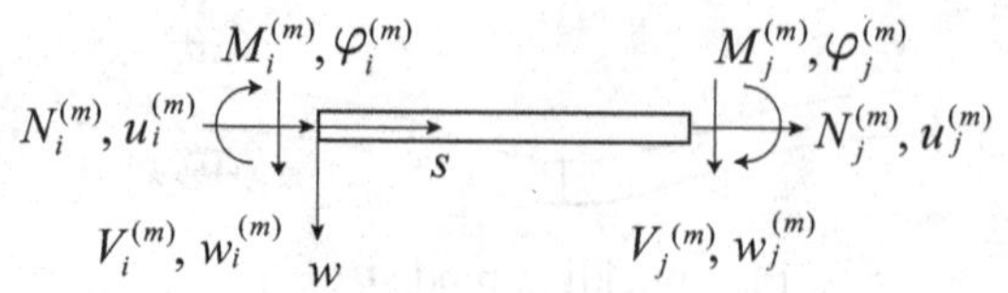

Figure 9.6 Coordinate system for a beam-column frame member

9.4.1 MEMBER STIFFNESS MATRIX

We consider the mth member, which is assumed to connect the nodes i and j. Using the coordinates in Fig. 9.6, we define the 6-dimensional member end displacement vector $\boldsymbol{u}^{(m)}$ and member end force vector $\boldsymbol{p}^{(m)}$, respectively, by

$$\boldsymbol{u}^{(m)} = \begin{pmatrix} \boldsymbol{u}_i^{(m)} \\ \boldsymbol{u}_j^{(m)} \end{pmatrix}, \qquad \boldsymbol{p}^{(m)} = \begin{pmatrix} \boldsymbol{p}_i^{(m)} \\ \boldsymbol{p}_j^{(m)} \end{pmatrix}, \tag{9.44}$$

where

$$\boldsymbol{u}_k^{(m)} = \begin{pmatrix} u_k^{(m)} \\ w_k^{(m)} \\ \varphi_k^{(m)} \end{pmatrix}, \qquad \boldsymbol{p}_k^{(m)} = \begin{pmatrix} N_k^{(m)} \\ V_k^{(m)} \\ M_k^{(m)} \end{pmatrix}, \quad k = i, j; \tag{9.45}$$

$u_k^{(m)}$ is the axial displacement, $w_k^{(m)}$ is the shear displacement, and $\varphi_k^{(m)}$ is the rotational displacement; $N_k^{(m)}$, $V_k^{(m)}$, and $M_k^{(m)}$ are the associated forces.[3] The axial member end forces $N_i^{(m)}$ and $N_j^{(m)}$ and the axial force $N^{(m)}$ are related as

$$N_i^{(m)} = -N^{(m)}, \qquad N_j^{(m)} = N^{(m)}. \tag{9.46}$$

Our goal is to derive the member stiffness equation

$$(K_{\mathrm{E}}^{(m)} + K_{\mathrm{G}}^{(m)})\,\boldsymbol{u}^{(m)} = \boldsymbol{p}^{(m)} \tag{9.47}$$

of the two-dimensional frame member, where the stiffness matrix $K_{\mathrm{E}}^{(m)}$ and the geometric stiffness matrix $K_{\mathrm{G}}^{(m)}$ are 6×6 symmetric matrices. The matrix $K_{\mathrm{S}}^{(m)}$ for the spring stiffness and the equivalent nodal load $\boldsymbol{q}^{(m)}$ of the distributed load in (9.14) are not included for simplicity, as with the corresponding equation (9.15) of the beam-column member.

The stiffness equation, expressing the axial-directional equilibrium, is given by

$$\begin{pmatrix} N_i^{(m)} \\ N_j^{(m)} \end{pmatrix} = \frac{EA}{L} \begin{pmatrix} 1 & -1 \\ -1 & 1 \end{pmatrix} \begin{pmatrix} u_i^{(m)} \\ u_j^{(m)} \end{pmatrix}, \tag{9.48}$$

[3]Note that the member end displacement vector and member end force vector in (9.44) are 6-dimensional by adding (u_i, u_j) and (N_i, N_j), respectively, to the 4-dimensional vectors in (9.2).

where $L = L^{(m)}$, $EA = EA^{(m)}$, and tensile axial force is defined to be positive.[4] By assembling the stiffness matrix in (9.48) for axial components and the 4×4 stiffness matrix in (9.12) for bending, we obtain the member stiffness matrix

$$K_{\mathrm{E}}^{(m)} = \begin{pmatrix} k_{\mathrm{E}ii}^{(m)} & k_{\mathrm{E}ij}^{(m)} \\ k_{\mathrm{E}ji}^{(m)} & k_{\mathrm{E}jj}^{(m)} \end{pmatrix} = \left(\begin{array}{ccc|ccc} \frac{EA}{L} & 0 & 0 & -\frac{EA}{L} & 0 & 0 \\ 0 & \frac{12EI}{L^3} & \frac{6EI}{L^2} & 0 & -\frac{12EI}{L^3} & \frac{6EI}{L^2} \\ 0 & \frac{6EI}{L^2} & \frac{4EI}{L} & 0 & -\frac{6EI}{L^2} & \frac{2EI}{L} \\ \hline -\frac{EA}{L} & 0 & 0 & \frac{EA}{L} & 0 & 0 \\ 0 & -\frac{12EI}{L^3} & -\frac{6EI}{L^2} & 0 & \frac{12EI}{L^3} & -\frac{6EI}{L^2} \\ 0 & \frac{6EI}{L^2} & \frac{2EI}{L} & 0 & -\frac{6EI}{L^2} & \frac{4EI}{L} \end{array}\right), \tag{9.49}$$

which is a 6×6 symmetric matrix consisting of

$$k_{\mathrm{E}ii}^{(m)} = \begin{pmatrix} \frac{EA}{L} & 0 & 0 \\ 0 & \frac{12EI}{L^3} & \frac{6EI}{L^2} \\ 0 & \frac{6EI}{L^2} & \frac{4EI}{L} \end{pmatrix} = \frac{2EI}{L^2} \begin{pmatrix} \beta/L & 0 & 0 \\ 0 & 6/L & 3 \\ 0 & 3 & 2L \end{pmatrix}, \tag{9.50}$$

$$k_{\mathrm{E}jj}^{(m)} = \begin{pmatrix} \frac{EA}{L} & 0 & 0 \\ 0 & \frac{12EI}{L^3} & -\frac{6EI}{L^2} \\ 0 & -\frac{6EI}{L^2} & \frac{4EI}{L} \end{pmatrix} = \frac{2EI}{L^2} \begin{pmatrix} \beta/L & 0 & 0 \\ 0 & 6/L & -3 \\ 0 & -3 & 2L \end{pmatrix}, \tag{9.51}$$

$$k_{\mathrm{E}ij}^{(m)} = k_{\mathrm{E}ji}^{(m)\top} = \begin{pmatrix} -\frac{EA}{L} & 0 & 0 \\ 0 & -\frac{12EI}{L^3} & \frac{6EI}{L^2} \\ 0 & -\frac{6EI}{L^2} & \frac{2EI}{L} \end{pmatrix} = \frac{2EI}{L^2} \begin{pmatrix} -\beta/L & 0 & 0 \\ 0 & -6/L & 3 \\ 0 & -3 & L \end{pmatrix}, \tag{9.52}$$

where $L = L^{(m)}$, $EA = EA^{(m)}$, $EI = EI^{(m)}$, and $\beta = \beta^{(m)}$ is defined as

$$\beta = \frac{EA}{L}\left(\frac{2EI}{L^3}\right)^{-1} = \frac{L^2}{2}\frac{EA}{EI} = \frac{L^2 A}{2I} = \frac{\lambda^2}{2}. \tag{9.53}$$

---

[4]The stiffness matrix in (9.48) is a part of the stiffness matrix in (8.52). Note that (9.48) implies $N_i^{(m)} = -N_j^{(m)}$.

Thus the parameter β is related to the slenderness ratio $\lambda = \lambda^{(m)}$ (cf., $\lambda = L\sqrt{A/I}$ in (7.43)).

By adding axial degrees-of-freedom to the 4×4 geometric stiffness matrix in (9.13), we obtain the geometric stiffness matrix

$$K_{\mathrm{G}}^{(m)} = \begin{pmatrix} k_{\mathrm{G}ii}^{(m)} & k_{\mathrm{G}ij}^{(m)} \\ k_{\mathrm{G}ji}^{(m)} & k_{\mathrm{G}jj}^{(m)} \end{pmatrix}, \tag{9.54}$$

which is a 6×6 symmetric matrix consisting of

$$k_{\mathrm{G}ii}^{(m)} = \frac{N}{30}\begin{pmatrix} 0 & 0 & 0 \\ 0 & 36/L & 3 \\ 0 & 3 & 4L \end{pmatrix}, \qquad k_{\mathrm{G}jj}^{(m)} = \frac{N}{30}\begin{pmatrix} 0 & 0 & 0 \\ 0 & 36/L & -3 \\ 0 & -3 & 4L \end{pmatrix}, \tag{9.55}$$

$$k_{\mathrm{G}ij}^{(m)} = k_{\mathrm{G}ji}^{(m)\top} = \frac{N}{30}\begin{pmatrix} 0 & 0 & 0 \\ 0 & -36/L & 3 \\ 0 & -3 & -L \end{pmatrix}, \tag{9.56}$$

where $N = N^{(m)}$.

Remark 9.3 When all members have a common length L, it is convenient, in both analytical study and numerical analysis, to employ the following normalized variables

$$\begin{aligned} \hat{\boldsymbol{u}}^{(m)} &= (u_i^{(m)}/L,\ w_i^{(m)}/L,\ \varphi_i^{(m)},\ u_j^{(m)}/L,\ w_j^{(m)}/L,\ \varphi_j^{(m)})^\top, \\ \hat{\boldsymbol{p}}^{(m)} &= (N_i^{(m)},\ V_i^{(m)},\ M_i^{(m)}/L,\ N_j^{(m)},\ V_j^{(m)},\ M_j^{(m)}/L)^\top, \end{aligned} \tag{9.57}$$

which are normalized versions of $\boldsymbol{u}^{(m)}$ and $\boldsymbol{p}^{(m)}$ in (9.44). Then we obtain the stiffness equation for these variables as

$$(\hat{K}_{\mathrm{E}}^{(m)} + \hat{K}_{\mathrm{G}}^{(m)})\,\hat{\boldsymbol{u}}^{(m)} = \hat{\boldsymbol{p}}^{(m)} \tag{9.58}$$

in place of (9.47). Here the matrices take simpler forms

$$\hat{K}_{\mathrm{E}}^{(m)} = \begin{pmatrix} \hat{k}_{\mathrm{E}ii}^{(m)} & \hat{k}_{\mathrm{E}ij}^{(m)} \\ \hat{k}_{\mathrm{E}ji}^{(m)} & \hat{k}_{\mathrm{E}jj}^{(m)} \end{pmatrix} = \frac{2EI}{L^2}\left(\begin{array}{ccc|ccc} \beta & 0 & 0 & -\beta & 0 & 0 \\ 0 & 6 & 3 & 0 & -6 & 3 \\ 0 & 3 & 2 & 0 & -3 & 1 \\ \hline -\beta & 0 & 0 & \beta & 0 & 0 \\ 0 & -6 & -3 & 0 & 6 & -3 \\ 0 & 3 & 1 & 0 & -3 & 2 \end{array}\right), \tag{9.59}$$

$$\hat{K}_{\mathrm{G}}^{(m)} = \begin{pmatrix} \hat{k}_{\mathrm{G}ii}^{(m)} & \hat{k}_{\mathrm{G}ij}^{(m)} \\ \hat{k}_{\mathrm{G}ji}^{(m)} & \hat{k}_{\mathrm{G}jj}^{(m)} \end{pmatrix} = \frac{N}{30}\left(\begin{array}{ccc|ccc} 0 & 0 & 0 & 0 & 0 & 0 \\ 0 & 36 & 3 & 0 & -36 & 3 \\ 0 & 3 & 4 & 0 & -3 & -1 \\ \hline 0 & 0 & 0 & 0 & 0 & 0 \\ 0 & -36 & -3 & 0 & 36 & -3 \\ 0 & 3 & -1 & 0 & -3 & 4 \end{array}\right), \tag{9.60}$$

where $\beta = \beta^{(m)} = L^2A/(2I) = \lambda^2/2$ (see (9.53)). □

9.4.2 PROCEDURE TO OBTAIN AXIAL FORCES

In the linear buckling analysis, the axial forces $N = N^{(m)}$ in (9.55), (9.56), and (9.60) are obtained by the small displacement analysis using the following procedure:

- By solving the equilibrium equation $K_{\rm E}\,\boldsymbol{u} = \boldsymbol{p}$ in (9.21), we obtain the unknown displacement $\boldsymbol{u}$.
- The member displacement vector $\boldsymbol{u}^{(m)}$ in (9.44) is obtained from $\boldsymbol{u}$.
- By the member stiffness equation $K_{\rm E}^{(m)}\,\boldsymbol{u}^{(m)} = \boldsymbol{p}^{(m)}$, which is obtained by omitting $K_{\rm G}^{(m)}$ in (9.47), we obtain the member end force vector $\boldsymbol{p}^{(m)}$.
- $N_i^{(m)}$ is obtained as the first component of $\boldsymbol{p}^{(m)}$ (see (9.44) and (9.45)) and the axial force is obtained as $N^{(m)} = -N_i^{(m)}$ by (9.46).

This procedure to obtain the axial forces in members is illustrated in Sections 9.5 and 9.6.2 for concrete structural examples.

9.5 STRUCTURAL EXAMPLE OF LINEAR BUCKLING ANALYSIS

The simple linear buckling analysis procedure introduced in Section 9.4 is illustrated for the L-shaped frame structure consisting of a vertical member and a horizontal member of the same length L in Fig. 9.7(a). This frame structure, which has only horizontal and vertical members, allows a simple formulation. A more general structure with tilted members is to be treated in Section 9.6.

9.5.1 DEFINITION OF VARIABLES

Node 2 of the L-shaped frame is free, while nodes 1 and 3 are pinned and are free to rotate but are restricted to translate in any direction. Then we can define the nodal

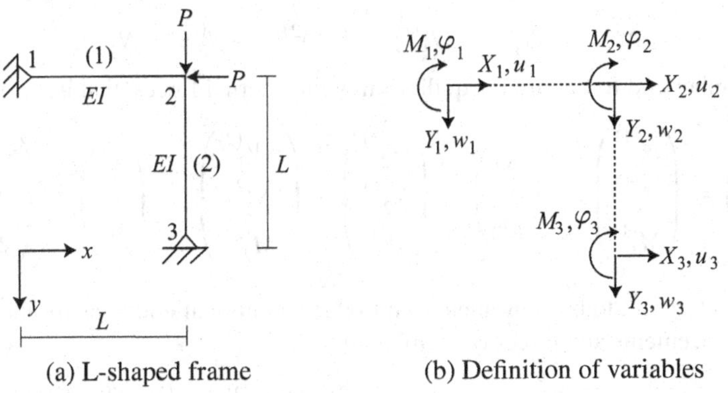

Figure 9.7 An L-shaped frame and the definition of nodal variables

displacements and forces as illustrated in Fig. 9.7(b) and define the displacement vector and the external force vector, respectively, by

$$\begin{aligned} \boldsymbol{u}^{\circ} &= (u_1, w_1, \varphi_1,\ u_2, w_2, \varphi_2,\ u_3, w_3, \varphi_3)^{\top}, \\ \boldsymbol{p}^{\circ} &= (X_1, Y_1, M_1,\ X_2, Y_2, M_2,\ X_3, Y_3, M_3)^{\top}. \end{aligned} \tag{9.61}$$

The vector $\boldsymbol{u}$ for unknown variables and the vector $\boldsymbol{u}^{\bullet}$ for the variables specified by boundary conditions are given, respectively, by

$$\boldsymbol{u} = (\varphi_1, u_2, w_2, \varphi_2, \varphi_3)^{\top}, \qquad \boldsymbol{u}^{\bullet} = (u_1, w_1, u_3, w_3)^{\top} = \boldsymbol{0}. \tag{9.62}$$

External forces are given by

$$X_2 = -P, \qquad Y_2 = P, \qquad M_1 = M_2 = M_3 = 0, \tag{9.63}$$

that is, $\boldsymbol{p} = (M_1, X_2, Y_2, M_2, M_3)^{\top} = (0, -P, P, 0, 0)^{\top}$ and unknown reaction forces are expressed by the vector $\boldsymbol{p}^{\bullet} = (X_1, Y_1, X_3, Y_3)^{\top}$.

For the horizontal member $m = 1$ with $(i, j) = (1, 2)$ the nodal variables are expressed by (9.44) with (9.45) as

$$\boldsymbol{u}^{(1)} = (u_1^{(1)}, w_1^{(1)}, \varphi_1^{(1)}, u_2^{(1)}, w_2^{(1)}, \varphi_2^{(1)})^{\top} = (u_1, w_1, \varphi_1, u_2, w_2, \varphi_2)^{\top},$$

whereas the vertical member $m = 2$ with $(i, j) = (2, 3)$ has

$$\boldsymbol{u}^{(2)} = (u_2^{(2)}, w_2^{(2)}, \varphi_2^{(2)}, u_3^{(2)}, w_3^{(2)}, \varphi_3^{(2)})^{\top} = (w_2, -u_2, \varphi_2, w_3, -u_3, \varphi_3)^{\top}.$$

The member end force vectors $\boldsymbol{p}^{(1)}$ and $\boldsymbol{p}^{(2)}$ are given by (9.44) with (9.45) as

$$\begin{aligned} \boldsymbol{p}^{(1)} &= (N_1^{(1)}, V_1^{(1)}, M_1^{(1)},\ N_2^{(1)}, V_2^{(1)}, M_2^{(1)}), \\ \boldsymbol{p}^{(2)} &= (N_2^{(2)}, V_2^{(2)}, M_2^{(2)},\ N_3^{(2)}, V_3^{(2)}, M_3^{(2)}). \end{aligned}$$

From (9.46), the axial forces $N^{(1)}$ and $N^{(2)}$ are given, using member end forces, as

$$N^{(1)} = -N_1^{(1)}\ (= N_2^{(1)}), \qquad N^{(2)} = -N_2^{(2)}\ (= N_3^{(2)}). \tag{9.64}$$

The member end forces are in equilibrium with external forces, that is,

$$\begin{pmatrix} X_1 \\ Y_1 \\ M_1 \end{pmatrix} = \begin{pmatrix} N_1^{(1)} \\ V_1^{(1)} \\ M_1^{(1)} \end{pmatrix}, \quad \begin{pmatrix} X_2 \\ Y_2 \\ M_2 \end{pmatrix} = \begin{pmatrix} N_2^{(1)} \\ V_2^{(1)} \\ M_2^{(1)} \end{pmatrix} + \begin{pmatrix} -V_2^{(2)} \\ N_2^{(2)} \\ M_2^{(2)} \end{pmatrix}, \quad \begin{pmatrix} X_3 \\ Y_3 \\ M_3 \end{pmatrix} = \begin{pmatrix} -V_3^{(2)} \\ N_3^{(2)} \\ M_3^{(2)} \end{pmatrix}. \tag{9.65}$$

It will turn out to be convenient to introduce additional notations for the member end displacements and forces of member $m = 2$:

$$\tilde{\boldsymbol{u}}^{(2)} = (u_2, w_2, \varphi_2, u_3, w_3, \varphi_3)^{\top} = (-w_2^{(2)}, u_2^{(2)}, \varphi_2^{(2)}, -w_3^{(2)}, u_3^{(2)}, \varphi_3^{(2)})^{\top}, \tag{9.66}$$

$$\tilde{\boldsymbol{p}}^{(2)} = (-V_2^{(2)}, N_2^{(2)}, M_2^{(2)},\ -V_3^{(2)}, N_3^{(2)}, M_3^{(2)}). \tag{9.67}$$

Note that $\tilde{\boldsymbol{u}}^{(2)}$ and $\tilde{\boldsymbol{p}}^{(2)}$ are obtained from $\boldsymbol{u}^{(2)}$ and $\boldsymbol{p}^{(2)}$, respectively, through permutation and sign inversion of some components.[5]

9.5.2 SMALL DISPLACEMENT ANALYSIS

We conduct the small displacement analysis to obtain member axial forces in (9.64).

The member stiffness matrix $K_{\mathrm{E}}^{(1)}$ for the member $m = 1$ is obtained from (9.49) with (9.50)–(9.52) as

$$K_{\mathrm{E}}^{(1)} = \frac{2EI}{L^2}\left(\begin{array}{ccc|ccc} \beta/L & 0 & 0 & -\beta/L & 0 & 0 \\ 0 & 6/L & 3 & 0 & -6/L & 3 \\ 0 & 3 & 2L & 0 & -3 & L \\ \hline -\beta/L & 0 & 0 & \beta/L & 0 & 0 \\ 0 & -6/L & -3 & 0 & 6/L & -3 \\ 0 & 3 & L & 0 & -3 & 2L \end{array}\right), \tag{9.68}$$

for which we have $K_{\mathrm{E}}^{(1)}\boldsymbol{u}^{(1)} = \boldsymbol{p}^{(1)}$. In the normalized variables, this matrix takes a simpler form

$$\hat{K}_{\mathrm{E}}^{(1)} = \frac{2EI}{L^2}\left(\begin{array}{ccc|ccc} \beta & 0 & 0 & -\beta & 0 & 0 \\ 0 & 6 & 3 & 0 & -6 & 3 \\ 0 & 3 & 2 & 0 & -3 & 1 \\ \hline -\beta & 0 & 0 & \beta & 0 & 0 \\ 0 & -6 & -3 & 0 & 6 & -3 \\ 0 & 3 & 1 & 0 & -3 & 2 \end{array}\right) \tag{9.69}$$

as in (9.59), where $\beta = L^2A/(2I) = \lambda^2/2$ by (9.53).

For the member $m = 2$, we similarly have $K_{\mathrm{E}}^{(2)}\boldsymbol{u}^{(2)} = \boldsymbol{p}^{(2)}$, where $K_{\mathrm{E}}^{(2)}$ is also equal to the matrix in (9.68). However, to carry out the assemblage of member stiffness matrices, it is convenient to use $\tilde{\boldsymbol{u}}^{(2)}$ and $\tilde{\boldsymbol{p}}^{(2)}$ in (9.66) and (9.67) to obtain an equivalent member stiffness equation $\tilde{K}_{\mathrm{E}}^{(2)}\tilde{\boldsymbol{u}}^{(2)} = \tilde{\boldsymbol{p}}^{(2)}$. Here the matrix $\tilde{K}_{\mathrm{E}}^{(2)}$ is obtained from $K_{\mathrm{E}}^{(2)}$ by permuting the first and second rows and columns as well as the fourth and fifth rows and columns and then changing the sign of the first and fourth rows and

[5]The transformation $\boldsymbol{u}^{(2)} \to \tilde{\boldsymbol{u}}^{(2)}$ corresponds to $\bar{\boldsymbol{u}}^{(m)} \to \boldsymbol{u}^{(m)}$ in (9.91) with $\theta^{(m)} = -\pi/2$ in (9.89) in Section 9.6.1. Similarly, $\boldsymbol{p}^{(2)} \to \tilde{\boldsymbol{p}}^{(2)}$ corresponds to $\bar{\boldsymbol{p}}^{(m)} \to \boldsymbol{p}^{(m)}$ in (9.91).

columns.[6] For the normalized variables, this matrix $\tilde{K}_{\mathrm{E}}^{(2)}$ is simplified to

$$\frac{2EI}{L^2}\left(\begin{array}{ccc|ccc} 6 & 0 & -3 & -6 & 0 & -3 \\ 0 & \beta & 0 & 0 & -\beta & 0 \\ -3 & 0 & 2 & 3 & 0 & 1 \\ \hline -6 & 0 & 3 & 6 & 0 & 3 \\ 0 & -\beta & 0 & 0 & \beta & 0 \\ -3 & 0 & 1 & 3 & 0 & 2 \end{array}\right). \tag{9.70}$$

We are now ready to obtain the equation for the whole structure

$$\hat{K}_{\mathrm{E}}^{\circ}\hat{\boldsymbol{u}}^{\circ} = \hat{\boldsymbol{p}}^{\circ} \tag{9.71}$$

for the normalized variables

$$\hat{\boldsymbol{u}}^{\circ} = (u_1/L, w_1/L, \varphi_1,\ u_2/L, w_2/L, \varphi_2,\ u_3/L, w_3/L, \varphi_3)^{\top},$$
$$\hat{\boldsymbol{p}}^{\circ} = (X_1, Y_1, M_1/L,\ \ X_2, Y_2, M_2/L,\ \ X_3, Y_3, M_3/L)^{\top}.$$

The matrix $\hat{K}_{\mathrm{E}}^{\circ}$ in (9.71), which is a 9×9 symmetric matrix, can be obtained by superposing the 6×6 member stiffness matrices in (9.69) and (9.70) appropriately. This superposition results in

$$\hat{K}_{\mathrm{E}}^{\circ} = \frac{2EI}{L^2}\left(\begin{array}{ccc|ccc|ccc} \beta & 0 & 0 & -\beta & 0 & 0 & 0 & 0 & 0 \\ 0 & 6 & 3 & 0 & -6 & 3 & 0 & 0 & 0 \\ 0 & 3 & 2 & 0 & -3 & 1 & 0 & 0 & 0 \\ \hline -\beta & 0 & 0 & \beta+6 & 0 & -3 & -6 & 0 & -3 \\ 0 & -6 & -3 & 0 & \beta+6 & -3 & 0 & -\beta & 0 \\ 0 & 3 & 1 & -3 & -3 & 4 & 3 & 0 & 1 \\ \hline 0 & 0 & 0 & -6 & 0 & 3 & 6 & 0 & 3 \\ 0 & 0 & 0 & 0 & -\beta & 0 & 0 & \beta & 0 \\ 0 & 0 & 0 & -3 & 0 & 1 & 3 & 0 & 2 \end{array}\right). \tag{9.72}$$

Using the displacement boundary conditions $u_1 = w_1 = u_3 = w_3 = 0$ in (9.62) and force boundary conditions $X_2 = -P$, $Y_2 = P$, and $M_1 = M_2 = M_3 = 0$ in (9.63), we obtain the small displacement equilibrium equation

$$\hat{K}_{\mathrm{E}}\,\hat{\boldsymbol{u}} = \hat{\boldsymbol{p}} \tag{9.73}$$

in $\hat{\boldsymbol{u}} = (\varphi_1, u_2/L, w_2/L, \varphi_2, \varphi_3)^{\top}$ with $\hat{\boldsymbol{p}} = (M_1/L, X_2, Y_2, M_2/L, M_3/L)^{\top} = (0, -P, P, 0, 0)^{\top}$ and

$$\hat{K}_{\mathrm{E}} = \frac{2EI}{L^2}\left(\begin{array}{c|ccc|c} 2 & 0 & -3 & 1 & 0 \\ \hline 0 & \beta+6 & 0 & -3 & -3 \\ -3 & 0 & \beta+6 & -3 & 0 \\ 1 & -3 & -3 & 4 & 1 \\ \hline 0 & -3 & 0 & 1 & 2 \end{array}\right). \tag{9.74}$$

[6]Since $\tilde{\boldsymbol{u}}^{(2)} = T\boldsymbol{u}^{(2)}$ and $\tilde{\boldsymbol{p}}^{(2)} = T\boldsymbol{p}^{(2)}$ for $T = \begin{pmatrix} S & O \\ O & S \end{pmatrix}$ with $S = \begin{pmatrix} 0 & -1 & 0 \\ 1 & 0 & 0 \\ 0 & 0 & 1 \end{pmatrix}$, we have $\tilde{K}_{\mathrm{E}}^{(2)} = TK_{\mathrm{E}}^{(2)}T^{\top}$. This corresponds to (9.97) with $T^{(m)} = S^{\top}$ in Section 9.6.1.

The equation (9.73) is solved for $\hat{\boldsymbol{u}}$ as

$$\hat{\boldsymbol{u}} = \begin{pmatrix} \varphi_1 \\ u_2/L \\ w_2/L \\ \varphi_2 \\ \varphi_3 \end{pmatrix} = \frac{1}{2\beta+3}\frac{PL^2}{2EI}\begin{pmatrix} 3 \\ -2 \\ 2 \\ 0 \\ -3 \end{pmatrix}. \tag{9.75}$$

We evaluate the axial forces in the members following the procedure given in Section 9.4.2. Using (9.75), member displacement vectors are obtained as

$$\hat{\boldsymbol{u}}^{(1)} = \begin{pmatrix} u_1/L \\ w_1/L \\ \varphi_1 \\ u_2/L \\ w_2/L \\ \varphi_2 \end{pmatrix} = \frac{1}{2\beta+3}\frac{PL^2}{2EI}\begin{pmatrix} 0 \\ 0 \\ 3 \\ -2 \\ 2 \\ 0 \end{pmatrix}, \quad \hat{\boldsymbol{u}}^{(2)} = \begin{pmatrix} w_2/L \\ -u_2/L \\ \varphi_2 \\ w_3/L \\ -u_3/L \\ \varphi_3 \end{pmatrix} = \frac{1}{2\beta+3}\frac{PL^2}{2EI}\begin{pmatrix} 2 \\ 2 \\ 0 \\ 0 \\ 0 \\ -3 \end{pmatrix}.$$

Then member end forces for $m = 1$ are evaluated from $\hat{\boldsymbol{p}}^{(1)} = \hat{K}_{\mathrm{E}}^{(1)}\hat{\boldsymbol{u}}^{(1)}$ using the matrix $\hat{K}_{\mathrm{E}}^{(1)}$ in (9.69) as

$$\begin{aligned}\hat{\boldsymbol{p}}^{(1)} &= (N_1^{(1)}, V_1^{(1)}, M_1^{(1)}/L, N_2^{(1)}, V_2^{(1)}, M_2^{(1)}/L)^\top \\ &= \frac{2EI}{L^2}\left(\begin{array}{ccc|ccc} \beta & 0 & 0 & -\beta & 0 & 0 \\ 0 & 6 & 3 & 0 & -6 & 3 \\ 0 & 3 & 2 & 0 & -3 & 1 \\ \hline -\beta & 0 & 0 & \beta & 0 & 0 \\ 0 & -6 & -3 & 0 & 6 & -3 \\ 0 & 3 & 1 & 0 & -3 & 2 \end{array}\right) \times \frac{1}{2\beta+3}\frac{PL^2}{2EI}\begin{pmatrix} 0 \\ 0 \\ 3 \\ -2 \\ 2 \\ 0 \end{pmatrix} = \frac{P}{2\beta+3}\begin{pmatrix} 2\beta \\ -3 \\ 0 \\ -2\beta \\ 3 \\ -3 \end{pmatrix}.\end{aligned} \tag{9.76}$$

The first component of this expression gives $N_1^{(1)} = [(2\beta)/(2\beta+3)]P$. Then from (9.64), we obtain the axial force of the member $m = 1$ as

$$N^{(1)} = -\frac{2\beta}{2\beta+3}P. \tag{9.77}$$

Similarly, we obtain the member end forces for $m = 2$ from $\hat{\boldsymbol{p}}^{(2)} = \hat{K}_{\mathrm{E}}^{(2)}\hat{\boldsymbol{u}}^{(2)}$, where $\hat{K}_{\mathrm{E}}^{(2)}$ is equal to the matrix in (9.69) (not (9.70)). Thus we obtain

$$\begin{aligned}\hat{\boldsymbol{p}}^{(2)} &= (N_2^{(2)}, V_2^{(2)}, M_2^{(2)}/L, N_3^{(2)}, V_3^{(2)}, M_3^{(2)}/L)^\top \\ &= \frac{2EI}{L^2}\left(\begin{array}{ccc|ccc} \beta & 0 & 0 & -\beta & 0 & 0 \\ 0 & 6 & 3 & 0 & -6 & 3 \\ 0 & 3 & 2 & 0 & -3 & 1 \\ \hline -\beta & 0 & 0 & \beta & 0 & 0 \\ 0 & -6 & -3 & 0 & 6 & -3 \\ 0 & 3 & 1 & 0 & -3 & 2 \end{array}\right) \times \frac{1}{2\beta+3}\frac{PL^2}{2EI}\begin{pmatrix} 2 \\ 2 \\ 0 \\ 0 \\ 0 \\ -3 \end{pmatrix} \frac{P}{2\beta+3}\begin{pmatrix} 2\beta \\ 3 \\ 3 \\ -2\beta \\ -3 \\ 0 \end{pmatrix}.\end{aligned} \tag{9.78}$$

The first component of this expression gives $N_2^{(2)} = [(2\beta)/(2\beta+3)]P$. Then from (9.64), we obtain the axial force of the member $m = 2$ as

$$N^{(2)} = -\frac{2\beta}{2\beta+3}P. \tag{9.79}$$

To sum up, we have

$$N^{(1)} = N^{(2)} = -\frac{2\beta}{2\beta+3}P. \tag{9.80}$$

Note that the axial forces tend to $-P$ as $\beta \to +\infty$ ($\lambda \to +\infty$), and to -0 as $\beta \to +0$ ($\lambda \to +0$) (cf., (9.53)).

9.5.3 LINEAR BUCKLING ANALYSIS

We conduct the linear buckling analysis to obtain buckling loads and buckling modes. Similarly to $\hat{K}_{\mathrm{E}}^{\circ}$ in (9.72), we can obtain the geometric stiffness matrix $\hat{K}_{\mathrm{G}}^{\circ}$ as

$$\hat{K}_{\mathrm{G}}^{\circ} = -\frac{\beta P}{15(2\beta+3)}\left(\begin{array}{ccc|ccc|ccc} 0 & 0 & 0 & 0 & 0 & 0 & 0 & 0 & 0 \\ 0 & 36 & 3 & 0 & -36 & 3 & 0 & 0 & 0 \\ 0 & 3 & 4 & 0 & -3 & -1 & 0 & 0 & 0 \\ \hline 0 & 0 & 0 & 36 & 0 & -3 & -36 & 0 & -3 \\ 0 & -36 & -3 & 0 & 36 & -3 & 0 & 0 & 0 \\ 0 & 3 & -1 & -3 & -3 & 8 & 3 & 0 & -1 \\ \hline 0 & 0 & 0 & -36 & 0 & 3 & 36 & 0 & 3 \\ 0 & 0 & 0 & 0 & 0 & 0 & 0 & 0 & 0 \\ 0 & 0 & 0 & -3 & 0 & -1 & 3 & 0 & 4 \end{array}\right)$$

using (9.60) and $N^{(1)} = N^{(2)} = -2\beta P/(2\beta+3)$ in (9.80). We can obtain the submatrix of $\hat{K}_{\mathrm{G}}^{\circ}$ corresponding to the unknown variables $\hat{\boldsymbol{u}}$ and the given external forces $\hat{\boldsymbol{p}}$ as

$$\hat{K}_{\mathrm{G}} = -\frac{\beta P}{15(2\beta+3)}\left(\begin{array}{c|ccc|c} 4 & 0 & -3 & -1 & 0 \\ \hline 0 & 36 & 0 & -3 & -3 \\ -3 & 0 & 36 & -3 & 0 \\ -1 & -3 & -3 & 8 & -1 \\ \hline 0 & -3 & 0 & -1 & 4 \end{array}\right). \tag{9.81}$$

With the use of $\hat{K}_{\mathrm{E}}$ in (9.74) and $\hat{K}_{\mathrm{G}}$ in (9.81), the buckling condition is given by $\det(\hat{K}_{\mathrm{E}} + \hat{K}_{\mathrm{G}}) = 0$ (cf., (9.32)), that is,

$$\det\left[\frac{2EI}{L^2}\left(\begin{array}{c|ccc|c} 2 & 0 & -3 & 1 & 0 \\ \hline 0 & \beta+6 & 0 & -3 & -3 \\ -3 & 0 & \beta+6 & -3 & 0 \\ 1 & -3 & -3 & 4 & 1 \\ \hline 0 & -3 & 0 & 1 & 2 \end{array}\right) - \frac{\beta P}{15(2\beta+3)}\left(\begin{array}{c|ccc|c} 4 & 0 & -3 & -1 & 0 \\ \hline 0 & 36 & 0 & -3 & -3 \\ -3 & 0 & 36 & -3 & 0 \\ -1 & -3 & -3 & 8 & -1 \\ \hline 0 & -3 & 0 & -1 & 4 \end{array}\right)\right]$$
$$= 0. \tag{9.82}$$

We hereafter solve this equation to obtain buckling loads. We consider the following three cases:

- The case of extremely small slenderness ratio $\lambda = +0$ represented by $\beta = +0$ (cf., (9.53)).
- A standard case of $\beta = 6$.
- The case of extremely large slenderness ratio $\lambda = +\infty$ represented by $\beta = +\infty$.

Case 1: Extremely small slenderness ratio

In the extreme case of $\beta = +0$ ($\lambda = +0$), we see that $\hat{K}_{\mathrm{G}} = O$ from (9.81) and no P satisfies the buckling condition (9.82). Owing to the absence of the geometric stiffness, no buckling takes place. We have $\hat{\boldsymbol{u}}$ in (9.75). From (9.65) with (9.76) and (9.78), we see that $(X_1, Y_1, X_3, Y_3) = (0, -P, P, 0)$ and there is no axial force.

Case 2: A standard case of $\beta = 6$

In this case, the buckling condition (9.82) becomes

$$\det\left[\frac{2EI}{L^2}\left(\begin{array}{c|ccc|c} 2 & 0 & -3 & 1 & 0 \\ \hline 0 & 12 & 0 & -3 & -3 \\ -3 & 0 & 12 & -3 & 0 \\ 1 & -3 & -3 & 4 & 1 \\ \hline 0 & -3 & 0 & 1 & 2 \end{array}\right) - \frac{2P}{75}\left(\begin{array}{c|ccc|c} 4 & 0 & -3 & -1 & 0 \\ \hline 0 & 36 & 0 & -3 & -3 \\ -3 & 0 & 36 & -3 & 0 \\ -1 & -3 & -3 & 8 & -1 \\ \hline 0 & -3 & 0 & -1 & 4 \end{array}\right)\right]$$
$$= \left(\frac{2EI}{L^2}\right)^5 108(1-5\alpha)^3(5-9\alpha)(1-\alpha) = 0,$$

where $\alpha = PL^2/(75EI)$. This buckling condition gives

$$\alpha = \frac{1}{5} \text{ (repeated 3 times) }, \quad \frac{5}{9}, \quad 1.$$

Then we have buckling loads

$$P_1 = 15\frac{EI}{L^2}, \qquad P_2 = \frac{125}{3}\frac{EI}{L^2}, \quad P_3 = 75\frac{EI}{L^2}.$$

The associated buckling modes in terms of $\hat{\boldsymbol{\eta}} = (\mathrm{d}\varphi_1, \mathrm{d}u_2/L, \mathrm{d}w_2/L, \mathrm{d}\varphi_2, \mathrm{d}\varphi_3)^\top$ are determined from $(\hat{K}_{\mathrm{E}} + \hat{K}_{\mathrm{G}})\hat{\boldsymbol{\eta}} = \mathbf{0}$ in (9.33) as

$$\hat{\boldsymbol{\eta}}_1 = c_1\begin{pmatrix} 1 \\ 0 \\ 0 \\ -1 \\ 1 \end{pmatrix} + c_2\begin{pmatrix} 1 \\ 1 \\ 1 \\ 1 \\ 1 \end{pmatrix} + c_3\begin{pmatrix} 2 \\ -1 \\ 1 \\ 0 \\ -2 \end{pmatrix}, \quad \hat{\boldsymbol{\eta}}_2 = \begin{pmatrix} 6 \\ 1 \\ -1 \\ 0 \\ -6 \end{pmatrix}, \quad \hat{\boldsymbol{\eta}}_3 = \begin{pmatrix} 1 \\ 0 \\ 0 \\ 1 \\ 1 \end{pmatrix} \tag{9.83}$$

with arbitrary constants c_1, c_2, and c_3. These buckling modes are depicted in Fig. 9.8(a), in which $c_1 = 1$, $c_2 = c_3 = 0$ are used for $\hat{\boldsymbol{\eta}}_1$ to be consistent with $\hat{\boldsymbol{\eta}}_1$ for Case 3 in (9.84).

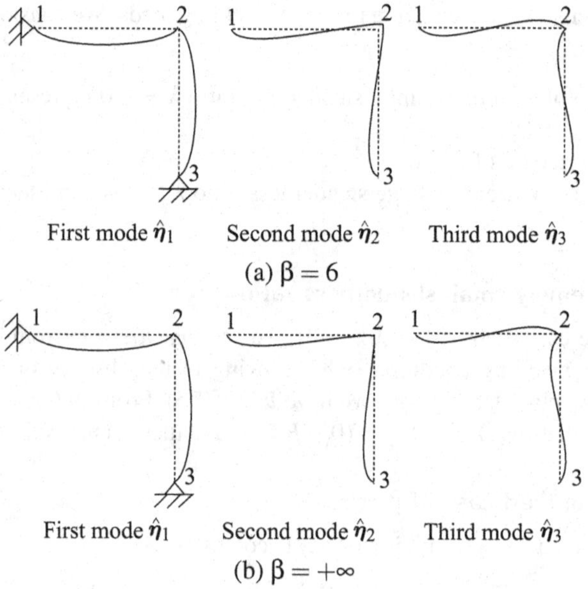

Figure 9.8 Buckling modes of an L-shaped frame

Case 3: Extremely large slenderness ratio

In the extreme case of $\beta = +\infty$ ($\lambda = +\infty$), the diagonal entries of $\beta + 6$ in the buckling condition (9.82) become $+\infty$. We, accordingly, delete the second and third rows and columns in the determinant in (9.82) to obtain

$$
\det \left[\frac{2EI}{L^2} \begin{pmatrix} 2 & 1 & 0 \\ 1 & 4 & 1 \\ 0 & 1 & 2 \end{pmatrix} - \frac{P}{30} \begin{pmatrix} 4 & -1 & 0 \\ -1 & 8 & -1 \\ 0 & -1 & 4 \end{pmatrix} \right]
= \left(\frac{2EI}{L^2} \right)^3 12(1-5\alpha)(1-2\alpha)(1-\alpha) = 0,
$$

where $\alpha = PL^2/(60EI)$. This buckling condition gives buckling loads

$$
P_1 = 12\frac{EI}{L^2}, \quad P_2 = 30\frac{EI}{L^2}, \quad P_3 = 60\frac{EI}{L^2}
$$

corresponding to $\alpha = 1/5,\ 1/2,\ 1$, respectively. The associated buckling modes in terms of $\hat{\boldsymbol{\eta}} = (\mathrm{d}\varphi_1, \mathrm{d}\varphi_2, \mathrm{d}\varphi_3)^\top$ are determined as

$$
\hat{\boldsymbol{\eta}}_1 = \begin{pmatrix} 1 \\ -1 \\ 1 \end{pmatrix}, \quad \hat{\boldsymbol{\eta}}_2 = \begin{pmatrix} 1 \\ 0 \\ -1 \end{pmatrix}, \quad \hat{\boldsymbol{\eta}}_3 = \begin{pmatrix} 1 \\ 1 \\ 1 \end{pmatrix}. \tag{9.84}
$$

These buckling modes are depicted in Fig. 9.8(b). It is to be noted that this case corresponds to the linear buckling analysis ignoring axial deformation, for which the

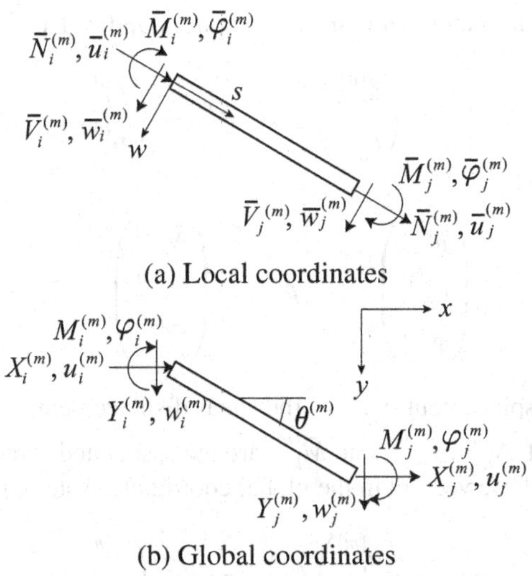

Figure 9.9 Two coordinate systems for a beam-column frame member

non-extensibility conditions $u_2 = w_2 = 0$ of members 1 and 2 are employed. Note that the modes $\hat{\boldsymbol{\eta}}_1$ and $\hat{\boldsymbol{\eta}}_3$ express the same deformation patterns with those for the Case 2 in (9.83). In contrast, the mode $\hat{\boldsymbol{\eta}}_2$ is different due to the restriction of the horizontal displacement of node 2 for the present case.

9.6 LINEAR BUCKLING ANALYSIS IN THE GLOBAL COORDINATES

In the derivation of the two-dimensional member stiffness (equilibrium) equation, it is convenient to consider the *local coordinate system* along the axis of the member in Fig. 9.9(a), whereas the *global coordinate system* in (b) is suitable for assembling member stiffness matrices. In this section, we describe the linear buckling analysis in the global coordinate system incorporating the axial deformation of beam-column members.

9.6.1 GLOBAL COORDINATE SYSTEM

We express the influence of the tilt of the member axis by transforming the member stiffness equation in (9.47) to the member stiffness equation in the global coordinate system in Fig. 9.9(b).

In this section, we use notations $\bar{\boldsymbol{u}}^{(m)}$ and $\bar{\boldsymbol{p}}^{(m)}$ for the nodal displacement vectors and nodal external force vectors in (9.44) expressed in the local coordinate system associated with the mth member connecting nodes i and j. That is,[7]

$$\bar{\boldsymbol{u}}^{(m)} = \begin{pmatrix} \bar{\boldsymbol{u}}_i^{(m)} \\ \bar{\boldsymbol{u}}_j^{(m)} \end{pmatrix}, \qquad \bar{\boldsymbol{p}}^{(m)} = \begin{pmatrix} \bar{\boldsymbol{p}}_i^{(m)} \\ \bar{\boldsymbol{p}}_j^{(m)} \end{pmatrix}, \tag{9.85}$$

where

$$\bar{\boldsymbol{u}}_k^{(m)} = \begin{pmatrix} \bar{u}_k^{(m)} \\ \bar{w}_k^{(m)} \\ \bar{\varphi}_k^{(m)} \end{pmatrix}, \qquad \bar{\boldsymbol{p}}_k^{(m)} = \begin{pmatrix} \bar{N}_k^{(m)} \\ \bar{V}_k^{(m)} \\ \bar{M}_k^{(m)} \end{pmatrix}, \quad k = i, j; \tag{9.86}$$

$\bar{u}_k^{(m)}$ is the axial displacement, $\bar{w}_k^{(m)}$ is the shear displacement, and $\bar{\varphi}_k^{(m)}$ is the rotational displacement; $\bar{N}_k^{(m)}$, $\bar{V}_k^{(m)}$, and $\bar{M}_k^{(m)}$ are the associated forces.

We now define these vectors in the global coordinate system, respectively, as

$$\boldsymbol{u}^{(m)} = \begin{pmatrix} \boldsymbol{u}_i^{(m)} \\ \boldsymbol{u}_j^{(m)} \end{pmatrix}, \qquad \boldsymbol{p}^{(m)} = \begin{pmatrix} \boldsymbol{p}_i^{(m)} \\ \boldsymbol{p}_j^{(m)} \end{pmatrix}, \tag{9.87}$$

where

$$\boldsymbol{u}_k^{(m)} = \begin{pmatrix} u_k^{(m)} \\ w_k^{(m)} \\ \varphi_k^{(m)} \end{pmatrix}, \qquad \boldsymbol{p}_k^{(m)} = \begin{pmatrix} N_k^{(m)} \\ V_k^{(m)} \\ M_k^{(m)} \end{pmatrix}, \quad k = i, j. \tag{9.88}$$

Using the tilt angle $\theta^{(m)}$ of the member axis in Fig. 9.9(b), we define the transformation matrix between the local coordinate system and the global coordinate system as

$$T^{(m)} = \begin{pmatrix} \cos\theta^{(m)} & \sin\theta^{(m)} & 0 \\ -\sin\theta^{(m)} & \cos\theta^{(m)} & 0 \\ 0 & 0 & 1 \end{pmatrix}. \tag{9.89}$$

Then the nodal displacement vectors and nodal external force vectors are transformed, respectively, as

$$\bar{\boldsymbol{u}}_k^{(m)} = T^{(m)} \boldsymbol{u}_k^{(m)}, \qquad \bar{\boldsymbol{p}}_k^{(m)} = T^{(m)} \boldsymbol{p}_k^{(m)}, \quad k = i, j, \tag{9.90}$$

which implies

$$\boldsymbol{u}^{(m)} = \begin{pmatrix} T^{(m)} & O \\ O & T^{(m)} \end{pmatrix}^{\top} \bar{\boldsymbol{u}}^{(m)}, \qquad \boldsymbol{p}^{(m)} = \begin{pmatrix} T^{(m)} & O \\ O & T^{(m)} \end{pmatrix}^{\top} \bar{\boldsymbol{p}}^{(m)}. \tag{9.91}$$

We recall the member stiffness equation (9.47), which reads

$$(\bar{K}_{\mathrm{E}}^{(m)} + \bar{K}_{\mathrm{G}}^{(m)})\, \bar{\boldsymbol{u}}^{(m)} = \bar{\boldsymbol{p}}^{(m)} \tag{9.92}$$

[7] We express by $(\bar{\cdot})$ the variables in the local coordinates, such as $\bar{u}_k^{(m)}$, $\bar{\varphi}_k^{(m)}$, and $\bar{N}_k^{(m)}$.

in the present notation. Here, the local stiffness matrix is given by (cf., (9.49)–(9.52))

$$\bar{K}_{\mathrm{E}}^{(m)} = \begin{pmatrix} \bar{k}_{\mathrm{E}ii}^{(m)} & \bar{k}_{\mathrm{E}ij}^{(m)} \\ \bar{k}_{\mathrm{E}ji}^{(m)} & \bar{k}_{\mathrm{E}jj}^{(m)} \end{pmatrix}, \tag{9.93}$$

which is a 6×6 symmetric matrix consisting of

$$\bar{k}_{\mathrm{E}ii}^{(m)} = \begin{pmatrix} \dfrac{EA}{L} & 0 & 0 \\ 0 & \dfrac{12EI}{L^3} & \dfrac{6EI}{L^2} \\ 0 & \dfrac{6EI}{L^2} & \dfrac{4EI}{L} \end{pmatrix}, \quad \bar{k}_{\mathrm{E}jj}^{(m)} = \begin{pmatrix} \dfrac{EA}{L} & 0 & 0 \\ 0 & \dfrac{12EI}{L^3} & -\dfrac{6EI}{L^2} \\ 0 & -\dfrac{6EI}{L^2} & \dfrac{4EI}{L} \end{pmatrix}, \tag{9.94}$$

$$\bar{k}_{\mathrm{E}ij}^{(m)} = \bar{k}_{\mathrm{E}ji}^{(m)\top} = \begin{pmatrix} -\dfrac{EA}{L} & 0 & 0 \\ 0 & -\dfrac{12EI}{L^3} & \dfrac{6EI}{L^2} \\ 0 & -\dfrac{6EI}{L^2} & \dfrac{2EI}{L} \end{pmatrix}, \tag{9.95}$$

where $L = L^{(m)}$, $EA = EA^{(m)}$, and $EI = EI^{(m)}$. Note that $\bar{K}_{\mathrm{G}}^{(m)}$ can be constructed similarly with reference to (9.54).

The member stiffness equation in the local coordinate system given in (9.92) is transformed to the member stiffness equation in the global coordinate system:

$$(K_{\mathrm{E}}^{(m)} + K_{\mathrm{G}}^{(m)})\,\boldsymbol{u}^{(m)} = \boldsymbol{p}^{(m)}, \tag{9.96}$$

where

$$K_{\mathrm{E}}^{(m)} = \begin{pmatrix} T^{(m)} & O \\ O & T^{(m)} \end{pmatrix}^{\top} \begin{pmatrix} \bar{k}_{\mathrm{E}ii}^{(m)} & \bar{k}_{\mathrm{E}ij}^{(m)} \\ \bar{k}_{\mathrm{E}ji}^{(m)} & \bar{k}_{\mathrm{E}jj}^{(m)} \end{pmatrix} \begin{pmatrix} T^{(m)} & O \\ O & T^{(m)} \end{pmatrix} \tag{9.97}$$

is the stiffness matrix in the global coordinate system and

$$K_{\mathrm{G}}^{(m)} = \begin{pmatrix} T^{(m)} & O \\ O & T^{(m)} \end{pmatrix}^{\top} \begin{pmatrix} \bar{k}_{\mathrm{G}ii}^{(m)} & \bar{k}_{\mathrm{G}ij}^{(m)} \\ \bar{k}_{\mathrm{G}ji}^{(m)} & \bar{k}_{\mathrm{G}jj}^{(m)} \end{pmatrix} \begin{pmatrix} T^{(m)} & O \\ O & T^{(m)} \end{pmatrix} \tag{9.98}$$

is the geometric stiffness matrix in the global coordinate system. We assemble these member matrices to obtain the matrices K_{E}° and K_{G}° for the whole structure.

By assembling the member stiffness equations and imposing the boundary conditions, we obtain the stiffness equation for the whole structure of the form

$$(K_{\mathrm{E}} + K_{\mathrm{G}})\,\boldsymbol{u} = \boldsymbol{p}, \tag{9.99}$$

in which $\boldsymbol{u}$ and $\boldsymbol{p}$ are obtained by assembling $\boldsymbol{u}^{(m)}$ and $\boldsymbol{p}^{(m)}$ and implementing boundary conditions. Then the linear buckling analysis proceeds as follows:

- The axial forces in the members are obtained by the small displacement analysis ignoring geometric stiffness, that is, by solving $K_E \boldsymbol{u} = \boldsymbol{p}$, and then by following the procedure in Section 9.4.2.
- The axial forces obtained in this manner are used in (9.55) and (9.56) to obtain K_G in (9.99).
- The buckling loads are obtained from the buckling condition $\det(K_E + K_G) = 0$ in (9.23). The buckling mode $\boldsymbol{\eta}$ is determined by $(K_E + K_G)\boldsymbol{\eta} = \boldsymbol{0}$ in (9.24).

This procedure for the linear buckling analysis is explained based on a concrete structural example in the following section.

9.6.2 STRUCTURAL EXAMPLE

The assemblage of the member stiffness equation using the transformation to the global coordinates is illustrated for the two-beam arch depicted in Fig. 9.10. Members 1 and 2 have the same cross-sectional rigidities EA and EI.

We assemble the member stiffness equations (9.96) for each member to arrive at

$$K_E^\circ = \begin{pmatrix} T^{(1)\top}\bar{k}_{E11}^{(1)}T^{(1)} & T^{(1)\top}\bar{k}_{E12}^{(1)}T^{(1)} & O \\ T^{(1)\top}\bar{k}_{E21}^{(1)}T^{(1)} & \sum_{m=1}^{2} T^{(m)\top}\bar{k}_{E22}^{(m)}T^{(m)} & T^{(2)\top}\bar{k}_{E23}^{(2)}T^{(2)} \\ O & T^{(2)\top}\bar{k}_{E32}^{(2)}T^{(2)} & T^{(2)\top}\bar{k}_{E33}^{(2)}T^{(2)} \end{pmatrix}, \tag{9.100}$$

$$K_G^\circ = \begin{pmatrix} T^{(1)\top}\bar{k}_{G11}^{(1)}T^{(1)} & T^{(1)\top}\bar{k}_{G12}^{(1)}T^{(1)} & O \\ T^{(1)\top}\bar{k}_{G21}^{(1)}T^{(1)} & \sum_{m=1}^{2} T^{(m)\top}\bar{k}_{G22}^{(m)}T^{(m)} & T^{(2)\top}\bar{k}_{G23}^{(2)}T^{(2)} \\ O & T^{(2)\top}\bar{k}_{G32}^{(2)}T^{(2)} & T^{(2)\top}\bar{k}_{G33}^{(2)}T^{(2)} \end{pmatrix} \tag{9.101}$$

for the whole structure. Here $\bar{k}_{Eij}^{(m)}$ are given by (9.94) and (9.95) and $\bar{k}_{Gij}^{(m)}$ by (9.55) and (9.56) $(i, j = 1, 2, 3)$, where $(i, j) = (1, 2)$ for $m = 1$ and $(i, j) = (2, 3)$ for $m = 2$.

Considering the displacement boundary conditions

$$\begin{pmatrix} w_1 \\ \varphi_1 \end{pmatrix} = \begin{pmatrix} w_3 \\ \varphi_3 \end{pmatrix} = \begin{pmatrix} 0 \\ 0 \end{pmatrix},$$

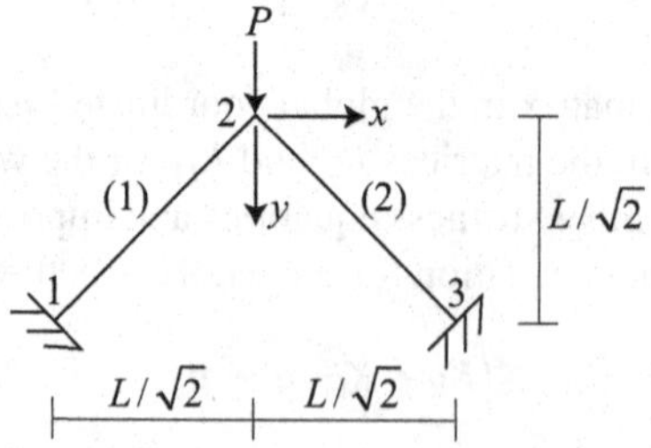

Figure 9.10 A two-beam arch

the structural stiffness equation is obtained as

$$(K_{\mathrm{E}}+K_{\mathrm{G}})\,\boldsymbol{u}=\boldsymbol{p}, \tag{9.102}$$

where $\boldsymbol{p}=(0,P,0)^{\top}$ is the known external force vector at node 2 and $\boldsymbol{u}=(u_2,w_2,\varphi_2)^{\top}$ is the unknown displacement vector of node 2, comprising x-directional displacement u_2, y-directional displacement w_2, and the rotational displacement φ_2. The stiffness matrixK_{E} and the geometric stiffness matrix K_{G} are obtained, respectively, from (9.100) and (9.101) as

$$K_{\mathrm{E}}=\sum_{m=1}^{2}T^{(m)\top}\bar{k}_{\mathrm{E}22}^{(m)}T^{(m)},\qquad K_{\mathrm{G}}=\sum_{m=1}^{2}T^{(m)\top}\bar{k}_{\mathrm{G}22}^{(m)}T^{(m)}. \tag{9.103}$$

Since $\theta^{(1)}=-\pi/4$ and $\theta^{(2)}=\pi/4$, the transformation matrices are given by

$$T^{(1)}=\begin{pmatrix}1/\sqrt{2} & -1/\sqrt{2} & 0\\ 1/\sqrt{2} & 1/\sqrt{2} & 0\\ 0 & 0 & 1\end{pmatrix},\qquad T^{(2)}=\begin{pmatrix}1/\sqrt{2} & 1/\sqrt{2} & 0\\ -1/\sqrt{2} & 1/\sqrt{2} & 0\\ 0 & 0 & 1\end{pmatrix}.$$

Then, from (9.103) with the definition of $\bar{k}_{\mathrm{E}22}^{(m)}$ in (9.94), we have

$$\begin{aligned}
K_{\mathrm{E}}&=\sum_{m=1}^{2}T^{(m)\top}\bar{k}_{\mathrm{E}22}^{(m)}T^{(m)}\\
&=\begin{pmatrix}1/\sqrt{2} & -1/\sqrt{2} & 0\\ 1/\sqrt{2} & 1/\sqrt{2} & 0\\ 0 & 0 & 1\end{pmatrix}^{\top}\begin{pmatrix}\frac{EA}{L} & 0 & 0\\ 0 & \frac{12EI}{L^3} & -\frac{6EI}{L^2}\\ 0 & -\frac{6EI}{L^2} & \frac{4EI}{L}\end{pmatrix}\begin{pmatrix}1/\sqrt{2} & -1/\sqrt{2} & 0\\ 1/\sqrt{2} & 1/\sqrt{2} & 0\\ 0 & 0 & 1\end{pmatrix}\\
&\quad+\begin{pmatrix}1/\sqrt{2} & 1/\sqrt{2} & 0\\ -1/\sqrt{2} & 1/\sqrt{2} & 0\\ 0 & 0 & 1\end{pmatrix}^{\top}\begin{pmatrix}\frac{EA}{L} & 0 & 0\\ 0 & \frac{12EI}{L^3} & \frac{6EI}{L^2}\\ 0 & \frac{6EI}{L^2} & \frac{4EI}{L}\end{pmatrix}\begin{pmatrix}1/\sqrt{2} & 1/\sqrt{2} & 0\\ -1/\sqrt{2} & 1/\sqrt{2} & 0\\ 0 & 0 & 1\end{pmatrix}\\
&=\frac{EA}{L}\begin{pmatrix}1 & -1 & 0\\ -1 & 1 & 0\\ 0 & 0 & 0\end{pmatrix}+\frac{2EI}{L^2}\begin{pmatrix}6/L & 0 & -3\sqrt{2}\\ 0 & 6/L & 0\\ -3\sqrt{2} & 0 & 4L\end{pmatrix}.
\end{aligned} \tag{9.104}$$

Similarly, we can obtain the geometric stiffness matrix as

$$K_{\mathrm{G}}=\frac{N}{30}\begin{pmatrix}36/L & 0 & -3\sqrt{2}\\ 0 & 36/L & 0\\ -3\sqrt{2} & 0 & 8L\end{pmatrix}. \tag{9.105}$$

Here we used $N^{(1)}=N^{(2)}=N$ due to the bilateral symmetry.

Small displacement analysis

For further analysis, we introduce a simplifying assumption

$$\frac{EA}{L} = \frac{12EI}{L^3}.$$

In this case, the concrete form of $K_{\rm E}$ in (9.104) is calculated as

$$K_{\rm E} = \frac{2EI}{L^2}\begin{pmatrix} 12/L & 0 & -3\sqrt{2} \\ 0 & 12/L & 0 \\ -3\sqrt{2} & 0 & 4L \end{pmatrix}. \tag{9.106}$$

As the first step of the linear buckling analysis, the axial forces in the members are obtained by the small displacement analysis ignoring geometric stiffness. By omitting $K_{\rm G}$ from (9.102) and using $K_{\rm E}$ in (9.106) and $\boldsymbol{p} = (0, P, 0)^\top$, we obtain

$$\boldsymbol{u} = (K_{\rm E})^{-1}\boldsymbol{p} = \frac{PL^3}{24EI}\begin{pmatrix} 0 \\ 1 \\ 0 \end{pmatrix}. \tag{9.107}$$

The member end forces of member 1 in the local coordinates are evaluated to

$$\bar{\boldsymbol{p}}^{(1)} = \begin{pmatrix} \bar{k}_{\rm E11}^{(1)} & \bar{k}_{\rm E12}^{(1)} \\ \bar{k}_{\rm E21}^{(1)} & \bar{k}_{\rm E22}^{(1)} \end{pmatrix}\begin{pmatrix} T^{(1)} & O \\ O & T^{(1)} \end{pmatrix}\begin{pmatrix} \boldsymbol{u}_1 \\ \boldsymbol{u}_2 \end{pmatrix} = \begin{pmatrix} \bar{k}_{\rm E12}^{(1)} T^{(1)}\boldsymbol{u} \\ \bar{k}_{\rm E22}^{(1)} T^{(1)}\boldsymbol{u} \end{pmatrix} = \frac{\sqrt{2}P}{8}\begin{pmatrix} 2 \\ 2 \\ -L \\ -2 \\ 2 \\ -L \end{pmatrix} \tag{9.108}$$

with $\boldsymbol{u}_1 = \boldsymbol{0}$ and $\boldsymbol{u}_2 = \boldsymbol{u}$ in (9.107). The fourth component of $\bar{\boldsymbol{p}}^{(1)}$ is nothing but $N^{(1)}$ by (9.46). Accordingly, the member axial forces are given by

$$N^{(1)} = N^{(2)} = N = -\frac{\sqrt{2}P}{4}. \tag{9.109}$$

Linear buckling analysis

The second step of the analysis uses the expression (9.105) of $K_{\rm G}$ with $N = -\sqrt{2}P/4$ in (9.109) to obtain the buckling condition. Let $\hat{K}_{\rm E}$ and $\hat{K}_{\rm G}$ denote $K_{\rm E}$ and $K_{\rm G}$ in (9.106) and (9.105), respectively, modified for the normalized variables. Then the buckling condition is given by

$$\begin{aligned}\det(\hat{K}_{\rm E} + \hat{K}_{\rm G}) &= \det\left[\frac{2EI}{L^2}\begin{pmatrix} 12 & 0 & -3\sqrt{2} \\ 0 & 12 & 0 \\ -3\sqrt{2} & 0 & 4 \end{pmatrix} - \frac{\sqrt{2}P}{120}\begin{pmatrix} 36 & 0 & -3\sqrt{2} \\ 0 & 36 & 0 \\ -3\sqrt{2} & 0 & 8 \end{pmatrix}\right] \\ &= \left(\frac{2EI}{L^2}\right)^3 72(1-5\alpha)(1-3\alpha)(5-9\alpha) = 0,\end{aligned}$$

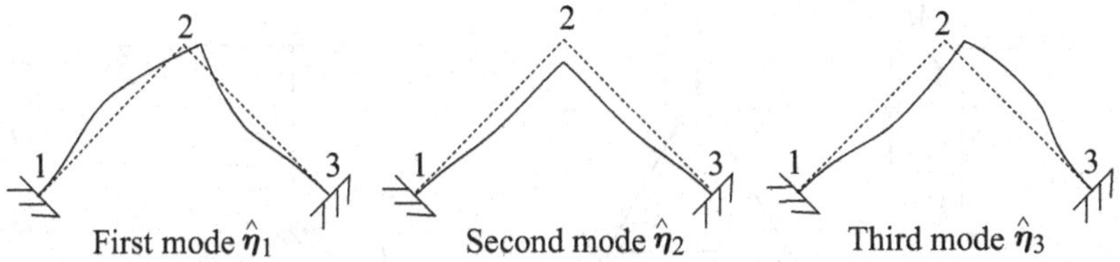

Figure 9.11 Buckling modes of the two-beam arch

where $\alpha = \sqrt{2}PL^2/(240EI)$. This condition yields $\alpha = 0.200$, 0.333, and 0.555. Then the buckling loads are

$$P_1 = 33.9\frac{EI}{L^2}, \qquad P_2 = 56.6\frac{EI}{L^2}, \qquad P_3 = 94.3\frac{EI}{L^2} \tag{9.110}$$

and the buckling modes in terms of $\hat{\eta} = (\mathrm{d}u_2/L, \mathrm{d}w_2/L, \mathrm{d}\varphi_2)^\top$ are

$$\hat{\eta}_1 = \begin{pmatrix} 1 \\ 0 \\ \sqrt{2} \end{pmatrix}, \qquad \hat{\eta}_2 = \begin{pmatrix} 0 \\ 1 \\ 0 \end{pmatrix}, \qquad \hat{\eta}_3 = \begin{pmatrix} 1 \\ 0 \\ -3\sqrt{2} \end{pmatrix}.$$

As can be seen from these buckling modes depicted in Fig. 9.11, the buckling loads P_1 and P_3 are associated with bifurcation and the buckling load P_2 with the snap-through buckling.

9.7 PROBLEMS

Problem 9.1 Prove (9.7).

Problem 9.2 Derive the (1,1), (1,2), (2,2) components of the matrix $K_{\mathrm{E}}^{(m)}$ in (9.12).

Problem 9.3 Derive the (1,1), (1,2), (2,2) components of the matrix $K_{\mathrm{G}}^{(m)}$ in (9.13).

Problem 9.4 Obtain the buckling loads of the two-element beam structure depicted in Fig. 9.12 by the linear buckling analysis ignoring the axial deformation of the members.

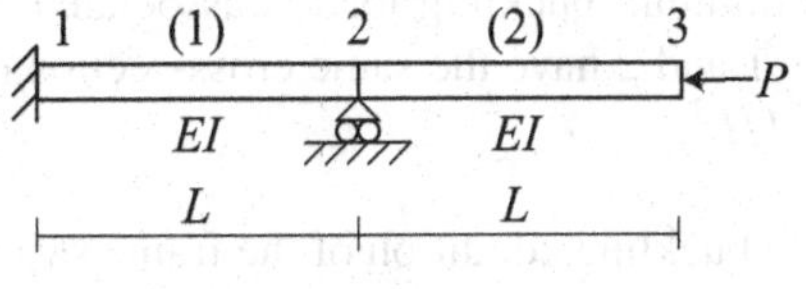

Figure 9.12 Problem 9.4: A two-element beam structure

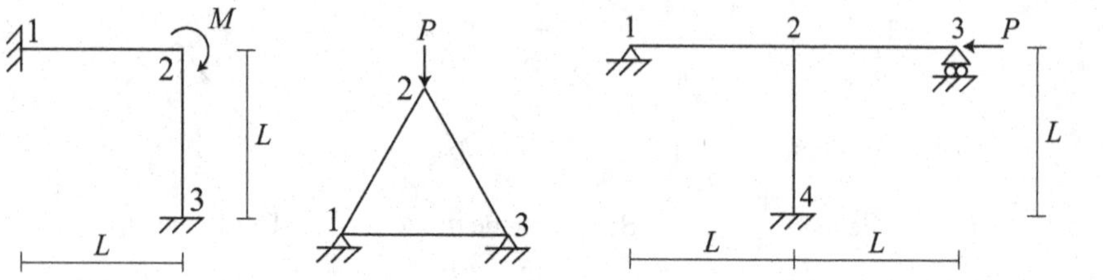

(a) (b) (c)

Figure 9.13 Problem 9.5: Frame structures

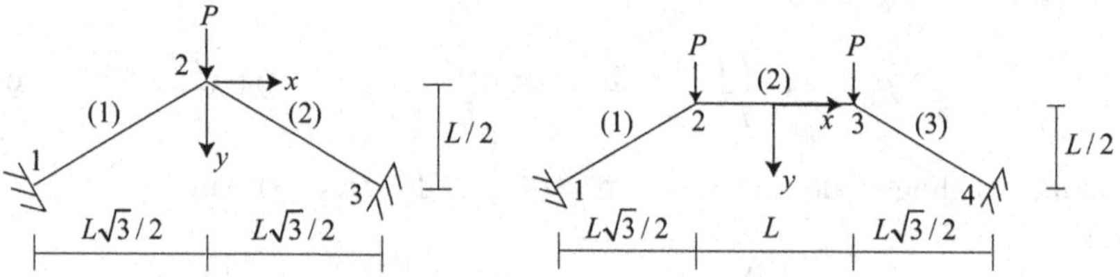

(a) Problem 9.8 (b) Problem 9.9

Figure 9.14 Frame structures

Problem 9.5 Judge if each of the three types of frame structures depicted in Fig. 9.13 undergoes bifurcation or not. Sketch the deformation or buckling mode of each structure (no calculation is needed). All members have the same length L and the same cross-sectional rigidities EA and EI.

Problem 9.6 For the frame structure in Fig. 9.13(a), conduct the linear buckling analysis ignoring the axial deformation of the members.

Problem 9.7 Obtain the buckling loads and buckling modes of the frame structure depicted in Fig. 9.13(c) by the linear buckling analysis ignoring the axial deformation of the members. Sketch these buckling modes.

Problem 9.8 Conduct the linear buckling analysis considering the axial deformation of the members (cf., Section 9.6.2) for the frame structure depicted in Fig. 9.14(a), and investigate if the minimum buckling load is associated with a snap-through or a bifurcation. Members 1 and 2 have the same cross-sectional rigidities EA and EI, satisfying $EA/L = 12EI/L^3$.

Problem 9.9 Derive the buckling condition of the frame structure in Fig. 9.14(b) by the linear buckling analysis considering the axial deformation of the members (cf., Section 9.6.2). All members have the same axial and flexural rigidities EA and EI, satisfying $EA/L = 12EI/L^3$.

REFERENCES

1. Bažant ZP, and Cedolin L (1991), *Stability of Structures*, Oxford University Press, New York.
2. Iyengar NGR (1988), *Structural Stability of Columns and Plates*, Ellis Horwood Series in Civil Engineering, Chichester.

10 Advanced Topics on Imperfect Systems

10.1 SUMMARY

Chapter 5 has introduced the bifurcation theory of the perfect system, which is an idealized modeling of a structure. While bifurcation of a structure is a mathematical concept for this idealized system, an actual structure does not undergo bifurcation in the strict mathematical sense, as it has inevitably some *initial imperfections* that denote deviations from the nominal values. A structure with imperfections is called an *imperfect system*. The presence of imperfections tends to reduce buckling loads. We extend the bifurcation equation for a perfect system to that for an imperfect system and investigate the sensitivity of buckling loads to imperfections based on this extended bifurcation equation. We present the theory of the worst imperfections that reduce the strength most rapidly and the theory of statistical variation of buckling loads for random imperfections.

Structural systems have imperfections of various structural properties, such as member length, modulus of elasticity, and structural shape. It is a vital issue in the design of structures to estimate the reduction of buckling loads due to the presence of imperfections. There are reviews of this development (Hutchinson Koiter [4]; Ikeda and Murota [8]) and books on the theory of imperfect structures (e.g., Thompson and Hunt [17, 18]; Ikeda and Murota [9]). Historical development of imperfect structures is to be reviewed in Chapter 11.

An example of a structure with an initial imperfection is given in Section 10.2. The formulation of imperfection sensitivity is presented in Section 10.3. The theory of the worst imperfection is introduced in Section 10.4 and the theory of buckling loads for random imperfection in Section 10.5. Imperfection sensitivity of elastic–plastic plate is presented in Section 10.6.

Keywords: • Bifurcation • Buckling load • Initial imperfection • Imperfect system • Random imperfection • Worst imperfection

10.2 STRUCTURAL EXAMPLE WITH IMPERFECTION

Structural systems have imperfections of various structural properties, such as member length, modulus of elasticity, and structural shape. A system with imperfections is called an imperfect system. An example of the buckling behavior of an imperfect structure is presented in this section as an introduction.

Consider an imperfect propped cantilever for which the rigid bar has a tilt by an angle of ε, as depicted in Fig. 10.1(a). When there is no tilt ($\varepsilon = 0$), this is identical with the propped cantilever in Fig. 1.8 in Section 1.5. Thus ε expresses the deviation

DOI: 10.1201/9781003112365-10

from the perfect state. Such deviation is called an *initial imperfection*. The system with $\varepsilon = 0$ is called the *perfect system*, while the system with $\varepsilon \neq 0$ is called an *imperfect system*.

An actual structural system is an imperfect system with some initial imperfections, whereas the perfect system is its idealized modeling. Note that there are various kinds of initial imperfections, such as a small horizontal load (see Problem 10.1).

The total potential energy U of an imperfect system is given by

$$U(u,f,\varepsilon) = kL^2 \left[\frac{1}{2}(\sin u - \sin \varepsilon)^2 - f(\cos \varepsilon - \cos u) \right]. \tag{10.1}$$

From this equation, the equilibrium equation and the tangent stiffness are obtained, respectively, as

$$F(u,f,\varepsilon) \equiv \frac{\partial U}{\partial u} = kL^2 \left[(\sin u - \sin \varepsilon) \cos u - f \sin u \right] = 0, \tag{10.2}$$

$$J(u,f,\varepsilon) \equiv \frac{\partial F}{\partial u} = kL^2 (\cos 2u + \sin \varepsilon \sin u - f \cos u). \tag{10.3}$$

The equilibrium paths for initial imperfections of $\varepsilon = 0.03$ and 0.1 are obtained from the equilibrium equation (10.2) and are plotted in Fig. 10.1(b). The thick lines

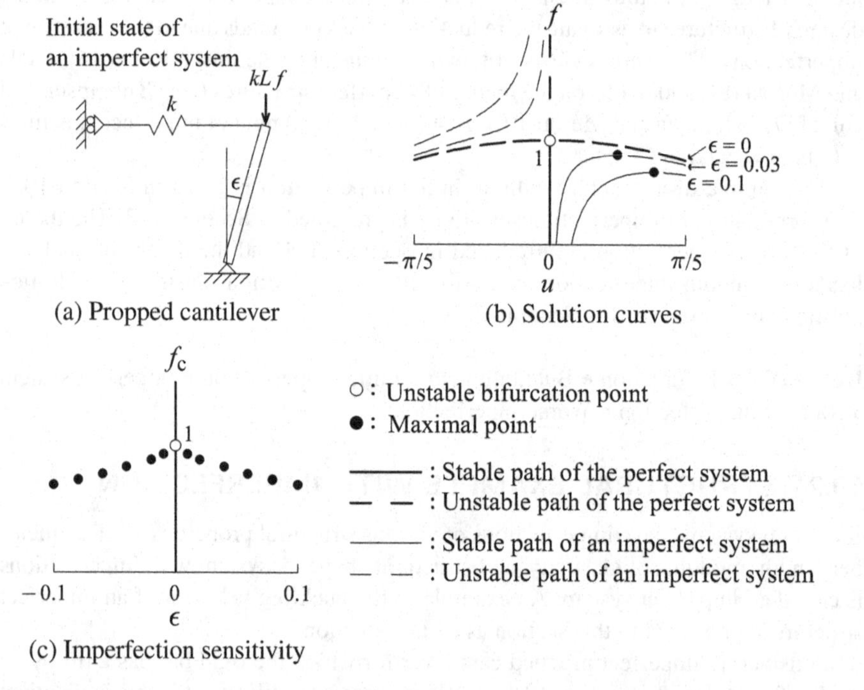

Figure 10.1 Solution curves and imperfection sensitivity of the propped cantilever with an initial tilt ε as an initial imperfection

denote the paths for the perfect system ($\varepsilon = 0$) and the thin lines denote those for imperfect ones ($\varepsilon \neq 0$). Each imperfect system has neither the trivial solution $u = 0$ nor a bifurcation point but instead has two separate paths. As the initial imperfection ε increases, the distance between the paths for the perfect system and those for imperfect systems increases.

On the equilibrium path of an imperfect system, there exists a maximal point of the load f, at which the tangent stiffness in (10.3) becomes zero. This imperfect system is stable until reaching this maximal point but becomes unstable beyond this point. The load at this point is the buckling load of this system.

The buckling load f_c of this system is plotted against the values of the initial imperfection $\varepsilon = 0, \pm 0.02, \ldots, \pm 0.1$ in Fig. 10.1(c). As $|\varepsilon|$ increases, f_c decreases rapidly. This kind of relation is called the *imperfection sensitivity* and is theoretically analyzed in Section 10.3.

10.3 FORMULATION OF IMPERFECTION SENSITIVITY

To describe the influence of initial imperfections, we extend the derivation of the bifurcation equation for the perfect system in Section 5.3 to an imperfect system with imperfections. We consider a nonlinear *equilibrium equation* (governing equation)

$$\boldsymbol{F}(\boldsymbol{u}, f, \boldsymbol{v}) = \boldsymbol{0} \tag{10.4}$$

with a p-dimensional *imperfection parameter vector* $\boldsymbol{v} = (v_1, \ldots, v_p)^\top$. Here $\boldsymbol{F} = (F_1, \ldots, F_n)^\top$ is a vector of sufficiently smooth nonlinear functions, $\boldsymbol{u} = (u_1, \ldots, u_n)^\top$ is an n-dimensional *displacement vector* to describe the state of the system, and f is a load parameter. The system is assumed to have potential, and the Jacobian matrix (tangent stiffness matrix) is a symmetric matrix.

We usually express the imperfection parameter vector $\boldsymbol{v}$ as the difference from the nominal value $\boldsymbol{v}^0$ for the perfect system as

$$\boldsymbol{v} = \boldsymbol{v}^0 + \varepsilon \boldsymbol{d}, \tag{10.5}$$

where ε is a scalar expressing the magnitude of the initial imperfection and $\boldsymbol{d}$ is an *imperfection pattern vector*. In this chapter, we investigate the influence of the imperfections when $\boldsymbol{d}$ is fixed and $|\varepsilon|$ is small (or $|\varepsilon| \to 0$).

We consider an equilibrium point $(\boldsymbol{u}, f)$ of the governing equation (10.4) and a simple critical point[1] $(\boldsymbol{u}_c^0, f_c^0)$ for the perfect system (the subscript $(\cdot)_c^0$ denotes that the variable therein is the value for a critical point of the perfect system). The incremental values $(\boldsymbol{w}, \tilde{f})$ between these two points are defined by[2]

$$\boldsymbol{u} - \boldsymbol{u}_c^0 = \sum_{i=1}^{n} \boldsymbol{\eta}_i w_i, \qquad \tilde{f} = f - f_c^0, \tag{10.6}$$

[1] We focus on simple critical points throughout this chapter for simplicity. The readers who are interested in multiple critical points may refer to the literature (Murota and Ikeda [13, 14] and Ikeda and Murota [9]).

[2] The definition (10.6) of the incremental variables is the same as the definition (5.18) for the perfect system.

where $\boldsymbol{\eta}_i$ are orthonormal eigenvectors of the Jacobian matrix J_{c}^0, as in (5.15). Recall the notation $\boldsymbol{\eta}_{\mathrm{c}} = \boldsymbol{\eta}_1$, for which we have $J_{\mathrm{c}}\boldsymbol{\eta}_{\mathrm{c}} = \mathbf{0}$ in (5.17).

Similarly to Section 5.3, by taking the inner product of the eigenvector $\boldsymbol{\eta}_i$ and the governing equation (10.4), using incremental values in (10.6), and eliminating passive coordinates $w_2, \ldots, w_n$, we obtain the bifurcation equation for an imperfect system as

$$\hat{F}(w, \tilde{f}, \varepsilon) = 0. \tag{10.7}$$

This bifurcation equation is expanded into a power series of w, $\tilde{f}$, and ε as

$$\hat{F}(w, \tilde{f}, \varepsilon) = \sum_{i=0}\sum_{j=0}\sum_{k=0} A_{ijk} w^i \tilde{f}^j \varepsilon^k, \tag{10.8}$$

where the coefficients A_{ijk} are

$$A_{ijk} = \frac{1}{i!j!k!}\frac{\partial^{i+j+k}\hat{F}}{\partial w^i \partial \tilde{f}^j \varepsilon^k}(0,0,0), \quad i,j,k = 0,1,2,\ldots. \tag{10.9}$$

These coefficients have a relation with the coefficients A_{ij} in (5.30) for the perfect system as

$$A_{ij0} = A_{ij}, \qquad i,j = 0,1,2,\ldots. \tag{10.10}$$

The coefficient A_{001}, which plays a pivotal role in the description of the influence of initial imperfection, is expressed as

$$A_{001} = \boldsymbol{\eta}_{\mathrm{c}}^\top \frac{\partial \boldsymbol{F}}{\partial \boldsymbol{v}}(\boldsymbol{u}_{\mathrm{c}}^0, f_{\mathrm{c}}^0, \boldsymbol{v}^0)\frac{\partial \boldsymbol{v}}{\partial \varepsilon} = \boldsymbol{\eta}_{\mathrm{c}}^\top \frac{\partial \boldsymbol{F}}{\partial \boldsymbol{v}}(\boldsymbol{u}_{\mathrm{c}}^0, f_{\mathrm{c}}^0, \boldsymbol{v}^0)\boldsymbol{d}. \tag{10.11}$$

Since $(w, \tilde{f}, \varepsilon) = (0,0,0)$ corresponds to the critical point of the perfect system, $A_{000} = A_{100} = 0$ holds.

We investigate in the sequel the asymptotic influence of the initial imperfection ε on a maximal/minimal point of load and a simple symmetric bifurcation point.[3] Initial imperfections degrade severely the buckling load of a structure at a simple symmetric bifurcation point but less severely at a maximal/minimal point.

10.3.1 MAXIMAL/MINIMAL POINT OF LOAD

For a maximal/minimal point of f, we have $A_{100} = 0$, $A_{010} \neq 0$, and $A_{200} \neq 0$. Then an asymptotic form of the bifurcation equation in (10.7) reads

$$\hat{F}(w, \tilde{f}, \varepsilon) \approx A_{200}w^2 + A_{010}\tilde{f} + A_{001}\varepsilon + A_{101}w\varepsilon + A_{110}w\tilde{f} = 0. \tag{10.12}$$

The criticality condition for maximality/minimality of $\tilde{f}$ determined by (10.12) becomes

$$\hat{J}(w, \tilde{f}, \varepsilon) \approx 2A_{200}w + A_{101}\varepsilon + A_{110}\tilde{f} = 0, \tag{10.13}$$

---

[3] The influence of the initial imperfection ε on a simple asymmetric bifurcation point is studied in Problem 10.4.

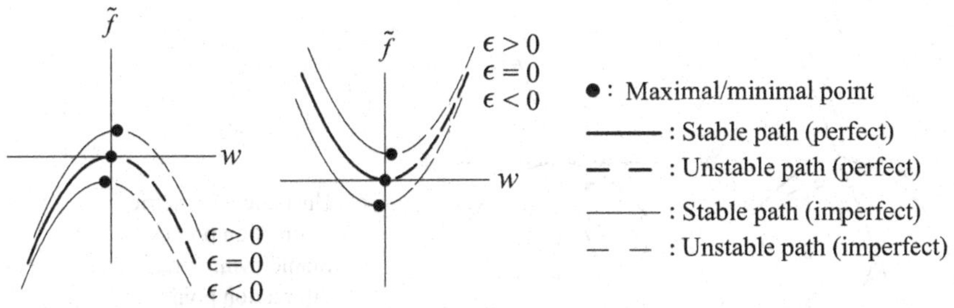

(a) Maximal point (b) Minimal point

Figure 10.2 Solution curves in the neighborhood of maximal/minimal points of f (stability is described for $A_{200} < 0$)

where $\hat{J} = \partial\hat{F}/\partial w$. The solution curves of (10.12) are depicted in Fig. 10.2. The buckling load increases or decreases according to $\varepsilon > 0$ or $\varepsilon < 0$ (when $A_{200} < 0$).

The simultaneous solution to (10.12) and (10.13) gives a maximal/minimal point $(w_c, \tilde{f}_c)$ of $\tilde{f}$ for an imperfect system as

$$\tilde{f}_c \approx -\frac{A_{001}}{A_{010}}\varepsilon, \tag{10.14}$$

$$w_c \approx -\frac{1}{2A_{200}A_{010}}(A_{101}A_{010} - A_{001}A_{110})\varepsilon. \tag{10.15}$$

We see from (10.14) that $\tilde{f}_c$ is linearly proportional to the magnitude ε of an initial imperfection. The relation expressing the influence of initial imperfection is called the *imperfection sensitivity law*.

10.3.2 SYMMETRIC BIFURCATION POINT

For a symmetric bifurcation point, we have $A_{100} = A_{010} = A_{200} = 0$ and $A_{300} \neq 0$. Then an asymptotic form of the bifurcation equation in (10.7) reads

$$\hat{F}(w, \tilde{f}, \varepsilon) \approx A_{300}w^3 + A_{110}w\tilde{f} + A_{020}\tilde{f}^2 + A_{001}\varepsilon = 0, \tag{10.16}$$

and the criticality condition becomes

$$\hat{J}(w, \tilde{f}, \varepsilon) \approx 3A_{300}w^2 + A_{110}\tilde{f} = 0. \tag{10.17}$$

The solution curves of (10.16) are depicted in Fig. 10.3. The curves for an imperfect system do not have a bifurcation point and split into two curves AB and CD. For an unstable-symmetric bifurcation point in Figs. 10.3(a) and (b) ($A_{300} < 0$), there is a maximal point of f on the curve AB. The buckling load of this system is governed by the maximal point shown by (•). The locations of the curve AB and the maximal point of f change according to the sign of the initial imperfection ε. In contrast, for a stable-symmetric bifurcation point in Figs. 10.3(c) and (d) ($A_{300} > 0$), there is no maximal point of f on the curve AB of an imperfect system and the load f on this

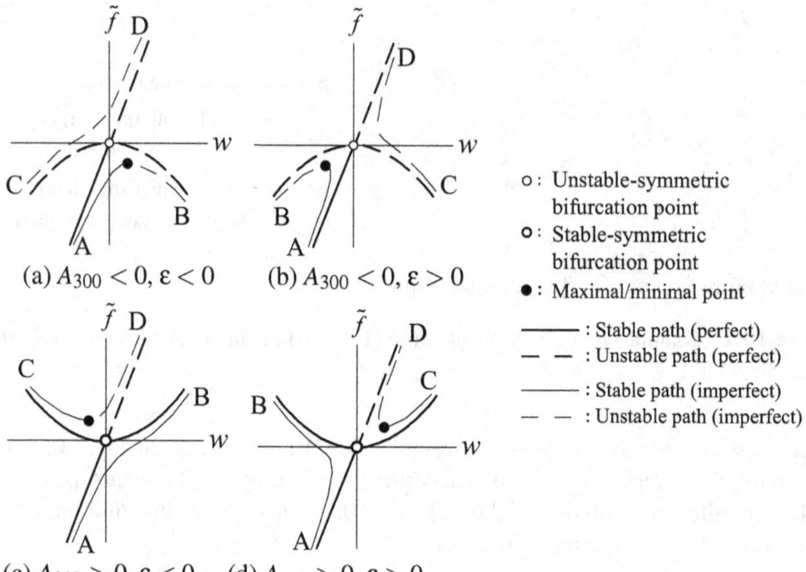

Figure 10.3 Solution curves in the neighborhood of symmetric bifurcation points ($A_{110} < 0$, $A_{020} > 0, A_{001} > 0$)

curve increases beyond the bifurcation point. There is a minimal point of f on the other curve CD.

The elimination of $\tilde{f}$ from (10.16) and (10.17) leads to an imperfection sensitivity law

$$w_c \approx \left(\frac{A_{001}}{2A_{300}}\right)^{1/3} \varepsilon^{1/3} \tag{10.18}$$

for the incremental displacement w_c. In the derivation of (10.18), it turns out that the term $A_{020}\tilde{f}^2$ in (10.16) can be omitted since it is of the order $\varepsilon^{4/3}$ and is a higher-order term. The substitution of (10.18) into (10.17) leads to the imperfection sensitivity law

$$\tilde{f}_c \approx -\frac{3A_{300}{}^{1/3}}{A_{110}}\left(\frac{A_{001}}{2}\right)^{2/3} \varepsilon^{2/3} \tag{10.19}$$

for the buckling load f_c. This is the so-called *two-thirds power law* (*Koiter law* [10]) of imperfection sensitivity, expressing that the buckling load f_c is reduced in proportion to $\varepsilon^{2/3}$. For an unstable-symmetric bifurcation point ($A_{300} < 0$), the sensitivity law (10.19) gives the information on the maximum load of the curve AB for an imperfect system (Figs. 10.3(a) and (b)) that is of engineering interest. In contrast, for a stable-symmetric bifurcation point ($A_{300} > 0$), the sensitivity law (10.19) gives information only on the minimal point of the curve CD (Figs. 10.3(c) and (d)) and is of no engineering interest.

The f_c versus ε relation given in Fig. 10.1(c) for the propped cantilever in Section 10.2 is an instance of the two-thirds power law (10.19).

10.3.3 STRUCTURAL EXAMPLE

As an example of imperfection sensitivity, we revive the truss arch structure depicted in Fig. 10.4(a) (see Section 3.4 for details of this structure). Several kinds of initial imperfections are considered simultaneously here, while the example in Section 10.2 has a single initial imperfection.

The equilibrium equation of this arch is given by (3.16) as

$$\boldsymbol{F} \equiv \begin{pmatrix} \sum_{m=1}^{2} EA^{(m)} \left(\dfrac{1}{L^{(m)}} - \dfrac{1}{\hat{L}^{(m)}} \right) (x - x_m) \\ \sum_{m=1}^{2} EA^{(m)} \left(\dfrac{1}{L^{(m)}} - \dfrac{1}{\hat{L}^{(m)}} \right) (y - y_m) - EAf \end{pmatrix} = \begin{pmatrix} 0 \\ 0 \end{pmatrix}. \tag{10.20}$$

Here, (x_i, y_i) $(i = 1,2,3)$ is the location of the ith node, $EA^{(m)}$ $(m = 1,2)$ is the cross-sectional rigidity of the m-th member and $L^{(m)}$ and $\hat{L}^{(m)}$ $(m = 1,2)$, respectively,

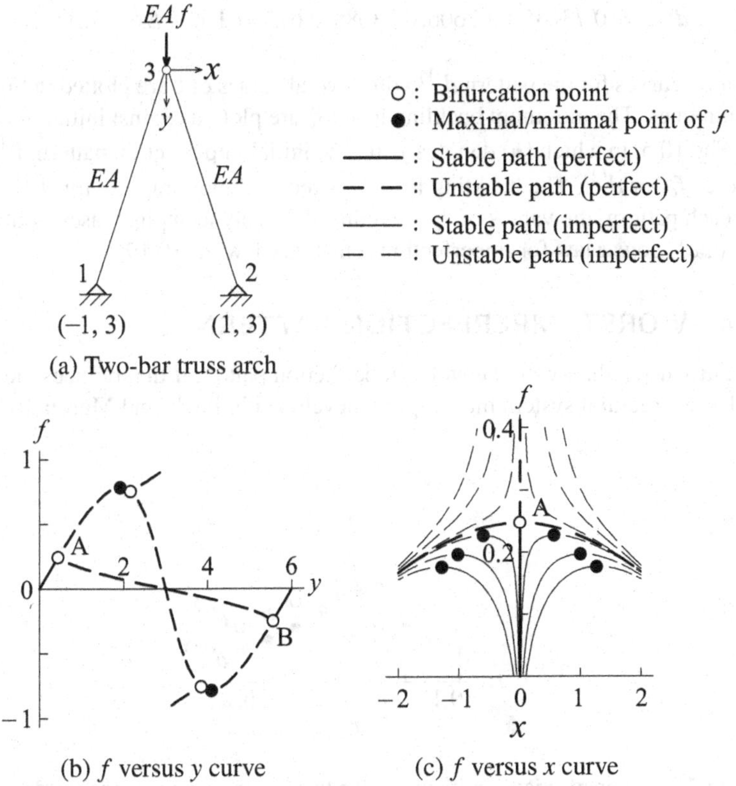

(a) Two-bar truss arch

(b) f versus y curve (c) f versus x curve

Figure 10.4 A truss arch and the f-y curves (perfect system) and f-x curves (Ikeda and Murota [6]; the perfect system shown by bold lines and imperfect systems shown by thin lines)

denote the initial and deformed member length of the m-th member.

The imperfection parameter vector is defined by

$$\boldsymbol{v} = (x_1, y_1, x_2, y_2, x_3, y_3, EA^{(1)}, EA^{(2)})^\top, \tag{10.21}$$

and the value of this vector for the perfect system is given by

$$\boldsymbol{v}^0 = (-1, 3, 1, 3, 0, 0, EA, EA)^\top. \tag{10.22}$$

Solution curves of the equilibrium equation (10.20) are plotted in Figs. 10.4(b) and (c). The minimum buckling load of the perfect system is governed by the unstable-symmetric bifurcation point A at $(x_c^0, y_c^0, f_c^0) = (0, 0.44735, 0.24776)$ (shown by ($\circ$)) on the fundamental path of this system shown by the bold curve. The critical eigenvector for the zero eigenvalue is $\boldsymbol{\eta}_c = (1, 0)^\top$.

We consider two imperfection patterns

$$\boldsymbol{d}^{(1)} = (-0.73685, -0.67606, -0.73685, 0.67606, 1, 0, EA, -EA)^\top,$$
$$\boldsymbol{d}^{(2)} = (0.73685, 0.67606, 0.73685, 0.67606, 1, 0, -EA, -EA)^\top.$$

Solution curves for the pattern $\boldsymbol{d}^{(1)}$ with several values of ε are plotted in Fig. 10.4(c) by thin lines. The minimum buckling loads f_c are plotted against initial imperfection ε in Fig. 10.5, in which ($\bullet$) denotes f_c for the initial imperfection pattern $\boldsymbol{d}^{(1)}$ and ($\circ$) denotes f_c for $\boldsymbol{d}^{(2)}$. The buckling load f_c is reduced more rapidly for $\boldsymbol{d}^{(1)}$ than $\boldsymbol{d}^{(2)}$. For each pattern, the values of f_c are reduced sharply as $|\varepsilon|$ increases, following the two-thirds power law for imperfection sensitivity law in (10.19).

10.4 WORST IMPERFECTION PATTERN

We introduce a theory on the worst imperfection pattern $\boldsymbol{d}$ that reduces the buckling load of a structural system most rapidly developed in Ikeda and Murota [6, 9].

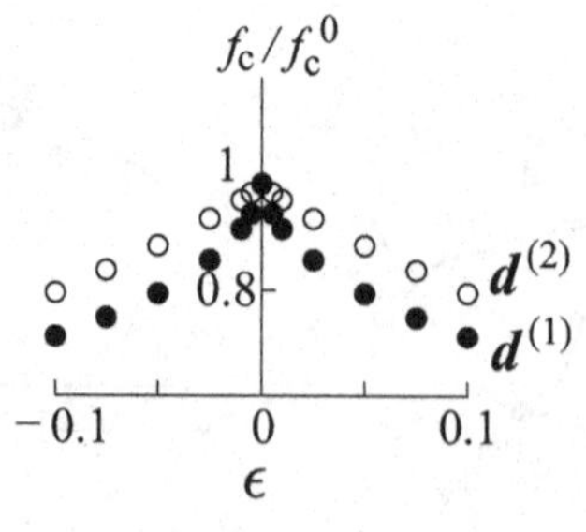

Figure 10.5 Imperfection sensitivity for the two imperfection patterns (Ikeda and Murota [9])

10.4.1 FORMULATION

As made clear in Section 10.3, the decrease of the buckling load for a small imperfection magnitude $|\varepsilon|$ is given asymptotically as

$$\tilde{f}_c = f_c - f_c^0 \approx \begin{cases} C_0\, a\varepsilon, & \text{maximal point,} \\ -C_0\,(a\varepsilon)^{2/3}, & \text{unstable-symmetric bifurcation point.} \end{cases} \tag{10.23}$$

Here C_0 is assumed to be a positive constant, and the variable $a \equiv A_{001}$ is expressed by (10.11) as a function in the imperfection pattern vector $\boldsymbol{d}$ as

$$a = \boldsymbol{\eta}_c^\top \frac{\partial \boldsymbol{F}}{\partial \boldsymbol{v}}(\boldsymbol{u}_c^0, f_c^0, \boldsymbol{v}^0)\boldsymbol{d} = \boldsymbol{\eta}_c^\top B\boldsymbol{d}, \tag{10.24}$$

where

$$B = \frac{\partial \boldsymbol{F}}{\partial \boldsymbol{v}}(\boldsymbol{u}_c^0, f_c^0, \boldsymbol{v}^0) \tag{10.25}$$

is an $n \times p$ constant matrix, called the *imperfection sensitivity matrix.*

We investigate the change of the buckling load f_c when the imperfection pattern vector $\boldsymbol{d}$ is changed under the condition of the unit weighted norm as

$$\boldsymbol{d}^\top W^{-1}\boldsymbol{d} = 1 \tag{10.26}$$

with respect to a positive definite matrix W. The variable a that governs the decrease $\tilde{f}_c$ of the buckling load in (10.23) is expressed as the inner product of the two vectors $B^\top\boldsymbol{\eta}_c$ and $\boldsymbol{d}$. If we choose $\boldsymbol{d}$ to be orthogonal to $B^\top\boldsymbol{\eta}_c$, $|\tilde{f}_c|$ becomes asymptotically zero. In contrast, if we choose $\boldsymbol{d}$ in the direction of

$$\boldsymbol{d}^* = \frac{WB^\top\boldsymbol{\eta}_c}{(\boldsymbol{\eta}_c^\top BWB^\top\boldsymbol{\eta}_c)^{1/2}}, \tag{10.27}$$

$|\tilde{f}_c|$ is maximized (note that $\boldsymbol{d}^{*\top} W^{-1}\boldsymbol{d}^* = 1$ is satisfied). The vector $\boldsymbol{d}^*$ in (10.27) is called the *worst imperfection pattern vector.*

10.4.2 STRUCTURAL EXAMPLE

As a structural example of the worst imperfection, we consider the truss arch in Section 10.3.3. The buckling load of this truss arch is governed by an unstable-symmetric bifurcation point (point A in Fig. 10.4).

We employ the same setting for initial imperfections as in (10.21) and (10.22):

$$\boldsymbol{v} = (x_1, y_1, x_2, y_2, x_3, y_3, EA^{(1)}, EA^{(2)})^\top,$$
$$\boldsymbol{v}^0 = (-1, 3, 1, 3, 0, 0, EA, EA)^\top.$$

The matrix W in (10.26) for the normalization of the imperfection pattern vector $\boldsymbol{d}$ is chosen as

$$W = \operatorname{diag}\left(1,1,1,1,1,1,(EA)^2,(EA)^2\right), \tag{10.28}$$

where diag$(\cdots)$ denotes a diagonal matrix with the values in the parentheses as diagonal entries. The imperfection sensitivity matrix B in (10.25) at the bifurcation point A is given by

$$B = \frac{EA}{10^2}\begin{pmatrix} 3.162 & 2.901 & 3.162 & -2.901 & -6.325 & 0 & -\frac{4.853}{EA} & \frac{4.853}{EA} \\ 4.316 & -2.553 & -4.316 & -2.553 & 0 & -4.843 & \frac{1.239}{EA} & \frac{1.239}{EA} \end{pmatrix}.$$

The use of this concrete form of B, the critical eigenvector $\boldsymbol{\eta}_{\rm c} = (1,0)^\top$ for the zero eigenvalue, and the concrete form (10.28) of W in the formula (10.27) leads to the worst imperfection pattern vector:

$$\boldsymbol{d}^* = (0.28404, 0.26061, 0.28404, -0.26061, -0.56812, 0, \\ -0.43592EA, 0.43592EA)^\top.$$

The worst patterns $-\boldsymbol{d}^*$ and $\boldsymbol{d}^*$ are depicted in Fig. 10.6(a). These patterns accelerate horizontal sway of the arch and, in turn, trigger the (bifurcation) buckling.

As depicted in Fig. 10.6(b), the buckling loads $f_{\rm c}$ for the worst imperfection pattern shown by (•) are the lower envelope of the buckling loads $f_{\rm c}$ for randomly given imperfection patterns. This ensures the validity of the theory.

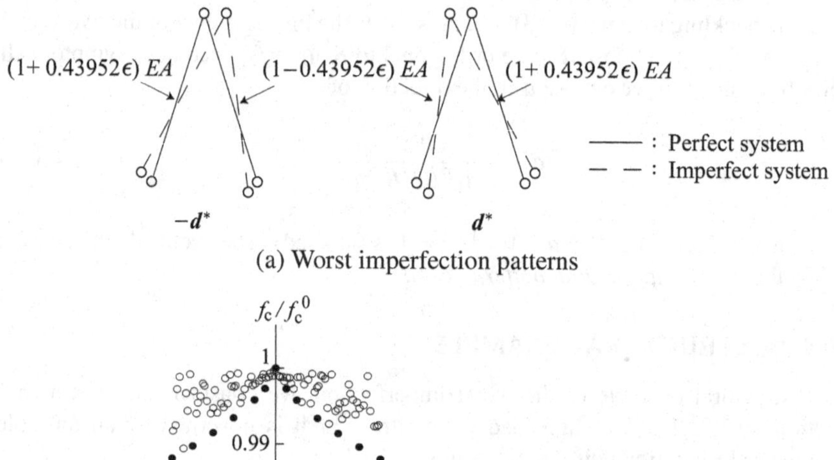

(a) Worst imperfection patterns

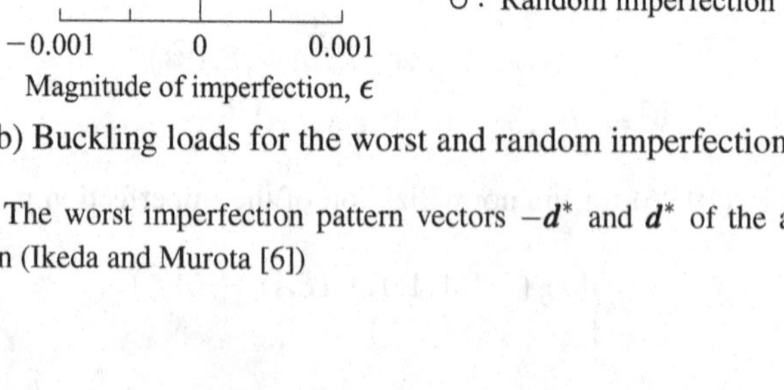

(b) Buckling loads for the worst and random imperfections

Figure 10.6 The worst imperfection pattern vectors $-\boldsymbol{d}^*$ and $\boldsymbol{d}^*$ of the arch and $f_{\rm c}/f_{\rm c}^0$ versus ε relation (Ikeda and Murota [6])

10.5 BUCKLING LOADS FOR RANDOM IMPERFECTIONS

We introduce a theory on the probabilistic variation of buckling loads by the probabilistic variation of initial imperfections developed in Ikeda and Murota [7, 9].

10.5.1 FORMULATION

The probabilistic variation of the buckling load $f_{\rm c}$ can be analyzed under the assumption that the imperfection pattern vector $\boldsymbol{d}$ is subject to a *multivariate normal distribution* with mean $\mathbf{0}$ and *variance–covariance matrix* W.

Recall the expression (10.23):

$$\tilde{f}_{\rm c} = f_{\rm c} - f_{\rm c}^0 \approx \begin{cases} C_0\, a\varepsilon & \text{at a maximal point,} \\ -C_0\,(a\varepsilon)^{2/3} & \text{at an unstable-symmetric bifurcation point} \end{cases} \tag{10.29}$$

for the influence of imperfection on (the decrease of) the buckling load $f_{\rm c}$. The variable $a = \boldsymbol{\eta}_{\rm c}^\top B\boldsymbol{d}$ in (10.29) is expressed as a linear combination of variables d_i as

$$a = \sum_{i=1}^{p} c_i d_i, \tag{10.30}$$

where

$$(c_1,\ldots,c_p) = \boldsymbol{\eta}_{\rm c}^\top B. \tag{10.31}$$

Because the variables d_i $(i = 1,\ldots,p)$ are subject to a multi-variate normal distribution $\mathrm{N}(\mathbf{0},W)$ with mean $\mathbf{0}$ and variance W, the variable a is subject to a normal distribution with mean 0 and variance

$$\tilde{\sigma}^2 = \boldsymbol{\eta}_{\rm c}^\top BWB^\top \boldsymbol{\eta}_{\rm c}$$

(see Problem 10.3 for the proof of this expression). Hence the probability density function of a is given by

$$\phi_a(a) = \frac{1}{\sqrt{2\pi}\tilde{\sigma}} \exp\left(\frac{-a^2}{2\tilde{\sigma}^2}\right). \tag{10.32}$$

Since a is subject to $\mathrm{N}(0,\tilde{\sigma}^2)$, a normalized variable

$$\tilde{a} = a/\tilde{\sigma} \tag{10.33}$$

is subject to the standard normal distribution $\mathrm{N}(0,1)$, whose probability density function is expressed as

$$\phi_{\rm N}(t) = \frac{1}{\sqrt{2\pi}} \exp\left(\frac{-t^2}{2}\right), \qquad -\infty < t < \infty. \tag{10.34}$$

It is convenient to introduce a *normalized critical load* (increment)

$$\zeta = \tilde{f}_{\rm c}/\hat{C} \tag{10.35}$$

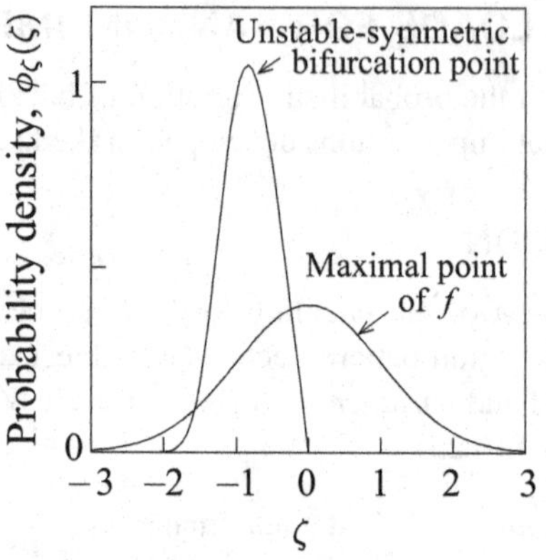

Figure 10.7 Probability density function $\phi_\zeta(\zeta)$ of the normalized buckling load ζ

with a scale factor

$$\hat{C} = \begin{cases} C_0\,\tilde{\sigma}\varepsilon & \text{at a maximal point,} \\ -C_0(\tilde{\sigma}\varepsilon)^{2/3} & \text{at an unstable-symmetric bifurcation point.} \end{cases} \tag{10.36}$$

From (10.29), (10.35), and (10.36), we obtain

$$\zeta = \begin{cases} \tilde{a} & \text{at a maximal point,} \\ -\tilde{a}^{2/3} & \text{at an unstable-symmetric bifurcation point.} \end{cases} \tag{10.37}$$

We already know that the normalized variable $\tilde{a}$ in (10.33) is subject to the standard normal distribution $\mathrm{N}(0,1)$. Then the distribution of ζ can be obtained easily from the relation (10.37) between $\tilde{a}$ and ζ. To be specific, the probability density function $\phi_\zeta(\zeta)$ of ζ can be obtained by the transformation of the variable as

$$\phi_\zeta(\zeta) = \begin{cases} \phi_{\mathrm{N}}(\tilde{a}) \left| \dfrac{d\tilde{a}}{d\zeta} \right| = \dfrac{1}{\sqrt{2\pi}} \exp\left(\dfrac{-\zeta^2}{2} \right), & -\infty < \zeta < \infty \text{ at a maximal point,} \\ 2\phi_{\mathrm{N}}(\tilde{a}) \left| \dfrac{d\tilde{a}}{d\zeta} \right| = \dfrac{3|\zeta|^{1/2}}{\sqrt{2\pi}} \exp\left(\dfrac{-|\zeta|^3}{2} \right), & -\infty < \zeta < 0 \\ & \text{at an unstable-symmetric bifurcation point.} \end{cases} \tag{10.38}$$

The probability density function $\phi_\zeta(\zeta)$ of the normalized buckling load obtained in this manner is plotted in Fig. 10.7.

From the relation

$$f_{\mathrm{c}} = f_{\mathrm{c}}^0 + \hat{C}\zeta \tag{10.39}$$

(cf., (10.35)), the probability density function of the critical load f_{c} can be obtained by a simple transformation of the probability density function in (10.38) of the normalized critical load ζ as follows:

$$\phi_{f_c}(f_c) = \begin{cases} \dfrac{1}{\sqrt{2\pi}\hat{C}} \exp\left[\dfrac{-1}{2}\left(\dfrac{f_c - f_c^0}{\hat{C}}\right)^2\right], & -\infty < f_c < \infty \\ & \text{at a maximal point,} \\ \dfrac{3|f_c - f_c^0|^{1/2}}{\sqrt{2\pi}\hat{C}^{3/2}} \exp\left[\dfrac{-1}{2}\left|\dfrac{f_c - f_c^0}{\hat{C}}\right|^3\right], & -\infty < f_c < f_c^0 \\ & \text{at an unstable-symmetric bifurcation point.} \end{cases} \quad (10.40)$$

The mean of f_c is computed as

$$\mathrm{E}[f_c] = f_c^0 + \mathrm{E}[\zeta]\hat{C} = \begin{cases} f_c^0 & \text{at a maximal point,} \\ f_c^0 - 0.802\hat{C} & \text{at an unstable-symmetric bifurcation point,} \end{cases} \quad (10.41)$$

and the variance of f_c as

$$\mathrm{Var}[f_c] = \mathrm{Var}[\zeta]\hat{C}^2 = \begin{cases} \hat{C}^2 & \text{at a maximal point,} \\ (0.432\hat{C})^2 & \text{at an unstable-symmetric bifurcation point.} \end{cases} \quad (10.42)$$

When a sample of buckling loads f_c is available, the probability density function of the buckling load f_c can be estimated as follows:

- Obtain the sample average $\mathrm{E}_{\mathrm{sample}}[f_c]$ and the sample variance $\mathrm{Var}_{\mathrm{sample}}[f_c]$.
- Substitute these values into the left-hand sides of (10.41) and (10.42) to estimate the buckling load f_c^0 of the perfect system and the scaling constant $\hat{C}$.
- Obtain the probability density function of the buckling load f_c from (10.40).

10.5.2 STRUCTURAL EXAMPLES

We illustrate the probabilistic variation of buckling loads due to the variation of initial imperfections for the regular-hexagonal truss dome in Fig. 10.8(a) subjected to the vertical load $f/2$ at the crown node and f at other free nodes. The buckling load of this dome is governed by an unstable-symmetric bifurcation point.

We consider imperfections of the initial locations of the nodes, each subject to a normal distribution with mean 0 and variance 10^{-6}. Buckling loads f_c are computed for 100 random imperfections subject to this normal distribution. As listed in Table 10.1, we obtain the sample mean $\mathrm{E}_{\mathrm{sample}}[f_c]$ and sample variance $\mathrm{Var}_{\mathrm{sample}}[f_c]$ of f_c and, in turn, f_c^0 and $\hat{C}$ from (10.41) and (10.42). The values of f_c^0 and $\hat{C}$ estimated in this manner are in good agreement with the true values obtained by the numerical analysis.

The histogram of the buckling load f_c for random imperfections is compared with the estimated probability density function shown by the dashed line and theoretical

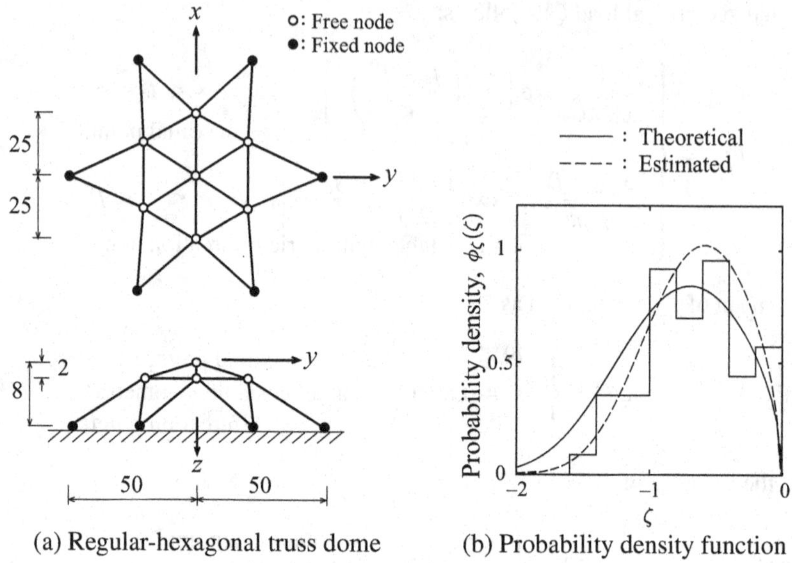

Figure 10.8 Regular-hexagonal truss dome and the probability density function of its buckling loads (Ikeda and Murota [7])

Table 10.1
Estimation of f_c^0 and $\hat{C}$ for the truss dome

	$\mathrm{E}_{\mathrm{sample}}[f_c]$	$(\mathrm{Var}_{\mathrm{sample}}[f_c])^{1/2}$	f_c^0	$\hat{C}$
Estimated value	7.648×10^{-4}	2.755×10^{-6}	7.699×10^{-4}	6.370×10^{-6}
True value	7.638×10^{-4}	3.346×10^{-6}	7.700×10^{-4}	7.736×10^{-6}

one shown by the solid line in Fig. 10.8(b).[4] The curves of the theoretical and the estimated probability density functions are in fair agreement, thereby demonstrating the validity and usefulness of the present theory.

10.6 IMPERFECTION SENSITIVITY OF ELASTIC–PLASTIC PLATES

Imperfect elastic–plastic plates under compression undergo elastic and/or plastic buckling, depending on their width–thickness ratio. The imperfection sensitivity for plastic buckling was first studied through extended use of Koiter's sensitivity law.[5]

[4]The estimated probability density function is obtained from (10.40) by estimating the values of f_c^0 and $\hat{C}$ from (10.41) and (10.42), while the theoretical one is obtained by computing the exact values of f_c^0 and $\hat{C}$.

[5]See Hutchinson [3], Needleman [15], Byskov [1], Ming and Wenda [12], and Su and Lu [16].

However, compressed plates in the plastic range often display complex buckling behaviors that are different from elastic buckling behaviors (Feldman and Abdoudi [2]; Lu, Obrecht, and Wunderlich [11]). The imperfection sensitivity law apparently needs some extension to express the diversity of buckling of plates in plastic range.

We hereafter introduce the results of a study of imperfection sensitivity of elastic–plastic plates (Ikeda et al. [5]). The finite-displacement, elastic–plastic analysis of simply-supported imperfect square steel plates was conducted by varying plate thickness to observe various kinds of imperfection sensitivity. Imperfections of sinusoidal initial deflections were considered, and the elastic–plastic material property modeling steel was employed (cf., Fig. 10.9).

Typical axial force versus out-of-plane displacement curves of a plate subjected to uniaxial compression are illustrated in Fig. 10.10(a). Three different kinds of buckling can be observed depending on the width-thickness ratio:

- For a thick plate with a small width–thickness ratio (Fig. 10.10(a)), plastic buckling takes place and the ultimate buckling strength is governed by yielding due to axial compression.
- For a plate with an intermediate thickness, elastic unstable bifurcation prevails and a sharp peak of the curve, which governs the ultimate buckling strength, is clearly seen.
- For a thin plate with a large width–thickness ratio, stable bifurcation occurs, and the load increases stably after the bifurcation. The ultimate buckling strength is not governed by bifurcation but by the peak of the post-bifurcation curve.

To sum up, the complexity of plate buckling arises from the mixed presence of elastic and plastic bifurcations.

The ultimate buckling strength, which is given either by the bifurcation load or by the maximum load, is plotted against the imperfection magnitude w_0 in Fig. 10.10(b) to observe imperfection sensitivity. For the three different kinds of buckling, the relations in this figure have different tangents (curvatures) for small imperfection magnitudes.

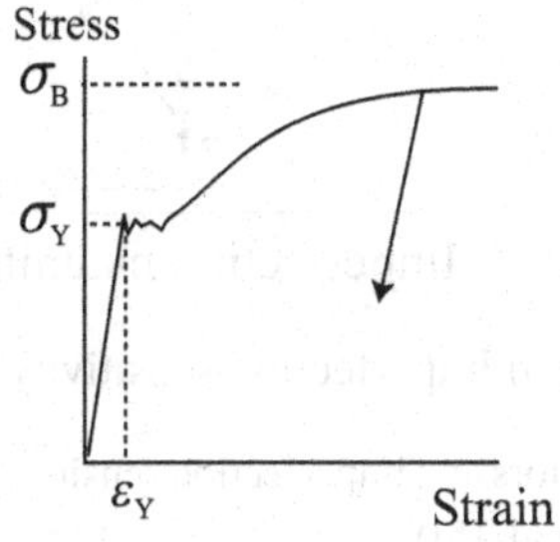

Figure 10.9 Stress versus strain curve of steel

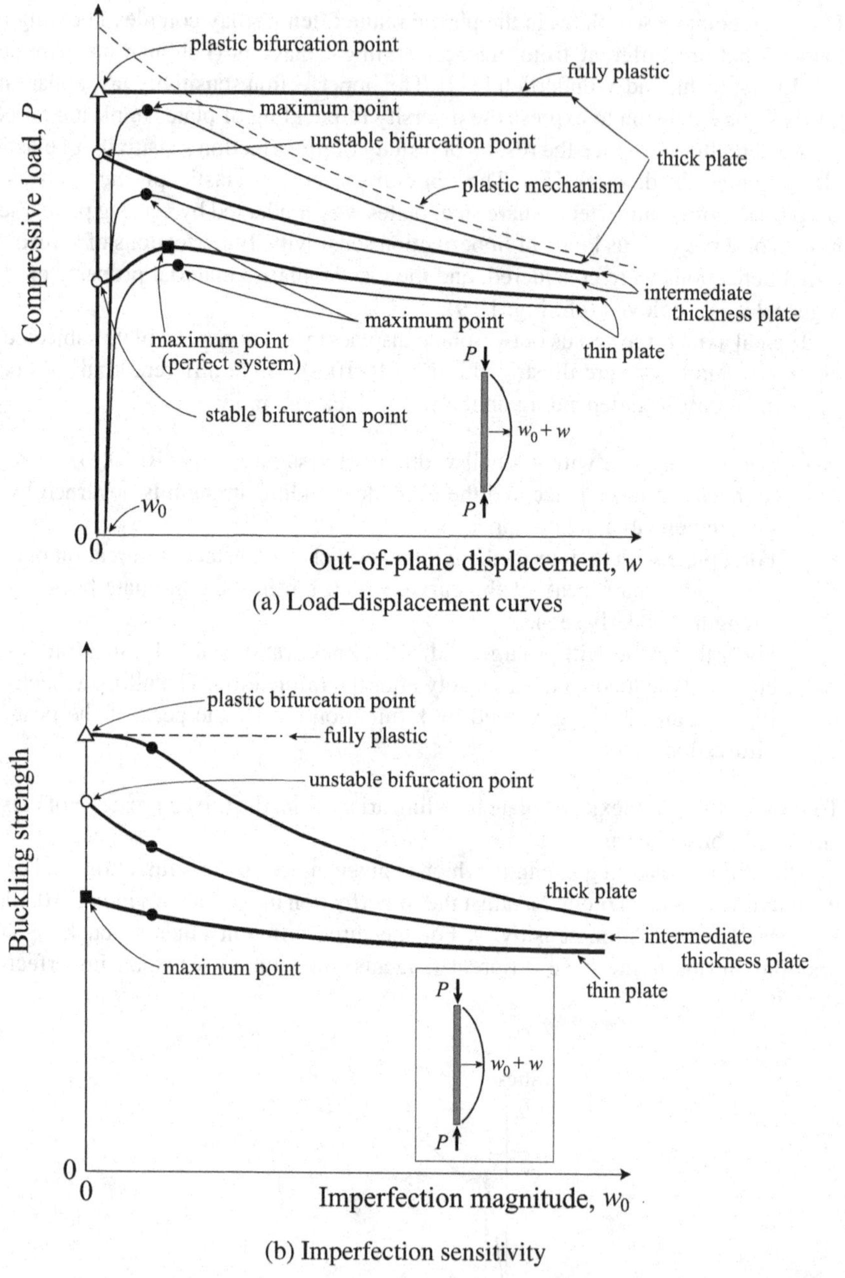

Figure 10.10 Buckling behaviors and imperfection sensitivities of plates under compression for various thicknesses (Ikeda et al. [5])

To describe the imperfection sensitivity of the plate, the imperfection sensitivity law of the following form may be employed:

$$\tilde{f}_c \approx C\varepsilon^{\rho},$$

where ρ is a constant that is to be determined based on the numerical results. The three different kinds of imperfection sensitivities in Fig. 10.10(b) can be described as follows:

- For a thick plate, the quadratic power law with $\rho = 2$ may be advanced, as an extension of the Koiter law, to describe imperfection-insensitive reduction of buckling strength in association with an increase of imperfection magnitude.
- For a plate with an intermediate thickness, the original Koiter law with $\rho = 2/3$ for unstable bifurcation seems adequate to express the sharp reduction of buckling strength.
- For a thin plate, the post-stable-bifurcation maximum load can be expressed by the original Koiter law with $\rho = 1$ for a maximum (limit) point.

For the Koiter laws, we have $\rho = 1$, 1/2, 2/3 and also $\rho = 2$ to explain the plastic buckling in view of the numerical results presented above.

10.7 PROBLEMS

Problem 10.1 Obtain the potential function and the governing equation of the propped cantilever in Fig. 10.11 with a linear spring when it is subjected to a rightward horizontal load $kL\varepsilon$ at the top in addition to the vertical load kLf.

Problem 10.2 Obtain the potential function and the governing equation of the propped cantilever with an initial tilt $\varepsilon \leq 0$ in Fig. 10.1(a) in Section 10.2 using an asymmetric spring property $F_s(x) = k\left(x + \frac{3}{2}\frac{x^2}{L}\right)$. Plot the solution curves for $\varepsilon = 0$, -0.03, and -0.1 and the imperfection sensitivity (f_c versus ε) relation.

Problem 10.3 Prove that the variable a in (10.30) is subject to a normal distribution with mean 0 and variance $\tilde{\sigma}^2 = \boldsymbol{\eta}_c^\top BWB^\top \boldsymbol{\eta}_c$.

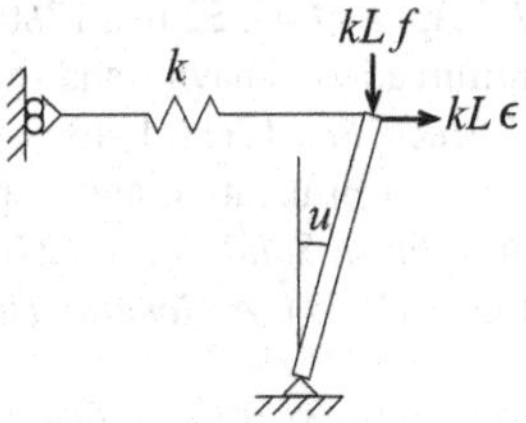

Figure 10.11 A tilted propped cantilever

Problem 10.4 Investigate the asymptotic influence of the initial imperfection ε on an asymmetric bifurcation point.

Problem 10.5 Suppose that the maximum load of a shell is 2.45 ton for an initial thickness imperfection of $\varepsilon = 10^{-1}$ and 2.60 ton for $\varepsilon = 10^{-3}$. (1) Obtain the buckling load f_c^0 for the perfect system with $\varepsilon = 0$ using the two-thirds power law. (2) Obtain the tolerance of the initial imperfection ε to ensure $f_c \geq 2.50$ ton.

REFERENCES

1. Byskov E (1982), Imperfection sensitivity of elastic-plastic truss columns, *AIAA J.* **20**(2), 263–267.
2. Feldman E, and Aboudi J (1993), Postbuckling analysis and imperfection sensitivity of viscoplastic plates and cylindrical panels, *Thin-Walled Struct.* **17**(4), 273–290.
3. Hutchinson JW (1972), On the postbuckling behavior of imperfection-sensitive structures in the plastic range, *ASME J. Appl. Mech.* **39**(1), 155–162.
4. Hutchinson JW, and Koiter WT (1970), Postbuckling theory, *Appl. Mech. Rev.* **23**(12), 1353–1366.
5. Ikeda K, Kitada T, Matsumura M, and Yamakawa Y (2007), Imperfection sensitivity of ultimate buckling strength of elastic–plastic square plates under compression, *Int. J. Non-Linear Mech.* **42**, 529–541.
6. Ikeda K, and Murota K (1990), Critical initial imperfection of structures, *Int. J. Solids Struct.* **26**(8), 865–886.
7. Ikeda K, and Murota K (1993), Statistics of normally distributed initial imperfections, *Int. J. Solids Struct.* **30**(18), 2445-2467.
8. Ikeda K, and Murota K (2008), Asymptotic and probabilistic approach to buckling of structures and materials, *Appl. Mech. Rev.* **61**, 040801-1–16.
9. Ikeda K, and Murota K (2019) *Imperfect Bifurcation in Structures and Materials: Engineering Use of Group-Theoretic Bifurcation Theory*, 3rd ed., Springer, New York.
10. Koiter WT (1945), *On the Stability of Elastic Equilibrium*, Dissertation, Delft Univ. Tech. (English translation: NASA Tech. Trans. F10:833, 1967).
11. Lu Z, Obrecht H, and Wunderlich W (1995), Imperfection sensitivity of elastic and elastic-plastic torispherical pressure vessel heads, *Thin-Walled Struct.* **23**(1-4), 21–39.
12. Ming S-X, and Wenda L (1990), Postbuckling and imperfection sensitivity analysis of structures in the plastic range—Part 1: Model analysis, *Thin-Walled Struct.* **10**(4), 263–275.
13. Murota K, and Ikeda K (1991), Critical imperfection of symmetric structures, *SIAM J. Appl. Math.* **51** (5) 1222–1254.
14. Murota K, and Ikeda K (1992), On random imperfections for structures of regular-polygonal symmetry, *SIAM J. Appl. Math.* **52** (6), 1780–1803.
15. Needleman A (1975), Postbifurcation behavior and imperfection sensitivity of elastic-plastic circular plates, *Int. J. Mech. Sci.* **17**(1), 1–13.
16. Su XM, and Lu WD (1991), Post-bifurcation and imperfection sensitivity analysis of plastic clamped circular plates, *Int. J. Solids Struct.* **27**(6), 769–782.
17. Thompson JMT, and Hunt GW (1973), *A General Theory of Elastic Stability*, Wiley, New York.
18. Thompson JMT, and Hunt GW (1984), *Elastic Instability Phenomena*, Wiley, Chichester.

11 History of Imperfect Buckling

11.1 SUMMARY

The experimental buckling loads of long thin-walled cylinders were found to fall markedly below the stability limit computed by the classical linearized theory of bifurcation. In order to resolve such inadequacy, *nonlinearity* and *initial imperfections* were implemented into the theory of elastic stability of shells.[1] Combined studies of imperfection-sensitive structures, experimental, analytical, and computational, were thus motivated.

The general theory of elastic stability invented by Koiter [109] played a pivotal role in dealing with the so-called initial post-buckling behavior, especially from an analytical standpoint, and led to excellent textbooks.[2] The concept of imperfection sensitivity was quite pertinent in gathering knowledge as to the behavior of imperfect structures. Thereafter research of initial post-buckling behavior and imperfection sensitivity of structures mushroomed worldwide.

We review, through the perspective of theoretical engineers, the historical development of buckling of imperfection-sensitive structures and materials. We deal with several topics, including initial post-buckling behaviors, search for prototype initial imperfections, probabilistic scatter of critical loads, generalized asymptotic method, and hilltop branching.

Keywords: • Bifurcation • Critical point • Initial imperfection • Imperfection sensitivity • Worst imperfection

11.2 INITIAL POST-BUCKLING BEHAVIORS

The importance of the general nonlinear theory was not recognized until the early 1960s, when research on *initial post-buckling behaviors* for imperfection-sensitive structures sprung up worldwide.

In the United States, Koiter's original work was reconstructed in a form suitable for elastic continua and was applied to imperfection-sensitive structures (Budiansky and Hutchinson [32]; Hutchinson and Amazigo [81]). Hutchinson and Koiter [82] and Ikeda and Murota [96] reviewed the early development of *post-buckling theory*.

In Europe, mainly in England, initial post-buckling behavior and related theories started to draw attention in the 1960s.[3] The direct experimental validation of imperfection sensitivity of structures was conducted (Roorda [155]). The perturbation

[1] See Flügge [65], Donnel [48], and von Kármán and Tsien [190].

[2] See Thompson and Hunt [178], Britvec [28], Budiansky [31], Huseyin [79], Pignataro et al. [150], and Godoy [67].

[3] See Britvec and Chilver [29], Thompson [175], Sewell [167], Supple [171], and Roorda [156].

DOI: 10.1201/9781003112365-11

technique was applied to the total potential energy function of a finite-dimensional system to derive asymptotic information on bifurcation buckling (Thompson and Hunt [178]).

In this connection, the optimal design of structures undergoing buckling was found to produce a coincidental critical point.[4] The danger of naive optimization of structures without due regard to imperfection sensitivity and the side effect of optimization by compound branching were pointed out.[5] For example, high imperfection-sensitivity was observed for shells when two or more bifurcation points nearly or strictly coincide (Hutchinson and Amazigo [81]) and for columns when they are subjected to interaction of local and global buckling modes (Koiter and Kuiken [112]). The study of coincidental buckling and the interaction between local and overall initial imperfections was thus motivated.[6] Imperfection sensitivity was applied to the optimization of imperfection-sensitive structures.[7]

11.3 SEARCH FOR PROTOTYPE INITIAL IMPERFECTIONS

Early studies of imperfect structures dealt with pre-specified imperfection modes. There arose a question, "What is the imperfection to be employed?" The geometrical imperfection in the shape of the buckling mode was used initially.[8] For example, classical axisymmetric buckling modes were used as *initial shape imperfections* for cylindrical shells (Koiter [110]), and *checkerboard imperfections* were assumed for spherical shells (Hutchinson [80]). *Dimple imperfections* were employed as realistic imperfections (Amazigo and Fraser [7] and Amazigo [4]).

In association with the development of techniques for measuring initial imperfections, detailed knowledge of geometric imperfections of shells was gathered.[9] Compilation and extensive analysis of the *international initial imperfection data banks* were pursued actively at the Delft University of Technology[10] and at the Israel Institute of Technology.[11] These data banks are useful in deriving characteristic initial imperfections that a given fabrication process is likely to produce. At Imperial College, information about measured imperfections was gathered to produce characteristic imperfection shapes.[12]

---

[4]See Thompson and Lewis [180], Tvergaard [186], and Ohsaki [132, 134].

[5]See, Koiter and Skaloud [113], Spunt [170], Thompson and Supple [182], and Thompson and Hunt [179].

[6]See Chilver [36], Supple [171], Ho [77], Byskov and Hutchinson [33], and Thompson [177].

[7]See Haug et al. [75], Palassopoulos [141, 142], Reitinger and Ramm [154], Ohsaki and Nakamura [138], Ohsaki [133], Ohsaki and Ikeda [136], Elishakoff and Ohsaki [61], and Kaveh [107].

[8]See Koiter [110], Hutchinson and Amazigo [81], and Hansen and Roorda [74].

[9]See Arbocz and Babcock [12, 13], Megson and Hallak [123], Chryssanthopoulos et al. [39], Bernard et al. [21], and Pircher and Wheeler [152].

[10]See Arbocz and Abramovich [11], Arbocz [10], Dancy and Jacobs [43], and de Vries [45].

[11]See Singer et al. [169] and Abramovich et al. [1].

[12]See Scott et al. [166], Chryssanthopoulos et al. [37, 38], and Chryssanthopoulos and Poggi [40].

The *worst imperfection* vector of an imperfect cubic potential system was proved to be in the direction of the bifurcated path of the largest slope by Ho [78], while Arbocz [9] carried out extensive studies on shape imperfections to identify "most rapidly growing imperfections." Thereafter study on the worst imperfection drew considerable attention.[13] The worst imperfection shape of structures was studied in a more general setting,[14] extended to imperfections other than structural shapes (Ikeda and Murota [89, 90] and Murota and Ikeda [124]), and implemented into the framework of finite-element analysis (Ikeda and Murota [90] and Deml and Wunderlich [46]).

11.4 PROBABILISTIC SCATTER OF BUCKLING LOADS

Through the search for the prototype imperfections, it came to be acknowledged that initial imperfections are subjected to probabilistic scatter and that the study of initial imperfections needs to be combined with probabilistic treatment to make them practical.

As first postulated by Bolotin [24], the critical load of a structure can be expressed as a function of a number of random variables representing initial imperfections. The straightforward evaluation of the probability density function of the critical load is divided into the following two stages:

1. Evaluate the probabilistic variation of the initial imperfections.
2. Compute the set of critical loads for a given set of initial imperfections and, in turn, to obtain the probabilistic variation of the critical loads.

The first and the most difficult stage was tackled through a series of attempts. For columns and bars, random imperfections were employed.[15] For shells, random axisymmetric imperfections[16] and general (non-symmetric) random initial imperfections[17] were employed. Initial imperfections were often assumed to be Gaussian (normally-distributed) random variables.[18] In addition to the initial imperfection of

[13]See Samuels [163], Sadovsky [160], Nishino and Hartono [131], Ikeda et al. [103], and Salerno and Uva [162].

[14]See Peek and Triantafyllidis [149], Triantafyllidis and Peek [184], and Peek [148].

[15]See Boyce [27], Fraser and Budiansky [66], Amazigo [3], Elishakoff [51], and Day et al. [44].

[16]See Amazigo [2], Tennyson et al. [173], Amazigo and Budiansky [5], Roorda and Hansen [159], and Elishakoff and Arbocz [56].

[17]See Amazigo [4], Arbocz and Babcock [12, 13], van Slooten and Soong [188], Hansen [72, 73], Elishakoff and Arbocz [57], Arbocz and Hol [15], Bielewicz et al. [22], and Schenk et al. [164].

[18]See Amazigo [2], Hansen and Roorda [74], Hansen [72, 73], and Elishakoff [50, 51].

structural shapes, various kinds of imperfections, such as loadings,[19] material properties (elastic moduli),[20] and thickness variation[21] were considered.

To tackle the second stage, several studies were conducted, as introduced below. The imperfection sensitivity law was used as a transfer function from an initial imperfection to the deterministic critical load and, in turn, to obtain the probabilistic variation of critical load for imperfections with a known probabilistic property.[22] Such use of imperfection sensitivity, however, was limited to a certain prototype imperfection. As a remedy of this, an initial imperfection was represented as a random process, and the method of stochastic differential equations was used to obtain an asymptotic relationship between the critical load and initial imperfection.[23] Sensitivity coefficients—*design sensitivity coefficients*—of linear buckling load factor with respect to design variables, such as stiffness and nodal locations, were employed in buckling design.

The probability of failure was employed to express the influence of the randomness of experimentally measured imperfections[24] and paved the way to the introduction of the results of statistical methods.[25] Several studies based on measured data were conducted[26] often with resort to the international initial imperfection data banks, from which stochastic imperfections with known average and autocorrelation can be produced. The *Monte Carlo simulation* came to be conducted to compute numerically the reliability of the buckling strength for measured or random initial imperfections.[27] An asymptotic approach was combined with statistical analysis (Palassopoulos [143, 144] and Trendafilova and Ivanova [183]). Koiter's special theory [110] for axisymmetric imperfections was combined with the Monte Carlo method (Elishakoff and Arbocz [56]) and, in turn, to introduce imperfection-sensitivity concept into design procedure (Elishakoff [54]). The Monte Carlo method was

---

[19]See Roorda [158], Li [116], Ikeda and Murota [90], Li et al. [118], Cederbaum and Arbocz [35], and Królak et al. [114].

[20]See Shinozuka [168], Ikeda and Murota [90], Elishakoff et al. [58], and Lagaros and Papadopoulos [115].

[21]See Tvergaard [187], Bielski [23], Koiter et al. [111], Li et al. [117], Gusic et al. [69], and Lagaros and Papadopoulos [115].

[22]See Bolotin [24], Roorda and Hansen [159], Thompson [176], Roorda [157], Ikeda and Ohsaki [100], and Ikeda et al. [102].

[23]See Fraser and Budiansky [66], Amazigo et al. [6], Amazigo [2, 3, 4], Amazigo and Budiansky [5], and van Slooten and Soong [188].

[24]See Hansen and Roorda [74] and Elishakoff [50, 51].

[25]See Bolotin [25, 26], Ang and Tang [8], Thoft-Christensen and Baker [174], Elishakoff [52], Augusti et al. [17], Ben-Haim [19], and Haldar and Mahadevan [71].

[26]See Elishakoff and Arbocz [56], Elishakoff [53], Arbocz and Hol [14, 15], Turčić [185], Ikeda et al. [97], Pircher et al. [151], and Lin and Teng [119].

[27]See Edlund and Leopoldson [49], Hansen [73], Elishakoff [50, 51], Elishakoff and Arbocz [56], Wang [191], and Palassopoulos [144].

replaced by the first-order second-moment method to reduce computational costs considerably.[28]

Stochastic finite element methods (SFEM)[29] or finite element methods for stochastic problems (FEMSP by Elishakoff and Ren [62]) were employed to numerically tackle the probabilistic properties of structures. The perturbation method was employed for most cases to deal with random quantities involved, and the second-moment analysis was often employed to compute the mean and the variance of the displacement or stress. The elastic modulus was often modeled as a random field in SFEM.

The *response surface approach* was used to evaluate the reliability of structures;[30] nonlinear finite element analyses, for example, were conducted for the parameter sweep of a few initial imperfections and/or design parameters to evaluate the reliability. A large number of imperfection modes were taken into account in the formulation by critical-imperfection-magnitude (CIM) method (Palassopoulos [140, 145]; Palassopoulos and Shinozuka [146]).

Possible limitations of probabilistic methods were pointed out (Elishakoff [55]), and a few attempts were conducted to arrive at a lower bound of buckling strength: *Knockdown factor* based on the so-called *lower bound design philosophy* is the most primitive but most robust way (Weingarten et al. [193]; NASA [129]). The knockdown factor was employed rather than the concept of imperfection sensitivity (Arbocz and Hol [14]; Elishakoff et al. [59]). The *reduced stiffness method* finds a lower bound of design strength of a shell through the identification of the components of the membrane energy that are eroded by imperfections and mode interactions (Croll [41]; Croll and Batista [42]). *Convex modeling of uncertainty,*[31] *robust reliability,*[32] *anti-optimization approach*[33] were developed to derive lower bounds based on the worst-case-scenario.

11.5 ASYMPTOTIC METHOD AND PLASTIC BIFURCATION OF MATERIALS

Koiter's asymptotic approach was combined with the finite element method to be consistent with computer-aided engineering environments.[34] As introduced in Chapter 10, the Koiter law was generalized to a number of initial imperfections as follows.

[28] See Elishakoff et al. [63] and Arbocz and Hol [15].

[29] An exhaustive review by Schuëller [165] is available. See also Astill et al. [16], Nakagiri and Hisada [128], der Kiureghian [47], and Liu et al. [121].

[30] See Faravelli [64], Bucher and Bourgund [30], Rajashekhar and Ellingwood [153], Myers and Montgomery [127], Kim and Na [108], Venter et al. [189], Zheng and Das [194], and Gomes and Awruch [68].

[31] See Ben-Haim and Elishakoff [20], Lindberg [120], Ben-Haim [18], and Elishakoff et al. [60].

[32] See Ben-Haim [19] and Papadimitriou et al. [147].

[33] See Elishakoff et al. [59], Lombardi and Haftka [122], and Zingales and Elishakoff [195].

[34] See Haftka et al. [70], Arbocz and Hol [14], Casciaro et al. [34], and Salerno and Lanzo [161].

Theoretical procedures to determine the worst initial imperfection (Ikeda and Murota [89, 90]) and to obtain the probability density function of critical loads for initial imperfections with known probabilistic characteristics were developed.[35] Weibull-like probability density functions of critical loads were derived for initial imperfections subjected to a multivariate normal distribution (Ikeda and Murota [92]).

Asymptotic methods were extended to experimentally observed bifurcation diagram and to a system with dihedral-group symmetry (Murota and Ikeda [124, 125]; Ikeda and Murota [95]) and, in turn, were applied to the description of the strength of structures and materials, such as shells, plates, steels, soils, and concretes (Ikeda et al. [84, 86, 87, 97]; Okazawa et al. [139]).

Plastic bifurcation, through the formation of a shear band, is acknowledged to govern the strength of materials (Hill and Hutchinson [76]). The horizon of the study of (elastic) bifurcation is extending towards the failure and deformation of materials. Elastic, diffuse mode bifurcation was shown to occur in soil specimens through the combination of asymptotic method, group-theoretic bifurcation theory, and numerical procedure.[36] This combined procedure was successfully applied to the stochastic description of the strength of steels and low-strength concretes (Ikeda et al. [87, 104]; Okazawa et al. [139]).

11.6 HILLTOP BRANCHING FOR MATERIALS AND STRUCTURES

Hilltop branching was found to take place for materials.[37] In mechanical instability of stressed atomic crystal lattices[38] and in numerical simulation of a long tensile steel specimen undergoing plastic instability,[39] a nearly coincidental pair of critical points of a bifurcation point and a limit point of loading parameter was found. Such a pair of points was approximated by a *hilltop branching (bifurcation) point*, at which the pair of points coincide strictly. This hilltop point was shown to enjoy piecewise linear imperfection sensitivity,[40] which is less severe than the two-thirds power-law for a simple pitchfork bifurcation point.

A structural form comprising n unlinked separate identical cells underwent explosive bifurcation at a hilltop bifurcation point with n-fold criticality and as many as 2^n bifurcating paths (Ikeda et al. [105]). Strictly coincidental hilltop branching points were found through the optimization of structural systems (Ohsaki [132, 134]). A

---

[35] See Ikeda and Murota [91, 92], Ikeda and Ohsaki [100], and Ikeda et al. [102].

[36] See Ikeda and Murota [93, 94, 95], Murota et al. [126], Ikeda et al. [84, 85, 88, 98, 99, 104, 106], and Tanaka et al. [172].

[37] The study of the scatter of strength of materials followed a completely different course of development from that of structures. Weibull [192] derived a famous distribution of the strength of materials governed by the fracture on the basis of the weakest link theory.

[38] See Thompson [177] and Thompson and Schorrock [181].

[39] See Okazawa et al. [139], Needleman [130], and Hutchinson and Miles [83].

[40] See Thompson [177], Thompson and Schorrock [181], Ikeda et al. [103], and Okazawa et al. [139].

hilltop point occurring as a coincidence of a limit point and a double bifurcation point of a system with dihedral-group symmetry also enjoys a piecewise linear law (Ikeda et al. [101]; Ohsaki and Ikeda [135, 137]).

REFERENCES

1. Abramovich H, Singer J, and Yaffe R (1981), Imperfection characteristics of stiffened shells: Group 1, *TAE Rep.* **406**, Dept. Aeron. Eng., Technion, Israel Inst. of Tech., Haifa.
2. Amazigo JC (1969), Buckling under axial compression of long cylindrical shells with random axisymmetric imperfections, *Quart. Appl. Math.* **26**(4), 537–566.
3. Amazigo JC (1971), Buckling of stochastically imperfect columns on nonlinear elastic foundations, *Quart. Appl. Math.* **29**(3), 403–409.
4. Amazigo JC (1974), Asymptotic analysis of the buckling of externally pressurized cylinders with random imperfections, *Quart. Appl. Math.* **31**(4), 429–442.
5. Amazigo JC, and Budiansky B (1972), Asymptotic formulas for the buckling stresses of axially compressed cylinders with localized or random axisymmetric imperfections, *ASME J. Appl. Mech.* **39**(1), 179–184.
6. Amazigo JC, Budiansky B, and Carrier GF (1970), Asymptotic analyses of the buckling of imperfect columns on nonlinear elastic foundations, *Int. J. Solids Struct.* **6**(10), 1341–1356.
7. Amazigo JC, and Fraser WB (1971), Buckling under external pressure of cylindrical shells with dimple shaped initial imperfections, *Int. J. Solids Struct.* **7**(8), 883–900.
8. Ang AH-S, and Tang WH (1975), *Probability Concepts in Engineering Planning and Design*, Wiley, New York.
9. Arbocz J (1974), The effect of initial imperfections on shell stability, *Thin Shell Structures: Theory, Experiment and Design*, YC Fung, and EE Sechler (eds), Prentice-Hall, Englewood Cliffs, NJ, 205–245.
10. Arbocz J (1982), The imperfection data bank: A means to obtain realistic buckling loads, *Buckling of Shells*, E Ramm (ed), Springer, Berlin, 535–567.
11. Arbocz J, and Abramovich H (1979), The initial imperfection data bank at the Delft University of Technology: Part I, *Tech. Rep.* LR-**290**, Dept. Aeron. Eng., Delft University of Tech., Delft.
12. Arbocz J, and Babcock CD Jr (1968), Experimental investigation of the effect of general imperfections on the buckling of cylindrical shells, *NASA* CR-**1163**.
13. Arbocz J, and Babcock CD Jr (1969), The effect of general imperfections on the buckling of cylindrical shells, *ASME J. Appl. Mech.* **36**(1), 28–38.
14. Arbocz J, and Hol JMAM (1990), Koiter's stability theory in a computer aided engineering (CAE) environment, *Int. J. Solids Struct.* **26**(9-10), 945–973.
15. Arbocz J, and Hol JMAM (1991), Collapse of axially compressed cylindrical shells with random imperfections, *AIAA J.* **29**(12), 2247–2256.
16. Astill J, Nosseir CJ, and Shinozuka M (1972), Impact loading on structures with random properties, *J. Struct. Mech.* **1**(1), 63–67.
17. Augusti G, Barratta A, and Casciati F (1984), *Probabilistic Methods in Structural Engineering*, Chapman and Hall, New York.
18. Ben-Haim Y (1993), Convex models of uncertainty in radial pulse buckling of shells, *ASME J. Appl. Mech.* **60**(3), 683–688.
19. Ben-Haim Y (1996), *Robust Reliability in the Mechanical Sciences*, Springer, Berlin.

20. Ben-Haim Y, and Elishakoff I (1990), *Convex Models of Uncertainty in Applied Mechanics*, Studies Appl. Mech. **25**, Elsevier, Amsterdam.
21. Bernard ES, Coleman R, and Bridge RQ (1999), Measurement and assessment of geometric imperfections in thin-walled panels, *Thin-Walled Struct.* **33**(2), 103–126.
22. Bielewicz E, Górski J, Schmidt R, and Walukiewicz H (1994), Random fields in the limit analysis of elastic-plastic shell structures, *Comput Struct.* **51**(3), 267–275.
23. Bielski J (1992), Influence of geometrical imperfections on buckling pressure and post-buckling behavior of elastic toroidal shells, *Mech. Struct. Machines* **20**(2), 145–154.
24. Bolotin VV (1958), *Statistical Methods in the Nonlinear Theory of Elastic Shells*, Izvestija Academii Nauk SSSR, Otdeleni Tekhnicheskikh Nauk **3** (English translation: NASA, 1962, TTF-**85**, 1–16).
25. Bolotin VV (1969), *Statistical Methods in Structural Mechanics*, Holden-Day Ser. Math. Phys. **7**, Holden-Day, San Francisco.
26. Bolotin VV (1984), *Random Vibrations of Elastic Systems*, Mech. Elastic Stability **7**, Martinus Nijhoff, The Hague.
27. Boyce WE (1961), Buckling of a column with random initial displacement, *J. Aeron. Sci.* **28**(4), 308–320.
28. Britvec SJ (1973), *The Stability of Elastic Systems*, Pergamon, Oxford.
29. Britvec SJ, and Chilver AH (1963), Elastic buckling of rigidly-jointed braced frames, *ASCE J. Eng. Mech. Div.* **89**(6), 217–255.
30. Bucher CG, and Bourgund U (1990), A fast and efficient response surface approach for structural reliability problems, *Struct. Safety* **7**(1), 57–66.
31. Budiansky B (1974), *Theory of Buckling and Postbuckling Behavior of Elastic Structures*, Advances Appl. Mech. **14**, C Yiu (ed), Academic Press, New York, 1–65.
32. Budiansky B, and Hutchinson JW (1964), Dynamic buckling of imperfection-sensitive structures, *Proc. Eleventh Int. Congress Appl. Mech.*, H Görtler (ed), Munich, 636–651.
33. Byskov E, and Hutchinson JW (1977), Mode interaction in axially stiffened cylindrical shells, *AIAA J.* **15**(7), 941–948.
34. Casciaro R, Salerno G, and Lanzo AD (1992), Finite element asymptotic analysis of slender elastic structures: A simple approach, *Int. J. Numer. Methods Eng.* **35**, 1397–1426.
35. Cederbaum G, and Arbocz J (1996), Reliability of shells via Koiter formulas, *Thin-Walled Struct.* **24**(2), 173–187.
36. Chilver AH (1967), Coupled modes of elastic buckling, *J. Mech. Phys. Solids* **15**(1), 15–28.
37. Chryssanthopoulos MK, Baker MJ, and Dowling PJ (1991), Statistical analysis of imperfections in stiffened cylinders, *ASCE J. Struct. Eng. Div.* **117**(7), 1979–1997.
38. Chryssanthopoulos MK, Baker MJ, and Dowling PJ (1991), Imperfection modeling for buckling analysis of stiffened cylinders, *ASCE J. Struct. Eng. Div.* **117**(7), 1998–2017.
39. Chryssanthopoulos MK, Giavotto V, and Poggi C (1995), Characterization of manufacturing effects for buckling-sensitive composite cylinders, *Composites Manuf.* **6**(2), 93–101.
40. Chryssanthopoulos MK, and Poggi C (1995), Stochastic imperfection modeling in shell buckling studies, *Thin-Walled Struct.* **23**(1-4), 179–200.
41. Croll JGA (1981), Lower bounds elasto-plastic buckling of cylinders, *Proc. Inst. Civ. Eng.* **2**(71), 235–261.
42. Croll JGA, and Batista RC (1981), Explicit lower bounds for the buckling of axially loaded cylinders, *Int. J. Mech. Sci.* **23**(6), 331–343.

43. Dancy R, and Jacobs D (1988), The initial imperfection data bank at the Delft University of Technology: Part II, *Tech. Rep.* LR-**559**, Dept. Aeron. Eng., Delft Univ. Tech., Delft.
44. Day W, Karwowski AJ, and Papanicolaou GC (1989), Buckling of randomly imperfect beams, *Acta Applicandae Mathematicae* **17**(3), 269–286.
45. de Vries J (2001), Imperfection Database, *CNES Conf.*, 3rd European Conf. Launcher Tech., Strasbourg, 323–332.
46. Deml M, and Wunderlich W (1997), Direct evaluation of the 'worst' imperfection shape in shell buckling, *Comput. Methods Appl. Mech. Eng.* **149**(1-4), 201–222.
47. der Kiureghian A (1985), Finite element methods in structural safety studies, *Proc. Symp. Struct. Safety Studies, ASCE*, Denver, 40–52.
48. Donnel LH (1934), A new theory for the buckling of thin cylinders under axial compression and bending, *Trans. Am. Soc. Mech. Eng.* **56**, 795–806.
49. Edlund BLO, and Leopoldson, ULC (1975), Computer simulation of the scatter in steel member strength, *Comput. & Struct.* **5**(4), 209–224.
50. Elishakoff I (1978), Impact buckling of thin bar via Monte Carlo method, *ASME J. Appl. Mech.* **45**(3), 586–590.
51. Elishakoff I (1979), Buckling of a stochastically imperfect finite column on a nonlinear elastic foundation: A reliability study, *ASME J. Appl. Mech.* **46**(2), 411–416.
52. Elishakoff I (1983), *Probabilistic Methods in the Theory of Structures*, Wiley, New York.
53. Elishakoff I (1988), Stochastic simulation of an initial imperfection data bank for isotropic shells with general imperfections, *Buckling of Struct.*, I Elishakoff (ed), Elsevier, Amsterdam.
54. Elishakoff I (1998), How to introduce initial-imperfection sensitivity concept into design 2, *M Stein Vol., NASA CP*, 206280.
55. Elishakoff I (2000), Possible limitations of probabilistic methods in engineering, *Appl. Mech. Rev.* **53**(2), 19–36.
56. Elishakoff I, and Arbocz J (1982), Reliability of axially compressed cylindrical shells with random axisymmetric imperfections, *Int. J. Solids Struct.* **18**(7), 563–585.
57. Elishakoff I, and Arbocz J (1985), Reliability of axially compressed cylindrical shells with general nonsymmetric imperfections, *ASME J. Appl. Mech.* **52**(1), 122–128.
58. Elishakoff I, Li YW, and Starnes JH Jr (1994), A deterministic method to predict the effect of unknown-but-bounded elastic moduli on the buckling of composite structures, *Comput. Methods Appl. Mech. Eng.* **111**(1-2), 155–167.
59. Elishakoff I, Li YW, and Starnes JH Jr (2001), *Non-Classical Problems in the Theory of Elastic Stability*, Cambridge University Press, Cambridge.
60. Elishakoff I, Lin YK, and Zhu, LP (1994), *Probabilistic and Convex Modelling of Acoustically Excited Structures*, Studies Appl. Mech. **39**, Elsevier, Amsterdam.
61. Elishakoff I, and Ohsaki M (2010), *Optimization and Anti-optimization of Structures under Uncertainty*, Imperial College Press, London.
62. Elishakoff I, and Ren YJ (2003), *Finite Element Methods for Structures with Large Stochastic Variations*, Oxford University Press, Oxford.
63. Elishakoff I, van Manen S, Vermeulen PG, and Arbocz J (1987), First-order second-moment analysis of the buckling of shells with random imperfections, *AIAA J.* **25**(8), 1113–1117.
64. Faravelli L (1989), Response surface approach for reliability analysis, *ASCE J. Eng. Mech. Div.* **115**(2), 2763–2781.
65. Flügge W (1932), Die Stabilität der Kreiszylinderschale, *Ing.-Arch.* **3**, 463.

66. Fraser WB, and Budiansky B (1969), The buckling of a column with random initial deflections, *ASME J. Appl. Mech.* **36**(2), 233–240.
67. Godoy LA (2000), *Theory of Elastic Stability: Analysis and Sensitivity*, Taylor and Francis, Philadelphia.
68. Gomes HM, and Awruch AM (2004), Comparison of response surface and neural network with other methods for structural reliability analysis, *Struct. Safety* **26**(1), 49–67.
69. Gusic G, Combescure A, and Jullien JF (2000), Influence of circumferential thickness variations on buckling of cylindrical shells under external pressure, *Comput. & Struct.* **74**(4), 461–477.
70. Haftka RT, Mallet RH, and Nachbar W (1971), Adaption of Koiter's method to finite element analysis of snap-through buckling behavior, *Int. J. Solids Struct.* **7**(10), 1427–1445.
71. Haldar A, and Mahadevan S (2000), *Probability, Reliability and Statistical Methods in Engineering Design*, Wiley, New York.
72. Hansen JS (1975), Influence of general imperfections in axially loaded cylindrical shells, *Int. J. Solids Struct.* **11**(11), 1223–1233.
73. Hansen JS (1977), General random imperfections in the buckling of axially loaded cylindrical shells, *AIAA J.* **15**(9), 1250–1256.
74. Hansen JS, and Roorda J (1974), On a probabilistic stability theory for imperfection sensitive structures, *Int. J. Solids Struct.* **10**(3), 341–359.
75. Haug EJ, Choi KK, and Komkov V (1986), *Design Sensitivity Analysis of Structural Systems*, Academic Press, Orlando FL.
76. Hill R, and Hutchinson JW (1975), Bifurcation phenomena in the plane tension test, *J. Mech. Phys. Solids* **23**(4-5), 239–264.
77. Ho D (1972), The influence of imperfections on systems with coincident buckling loads, *Int. J. Non-Linear Mech.* **7**, 311–321.
78. Ho D (1974), Buckling load of non-linear systems with multiple eigenvalues, *Int. J. Solids Struct.* **10**(11), 1315–1330.
79. Huseyin K (1975), *Non-Linear Theory of Elastic Stability*, Noordhoff, Leyden.
80. Hutchinson JW (1967), Imperfection sensitivity of externally pressurized spherical shells, *ASME J. Appl. Mech.* **34**(1), 49–55.
81. Hutchinson JW, and Amazigo C (1967), Imperfection-sensitivity of eccentrically stiffened cylindrical shells, *AIAA J.* **5**(3), 392–401.
82. Hutchinson JW, and Koiter WT (1970), Postbuckling theory, *Appl. Mech. Rev.* **23**(12), 1353–1366.
83. Hutchinson JW, and Miles JP (1974), Bifurcation analysis of the onset of necking in an elastic/plastic cylinder under uniaxial tension, *J. Mech. Phys. Solids* **22**(1), 61–71.
84. Ikeda K, Chida T, and Yanagisawa E (1997), Imperfection sensitive strength variation of soil specimens, *J. Mech. Phys. Solids* **45**(2), 293–315.
85. Ikeda K, and Goto S (1993), Imperfection sensitivity for size effect of granular materials, *Soils & Foundations* **33**(2), 157-170.
86. Ikeda K, Kitada T, Matsumura M, and Yamakawa Y (2007), Imperfection sensitivity of ultimate buckling strength of elastic–plastic square plates under compression, *Int. J. Non-Linear Mech.* **42**, 529–541.
87. Ikeda K, Maruyama K, Ishida H, and Kagawa S (1997), Bifurcation in compressive behavior of concrete, *ACI Mater. J.* **94**(6), 484-491.
88. Ikeda K, Murakami S, Saiki I, Sano I, and Oguma N (2001), Image simulation of uniform materials subjected to recursive bifurcation, *Int. J. Eng. Sci.* **39**(17), 1963–1999.

89. Ikeda K, and Murota K (1990), Critical initial imperfection of structures, *Int. J. Solids Struct.* **26**(8), 865–886.
90. Ikeda K, and Murota K (1990), Computation of critical initial imperfection of truss structures, *ASCE J. Eng. Mech. Div.* **116**(10), 2101–2117.
91. Ikeda K, and Murota K (1991), Random initial imperfections of structures, *Int. J. Solids Struct.* **28**(8), 1003–1021.
92. Ikeda K, and Murota K (1993), Statistics of normally distributed initial imperfections, *Int. J. Solids Struct.* **30**(18), 2445–2467.
93. Ikeda K, and Murota K (1996), Bifurcation as sources of uncertainty in soil shearing behavior, *Soils & Foundations* **36**(1), 73–84.
94. Ikeda K, and Murota K (1997), Recursive bifurcation as sources of complexity in soil shearing behavior, *Soils & Foundations* **37**(3), 17–29.
95. Ikeda K, and Murota K (1999), Systematic description of imperfect bifurcation behavior of symmetric systems, *Int. J. Solids Struct.* **36**(11), 1561–1596.
96. Ikeda K, Murota K (2007), Asymptotic and probabilistic approach to buckling of structures and materials, *Appl. Mech. Rev.* **61**(4), 040801.
97. Ikeda K, Murota K, and Elishakoff I (1996), Reliability of structures subject to normally distributed initial imperfections, *Comput. & Struct.* **59**(3), 463–469.
98. Ikeda K, Murota K, and Nakano, M. (1994), Echelon modes in uniform materials, *Int. J. Solids Struct.* **31**(19), 2709–2733.
99. Ikeda K, Murota K, Yamakawa Y, and Yanagisawa E (1997), Mode switching and recursive bifurcation in granular materials, *J. Mech. Phys. Solids* **45**(11–12), 1929–1953.
100. Ikeda K, and Ohsaki M (2006), Generalized sensitivity and probabilistic analysis of buckling loads of structures, *Int. J. Non-linear Mech.* **42**(5), 733–743.
101. Ikeda K, Ohsaki M, and Kanno Y (2005), Imperfection sensitivity of hilltop branching points of systems with dihedral group symmetry, *Int. J. Non-linear Mech.* **40**, 755–774.
102. Ikeda K, Ohsaki M, Sudo K, and Kitada T (2009), Probabilistic analysis of buckling loads of structures via extended Koiter law, *Struct. Eng. Mech.* **32**(1), 167–178.
103. Ikeda K, Oide K, and Terada K (2002), Imperfection sensitive variation of critical loads at hilltop bifurcation point,. *Int. J. Eng. Sci.* **40**(7), 743–772.
104. Ikeda K, Okazawa S, Terada K, Noguchi H, and Usami T (2001), Recursive bifurcation of tensile steel specimens, *Int. J. Eng. Sci.* **39**(17), 1913–1934.
105. Ikeda K, Providéncia P, and Hunt GW (1993), Multiple equilibria for unlinked and weakly-linked cellular forms, *Int. J. Solids Struct.* **30**(3), 371–384.
106. Ikeda K, Yamakawa Y, and Tsutsumi S (2003), Simulation and interpretation of diffuse mode bifurcation of elastoplastic solids, *J. Mech. Phys. Solids* **51**(9), 1649-1673.
107. Kaveh A (2013), *Optimal Analysis of Structures by Concepts of Symmetry and Regularity*, Springer, Wien.
108. Kim S-H, and Na S-W (1997), Response surface method using vector projected sampling points, *Struct. Safety* **19**(1), 3–19.
109. Koiter WT (1945), *On the Stability of Elastic Equilibrium*, Dissertation, Delft Univ. Tech. (English translation: NASA technical translation F **10: 833**, 1967).
110. Koiter WT (1963), The effects of axisymmetric imperfections on the buckling of cylindrical shells under axial compression, *Proc. Kon. Ned. Akad. Wet., Ser. B* **66**, 265–279.
111. Koiter WT, Elishakoff I, Li YW, and Starnes JH Jr (1994), Buckling of an axially compressed cylindrical shells of variable thickness, *Int. J. Solids Struct.* **31**(6), 797–805.
112. Koiter WT, and Kuiken GDC (1971), The interaction between local buckling and overall buckling in the behavior of built-up columns, *Rep.* WTHD-**23**, Delft Univ. Tech., Delft.

113. Koiter WT, and Skaloud M (1963), Interventions, comportment postcritique des plaques utilisees en construction metallique, *Mém. Soc. Sci. Liege*, 5 Serte, **8**(5), 64–68, 103–104.
114. Królak M, Kołakowski Z, and Kotełko M (2001), Influence of load-non-uniformity and eccentricity on the stability and load carrying capacity of orthotropic tubular columns of regular hexagonal cross-section, *Thin-Walled Struct.* **39**(6), 483–498.
115. Lagaros ND, and Papadopoulos V (2006), Optimum design of shell structures with random geometric, material and thickness imperfections, *Int. J. Solids Struct.* **43**(22-23), 6948–6964.
116. Li L-Y (1990), Influence of loading imperfections on the stability of an axially compressed cylindrical shell, *Thin-Walled Struct.* **10**(3), 215–220.
117. Li YW, Elishakoff I, and Starnes JH Jr (1995), Axial buckling of composite cylindrical shells with periodic thickness variation, *Comput. & Struct.* **56**(1), 65–74.
118. Li YW, Elishakoff I, Starnes JH Jr, and Shinozuka M (1995), Nonlinear buckling of a structure with random imperfection and random axial compression by a conditional simulation technique, *Comput. & Struct.* **56**(1), 59–64.
119. Lin X, and Teng JG (2003), Iterative Fourier decomposition of imperfection measurements at non-uniformly distributed sampling points, *Thin-Walled Struct.* **41**(10), 901–924.
120. Lindberg HE (1992), Convex models of uncertain imperfection control in multimode dynamic buckling, *ASME J. Appl. Mech.* **59**(4), 937–945.
121. Liu WK, Belytschko T, and Mani A (1986), Probabilistic finite elements for nonlinear structural dynamics, *Comput. Methods Appl. Mech. Eng.* **56**(1), 61–81.
122. Lombardi M, and Haftka RT (1998), Anti-optimization technique for strutural design under load uncertainties, *Comput. Methods Appl. Mech. Eng.* **157**(1-2), 19–31.
123. Megson THG, and Hallak G (1992), Measurement of the geometric initial imperfections in diaphragms, *Thin-Walled Struct.* **14**(5), 381–394.
124. Murota K, and Ikeda K (1991), Critical imperfection of symmetric structures, *SIAM J. Appl. Math.* **51**(5), 1222–1254.
125. Murota K, and Ikeda K (1992), On random imperfections for structures of regular-polygonal symmetry, *SIAM J. Appl. Math.* **52**(6), 1780–1803.
126. Murota K, Ikeda K, and Terada K (1999), Bifurcation mechanism underlying echelon mode formation, *Comp. Methods Appl. Mech. Eng.* **170**(3–4), 423–448.
127. Myers RH, and Montgomery DC (1995), *Response Surface Methodology: Process and Product Optimization using Design Experiments*, Wiley, New York.
128. Nakagiri S, and Hisada T (1980), A note on stochastic finite element method (Part 1): Variation of stress and strain caused by shape fluctuation, *Monthly J. Inst. Industrial Sci., Univ. Tokyo* **32**, 39–42.
129. NASA (1968), Buckling of thin-walled circular cylinders, *NASA* SP **8007**.
130. Needleman A (1972), A numerical study of necking in circular cylindrical bars, *J. Mech. Phys. Solids* **20**(2), 111–127.
131. Nishino F, and Hartono W (1989), Influential mode of imperfection on carrying capacity of structures, *ASCE J. Eng. Mech. Div.* **115**(10), 2150–2165.
132. Ohsaki M (2000), Optimization of geometrically non-linear symmetric systems with coincident critical points, *Int. J. Numer. Methods Eng.* **48**(9), 1345–1357.
133. Ohsaki M (2001), Sensitivity analysis and optimization corresponding to a degenerate critical point, *Int. J. Solids Struct.* **38**(19), 4955–4967.

134. Ohsaki M (2003), Sensitivity analysis of an optimized bar-spring model with hill-top branching, *Archive Appl. Mech.* **73**, 241–251.
135. Ohsaki M, and Ikeda K (2006), Imperfection sensitivity analysis of hill-top branching with many symmetric bifurcation points, *Int. J. Solids Struct.* **43**(16), 4704–4719.
136. Ohsaki M, and Ikeda M (2007), *Stability and Optimization of Structures: Generalized Sensitivity Analysis*, Springer, New York.
137. Ohsaki M, and Ikeda K (2009), Imperfection sensitivity of degenerate hilltop branching points, *Int. J. Non-Linear Mech.* **44**(3), 324–336.
138. Ohsaki M, and Nakamura T (1994), Optimum design with imperfection sensitivity coefficients for limit point loads, *Struct. Opt.* **8**, 131–137.
139. Okazawa S, Oide K, Ikeda K, and Terada K (2002), Imperfection sensitivity and probabilistic variation of tensile strength of steel members, *Int. J. Solids Struct.* **39**(6), 1651–1671.
140. Palassopoulos GV (1973), On the buckling of axially compressed thin cylindrical shells, *J. Struct. Mech.* **2**(3), 177–193.
141. Palassopoulos GV (1989), Optimization of imperfection-sensitive structures, *ASCE J. Eng. Mech.* **115**(8), 1663–1682.
142. Palassopoulos GV (1991), On the optimization of imperfection-sensitive structures with buckling constraints, *Eng. Opt.* **17**, 219–227.
143. Palassopoulos GV (1991), Reliability-based design of imperfection sensitive structures, *ASCE J. Eng. Mech. Div.* **117**(6), 1220–1240.
144. Palassopoulos GV (1992), Response variability of structures subjected to bifurcation buckling, *ASCE J. Eng. Mech. Div.* **118**(6), 1164–1183.
145. Palassopoulos GV (1993), New approach to buckling of imperfection-sensitive structures, *ASCE J. Eng. Mech.* **119**(4), 850–869.
146. Palassopoulos GV, and Shinozuka M (1973), On the elastic stability of thin shells, *J. Struct. Mech.* **1**(4), 439–449.
147. Papadimitriou C, Beck JL, and Katafygiotis LS (2001), Updating robust reliability using structural test data, *Prob. Eng. Mech.* **16**(2), 103–113.
148. Peek R (1993), Worst shapes of imperfections for space trusses with multiple global and local buckling modes, *Int. J. Solids Struct.* **30**(16), 2243–2260.
149. Peek R, and Triantafyllidis N (1992), Worst shapes of imperfections for space trusses with many simultaneously buckling modes, *Int. J. Solids Struct.* **29**(19), 2385–2402.
150. Pignataro M, Rizzi N, and Luongo A (1991), *Stability, Bifurcation and Postcritical Behaviour of Elastic Structures*, Elsevier, Amsterdam.
151. Pircher M, Berry PA, Ding X, and Bridge RQ (2001), The shape of circumferential weld-induced imperfections in thin-walled steel silos and tanks, *Thin-Walled Struct.* **39**(12), 999–1014.
152. Pircher M, and Wheeler A (2003), The measurement of imperfections in cylindrical thin-walled members, *Thin-Walled Struct.* **41**(5), 419–433.
153. Rajashekhar MR, and Ellingwood BR (1993), A new look at the response surface approach for reliability analysis, *Struct. Safety* **12**(3), 205–220.
154. Reitinger R, and Ramm E (1995), Buckling and imperfection sensitivity in the optimization of shell structures, *Thin-Walled Struct.* **23**(1-4), 159–177.
155. Roorda J (1965), Stability of structures with small imperfections, *ASCE J. Eng. Mech. Div.* **91**, 87–106.
156. Roorda J (1968), On the buckling of symmetric structural systems with first and second order imperfections, *Int. J. Solids Struct.* **4**(12), 1137–1148.

157. Roorda J (1969), Some statistical aspects of the buckling of imperfection-sensitive structures, *J. Mech. Phys. Solids* **17**(2), 111–123.
158. Roorda J (1980), *Buckling of Elastic Structures*, University of Waterloo Press, Waterloo.
159. Roorda J, and Hansen JS (1972), Random buckling behavior in axially loaded cylindrical shells with axisymmetric imperfections, *J. Spacecraft* **9**(2), 88–91.
160. Sadovsky Z (1978), A theoretical approach to the problem of the most dangerous initial deflection shape in stability type structural problems, *Aplikace Mathematiky* **23**(4), 248–266.
161. Salerno G, and Lanzo AD (1997), A nonlinear beam finite element for the post-buckling analysis of plane frames by Koiter's perturbation approach, *Comput. Methods Appl. Mech. Eng.* **146**, 325–349.
162. Salerno G, and Uva G (2006), Ho's theorem in global–local mode interaction of pin-jointed bar structures, *Int. J. Non-Linear Mech.* **41**(3), 359–376.
163. Samuels P (1980), Bifurcation and limit point instability of dual eigenvalue third order system, *Int. J. Solids Struct.* **16**(8), 743–756.
164. Schenk CA, Schuëller GI, and Arbocz J, (2000), Buckling analysis of cylindrical shells with random imperfections, *Int. Conf. Monte Carlo Simulation*, Monte Carlo, 18–21.
165. Schuëller GI (ed) (1997), A state-of-the-art report on computational stochastic mechanics, *Prob. Eng. Mech.* **12**(4), 197–321.
166. Scott ND, Harding JE, and Dowling PJ (1987), Fabrication of small scale stiffened cylindrical shells, *J. Strain Anal.* **22**(2), 97–106.
167. Sewell MJ (1965), The static perturbation technique in buckling problems, *J. Mech. Phys. Solids* **13**(4), 247–265.
168. Shinozuka M (1987), Structural response variability, *ASCE J. Eng. Mech. Div.* **113**(6), 825–843.
169. Singer J, Abramovich H, and Yaffe R (1978), Initial imperfection measurements of integrally stringer-stiffened cylindrical shells, *TAE Rep.* **330**, Dept. Aeron. Eng., Technion, Israel Inst. of Tech., Haifa.
170. Spunt L (1971), *Optimum Structural Design*, Prentice-Hall, Englewood Cliffs NJ.
171. Supple WJ (1967), Coupled branching configurations in the elastic buckling of symmetric structural system, *Int. J. Mech. Sci.* **9**(2), 97–112.
172. Tanaka R, Saiki I, and Ikeda K (2002), Group-theoretic bifurcation mechanism for pattern formation in three-dimensional uniform materials, *Int. J. Bifurcation Chaos* **12**(12), 2767–2797.
173. Tennyson RC, Muggeridge DB, and Caswell RD (1971), Buckling of circular cylindrical shells having axisymmetric imperfection distributions, *AIAA J.* **9**(5), 924–930.
174. Thoft-Christensen P, and Baker MJ (1982), *Structural Reliability Theory and its Applications*, Springer, Berlin.
175. Thompson JMT (1965), Discrete branching points in the general theory of elastic stability, *J. Mech. Phys. Solids* **13**(5), 295–310.
176. Thompson JMT (1967), Towards a general statistical theory of imperfection-sensitivity in elastic post-buckling, *J. Mech. Phys. Solids* **15**(6), 413–417.
177. Thompson JMT (1982), *Instabilities and Catastrophes in Science and Engineering*, Wiley, Chichester.
178. Thompson JMT, and Hunt GW (1973), *A General Theory of Elastic Stability*, Wiley, New York.
179. Thompson JMT, and Hunt GW (1974), Dangers of structural optimization, *Eng. Opt.* **1**, 99–110.

180. Thompson JMT, and Lewis GM (1972), On the optimum design of thin-walled compression members, *J. Mech. Phys. Solids* **20**(2), 101–109.
181. Thompson JMT, and Schorrock PA (1975), Bifurcation instability of an atomic lattice, *J. Mech. Phys. Solids* **23**(1), 21–37.
182. Thompson JMT, and Supple WJ (1973), Erosion of optimum designs by compound branching phenomena, *J. Mech. Phys. Solids* **21**(3), 135–144.
183. Trendafilova I, and Ivanova J (1995), Loss of stability of thin, elastic, strongly convex shells of revolution with initial imperfections, subjected to uniform pressure: A probabilistic approach, *Thin-Walled Struct.* **23**(1-4), 201–214.
184. Triantafyllidis N, and Peek R (1992), On stability and the worst imperfection shape in solids with nearly simultaneous eigenmodes, *Int. J. Solids Struct.* **29**(18), 2281–2299.
185. Turčić F (1991), Resistance of axially compressed cylindrical shells determined for measured geometrical imperfections, *J. Construct. Steel Res.* **19**(3), 225–234.
186. Tvergaard V (1973), Influence of post-buckling behaviour on optimum design of stiffened panels, *Int. J. Solids Struct.* **9**(12), 1519–1534.
187. Tvergaard V (1976), Effect of thickness inhomogeneities in internally pressurized elastic-plastic spherical shells, *J. Mech. Phys. Solids* **24**(5), 291–304.
188. van Slooten RA, and Soong TT (1972), Buckling of a long, axially compressed, thin cylindrical shell with random initial imperfections, *ASME J. Appl. Mech.* **39**(4), 1066–1071.
189. Venter G, Haftka RT, and Starnes JH Jr (1998), Construction of response surface approximation for design optimization, *AIAA J.* **36**(12), 2242–2249.
190. von Kármán T, and Tsien H-S (1939), The buckling of spherical shells by external pressure, *J. Aeron. Sci.* **1**, 43–50.
191. Wang F-Y (1990), Monte Carlo analysis of nonlinear vibration of rectangular plates with random geometric imperfections, *Int. J. Solids Struct.* **26**(1), 99–109.
192. Weibull W (1939), A statistical theory on the strength of materials, *Roy. Swedish Inst. Eng. Research, Proc.* NR-**151**, 1–52.
193. Weingarten VI, Morgan EJ, and Seide P (1965), Elastic stability of thin-walled cylindrical and conical shells under axial compression, *AIAA J.* **3**(3), 500–505.
194. Zheng Y, and Das PK (2000), Improved response surface method and its application to stiffened plate reliability analysis, *Eng. Struct.* **22**(5), 544–551.
195. Zingales M, and Elishakoff I (2000), Anti-optimization versus probability in an applied mechanics problem: Vector uncertainty, *ASME J. Appl. Mech.* **67**(3), 472–484.

A Answers to Problems

Chapter 1

Problem 1.1: (1) From (1.4), the Jacobian $J(u,f) = \partial^2 U/\partial u^2 = k$ is always positive. Accordingly, this system is stable. (2) For a solution (u_*, f_*) satisfying the governing equation, we have $f_* = ku_*$. Then (1.4) gives

$$U(u, f_*) = \frac{ku^2}{2} - ku_* u = \frac{k}{2}\left[(u-u_*)^2 - {u_*}^2\right].$$

Thus this potential U has a local minimum at $u = u_*$ for a constant value f_* of f; accordingly, this system is stable.

Problem 1.2: (1) For the nonlinear spring, we have

$$U(u,f) = k\left(\frac{1}{2}u^2 - \frac{1}{3}u^3\right) - fu, \quad F(u,f) = k(u - u^2 - f), \quad J(u,f) = k(1-2u).$$

(2) This system is stable for $u < 1/2$ and unstable for $u > 1/2$, where $(u_c, f_c) = (1/2, k/4)$ is the bifurcation point.

Problem 1.3: For the rigid-bar-spring system in Fig. 1.10, we have

$$\begin{aligned}
U(u,f) &= \frac{1}{2}k_1(2L\sin u)^2 + \frac{1}{2}k_2(L\sin u)^2 + \frac{1}{2}k_\theta u^2 - f\cdot 2L(1-\cos u) \\
&= kL^2\left(7\sin^2 u + \frac{1}{2}u^2\right) - 2Lf(1-\cos u), \\
F(u,f) &= \frac{\partial U}{\partial u} = kL^2(7\sin 2u + u) - 2Lf\sin u, \\
J(u,f) &= \frac{\partial F}{\partial u} = kL^2(14\cos 2u + 1) - 2Lf\cos u.
\end{aligned}$$

Problem 1.4: $F(x,f) = \partial U/\partial x = 3x^2 + 2xf = x(3x+2f) = 0.$

Problem 1.5: $F(x,f) = \partial U/\partial x = 3x^2 - 2xf - f^2 = (3x+f)(x-f) = 0.$

Chapter 2

Problem 2.1:

(1) $\boldsymbol{F}(x,y) = (\partial U/\partial x, \partial U/\partial y)^\top = (4x^3 + 2xy^2, 2x^2y + 4y^3)^\top.$

(2) $\boldsymbol{F}(x,y,z) = (4x^3 + 4x\cos(x^2)\sin(y^2), 4y\sin(x^2)\cos(y^2), -2z\sin(z^2))^\top.$

(3) $\boldsymbol{F}(x,y,f) = (4x^3 + 2xf, 2y + f - 1)^\top.$

DOI: 10.1201/9781003112365-A

Problem 2.2: (1) The governing equation is $\left(\frac{\partial U}{\partial x}, \frac{\partial U}{\partial y}\right)^\top = (2x-2-y, 2y-x)^\top = \mathbf{0}$ and its solution is $x = 4/3$, $y = 2/3$. (2) The governing equation is

$$\left(\frac{\partial U}{\partial x}, \frac{\partial U}{\partial y}, \frac{\partial U}{\partial z}\right)^\top = (2(x+y+z-2), 2(x-y+z), 2(x+y-z))^\top = \mathbf{0}$$

and its solution is $x = 0$, $y = z = 1$.

Problem 2.3: The governing equation is $\left(\frac{\partial U}{\partial x}, \frac{\partial U}{\partial y}\right)^\top = (3x^2 - 2xf, f - 2y) = \mathbf{0}$. By eliminating f, we can obtain the x-y curve, $x(3x-4y) = 0$.

Problem 2.4: (1) The governing equation is

$$\frac{\partial U}{\partial x} = 4x^3 + 4xy^2 + 4x + y = 0, \qquad \frac{\partial U}{\partial y} = 4y^3 + 4x^2 y + 2y + x = 0,$$

for which the origin $(0,0)$ is a solution. We have

$$J(0,0) = \begin{pmatrix} 4 & 1 \\ 1 & 2 \end{pmatrix}, \qquad \det\begin{pmatrix} 4-\lambda & 1 \\ 1 & 2-\lambda \end{pmatrix} = \lambda^2 - 6\lambda + 7$$

and the eigenvalues of $J(0,0)$ are $\lambda = 3 \pm \sqrt{2} > 0$. Accordingly, the origin is stable. (2) The governing equation is

$$\frac{\partial U}{\partial x} = 4x^3 + 2x + 2xy^2 + 3y = 0, \qquad \frac{\partial U}{\partial y} = 2x^2 y + 3x + 2y = 0,$$

for which the origin $(0,0)$ is a solution. The Jacobian matrix at the origin $J(0,0) = \begin{pmatrix} 2 & 3 \\ 3 & 2 \end{pmatrix}$ has eigenvalues $\lambda = -1, 5$, one of which is negative. Accordingly, the origin is unstable. (3) The governing equation is

$$\frac{\partial U}{\partial x} = 4x^3 + 2x + 2y = 0, \qquad \frac{\partial U}{\partial y} = 2y + 2x + 2z = 0, \qquad \frac{\partial U}{\partial z} = 2z + 2y = 0,$$

for which the origin $(0,0,0)$ is a solution. The Jacobian matrix at the origin

$$J(0,0,0) = \begin{pmatrix} 2 & 2 & 0 \\ 2 & 2 & 2 \\ 0 & 2 & 2 \end{pmatrix}$$

has eigenvalues $\lambda = 2, 2 \pm 2\sqrt{2}$. Accordingly, the origin is unstable.

Problem 2.5: The governing equation is

$$\frac{\partial U}{\partial x} = 2f(x + 2x^3) + 2xy^2 + 3y = 0, \qquad \frac{\partial U}{\partial y} = 2fy + 2x^2 y + 3x,$$

for which the origin $(0,0)$ is a solution. The Jacobian matrix $J(0,0,f) = \begin{pmatrix} 2f & 3 \\ 3 & 2f \end{pmatrix}$ has eigenvalues $\lambda = 2f \pm 3$. Hence the origin is stable for $f > 3/2$ and is unstable for $f < 3/2$. The potential for $f = 3/2$,

$$U(x,y,3/2) = \frac{3}{2}(x+y)^2 + \frac{3}{2}x^4 + x^2y^2,$$

has a local minimum at the origin $(x,y) = (0,0)$. Accordingly, the origin is stable for $f = 3/2$.

Problem 2.6: (1) The reciprocity holds from $\partial F_x/\partial y = \partial F_y/\partial x = 8xy$. The integration of F_x with respect to x leads to a potential

$$U(x,y) = \int F_x \mathrm{d}x = 2x^2y^2 + x^2 + C(y) \tag{A.1}$$

containing a function $C(y)$ independent of x. The partial differentiation of this potential function with respect to y leads to $F_y = \partial U/\partial y = 4x^2y + C'(y)$. By comparing this equation and $F_y = 4x^2y + 2y$, we obtain $C'(y) = 2y$, that is, $C(y) = y^2 + c$. Then from (A.1), we have $U(x,y) = 2x^2y^2 + x^2 + y^2 + c$. (2) The reciprocity holds from

$$\frac{\partial F_x}{\partial y} = \frac{\partial F_y}{\partial x} = 4xyz^2, \qquad \frac{\partial F_y}{\partial z} = \frac{\partial F_z}{\partial y} = 4x^2yz, \qquad \frac{\partial F_z}{\partial x} = \frac{\partial F_x}{\partial z} = 4xy^2z.$$

The integration of F_x with respect to x leads to a potential

$$U(x,y,z) = \int F_x \mathrm{d}x = x^4 + x^2y^2z^2 + C(y,z) \tag{A.2}$$

containing a function $C(y,z)$ independent of x. The partial differentiation of this potential function with respect to y leads to $F_y = \partial U/\partial y = 2x^2yz^2 + \partial C/\partial y$. By comparing this equation and $F_y = 2x^2yz^2$, we obtain $\partial C/\partial y = 0$, that is, $C(y,z) = C(z)$. Substituting this equation into (A.2), partially differentiating with respect to z, and comparing with $F_z = 2x^2y^2z + 4z^3$, we obtain $C'(z) = 4z^3$, that is, $C(z) = z^4 + c$. Then from (A.2), we have $U(x,y,z) = x^4 + x^2y^2z^2 + z^4 + c$.

Problem 2.7: For the structural system in Fig. 2.10(a), we have

$$\begin{aligned} U(u,f) &= \frac{1}{2}k\{L(1-\cos u) - L\sin u\}^2 + \frac{1}{2}k\{L(1-\cos u) + L\sin u\}^2 \\ &\quad + \frac{1}{2}2k(2L\sin u)^2 + \frac{1}{2}k_\theta u^2 - f\,2L(1-\cos u) \\ &= kL^2\left(2 - 2\cos u + 4\sin^2 u + \frac{u^2}{2}\right) - 2Lf(1-\cos u), \\ F(u,f) &= kL^2(2\sin u + 4\sin 2u + u) - 2Lf\sin u. \end{aligned}$$

The bifurcated solution is $f = kL[1 + 4\cos u + u/(2\sin u)]$. By setting $u \to 0$ on this bifurcated solution, we can obtain the bifurcation load $f_c = 11kL/2$.

Problem 2.8: For the structural system in Fig. 2.10(b), we have

$$
\begin{aligned}
U(u,f) &= \frac{1}{2}k_\theta u^2 + \frac{1}{2}2k(2L\sin u + L\cos u - L)^2 \\
&\quad + \frac{1}{2}k[\{2L(1-\cos u)+L\sin u\}^2 + \{2L(1-\cos u) - L\sin u\}^2] \\
&\quad - f[\{2L(1-\cos u)+L\sin u\} + \{2L(1-\cos u) - L\sin u\}] \\
&= kL^2\left[\frac{1}{2}u^2 + (2\sin u + \cos u - 1)^2 + 4(1-\cos u)^2 + \sin^2 u\right] \\
&\quad - 4Lf(1-\cos u), \\
&= kL^2\left[\frac{1}{2}u^2 + 10(1-\cos u) + 2\sin 2u - 4\sin u\right] - 4Lf(1-\cos u),
\end{aligned}
$$

$$
F(u,f) = kL^2[u + 10\sin u + 4\cos 2u - 4\cos u] - 4Lf\sin u,
$$

$$
\text{bifurcated solution: } f = \frac{kL}{4}\left[\frac{u}{\sin u} + 10 + 4\frac{\cos 2u - \cos u}{\sin u}\right].
$$

By setting $u \to 0$ on this bifurcated solution, we can obtain the bifurcation load $f_c = 11kL/4$.

Problem 2.9: The potential is given by

$$
U(u_1,u_2,f) = \frac{k}{2}\left({u_1}^2 + {u_2}^2\right) - fkL\left(2L - \sqrt{L^2 - {u_1}^2} - \sqrt{L^2 - (u_1-u_2)^2}\right).
$$

The governing equation

$$
\begin{aligned}
\frac{\partial U}{\partial u_1}(u_1,u_2,f) &= ku_1 + fkL\left(\frac{-u_1}{\sqrt{L^2-{u_1}^2}} + \frac{u_2-u_1}{\sqrt{L^2-(u_1-u_2)^2}}\right) = 0, \\
\frac{\partial U}{\partial u_2}(u_1,u_2,f) &= ku_2 + fkL\frac{u_1-u_2}{\sqrt{L^2-(u_1-u_2)^2}} = 0
\end{aligned}
$$

has the trivial solution path $u_1 = u_2 = 0$ and the Jacobian matrix on this path is given by

$$
J(0,0,f) = k\begin{pmatrix} 1-2f & f \\ f & 1-f \end{pmatrix}.
$$

Then the buckling condition is given by $\det J(0,0,f) = k^2(f^2 - 3f + 1) = 0$ and the buckling loads are evaluated to $f_c = (3 \pm \sqrt{5})/2$, which are identical with those obtained in Section 2.8.1.

Problem 2.10: (1) For $n = 2$ in (2.40), we obtain

$$
J(\mathbf{0},f) = k\begin{pmatrix} 1-2f & f \\ f & 1-2f \end{pmatrix}.
$$

The buckling condition

$$\det J(\mathbf{0},f) = k^2\,[(1-2f)^2 - f^2] = k^2(3f-1)(f-1) = 0$$

gives the buckling loads $f_A = 1/3$ and $f_B = 1$, and associated buckling modes are $\boldsymbol{\eta}_A = (1,-1)^\top$ and $\boldsymbol{\eta}_B = (1,1)^\top$. (2) For $n = 3$ in (2.40), we obtain

$$J(\mathbf{0},f) = k\begin{pmatrix} 1-2f & f & 0 \\ f & 1-2f & f \\ 0 & f & 1-2f \end{pmatrix}.$$

The buckling condition

$$\det J(\mathbf{0},f) = k^3(1-2f)(2f^2-4f+1) = 0$$

gives the buckling loads

$$f_A = 1-\sqrt{2}/2 \approx 0.293, \qquad f_B = 0.5, \qquad f_C = 1+\sqrt{2}/2 \approx 1.707,$$

and the associated (normalized) buckling modes are

$$\boldsymbol{\eta}_A = \frac{1}{2}\begin{pmatrix} 1 \\ -\sqrt{2} \\ 1 \end{pmatrix}, \qquad \boldsymbol{\eta}_B = \frac{1}{\sqrt{2}}\begin{pmatrix} 1 \\ 0 \\ -1 \end{pmatrix}, \qquad \boldsymbol{\eta}_C = \frac{1}{2}\begin{pmatrix} 1 \\ \sqrt{2} \\ 1 \end{pmatrix}.$$

Problem 2.11: The potential is given by

$$U(u_1,u_2,f) = L^2\left[\frac{1}{2}k_1\sin^2 u_1 + \frac{1}{2}k_2(\sin u_1 + \sin u_2)^2 - (2-\cos u_1 - \cos u_2)fk\right].$$

The governing equation

$$\frac{\partial U}{\partial u_1}(u_1,u_2,f) = L^2[(k_1\sin u_1 + k_2(\sin u_1+\sin u_2))\cos u_1 - fk\cdot\sin u_1] = 0,$$
$$\frac{\partial U}{\partial u_2}(u_1,u_2,f) = L^2[k_2(\sin u_1+\sin u_2)\cos u_2 - fk\cdot\sin u_2] = 0$$

has the trivial solution $u_1 = u_2 = 0$ and the Jacobian matrix on this solution path is given by

$$J(0,0,f) = kL^2\begin{pmatrix} (k_1+k_2)/k - f & k_2/k \\ k_2/k & k_2/k - f \end{pmatrix}.$$

Then the buckling condition is given by

$$\det J(0,0,f) = (kL^2)^2\left(f^2 - \frac{k_1+2k_2}{k}f + \frac{k_1k_2}{k^2}\right) = 0. \qquad \text{(A.3)}$$

For $k_1 = 2k$ and $k_2 = k$, the buckling condition becomes

$$\det J(0,0,f) = (kL^2)^2(f^2-4f+2) = 0,$$

and the bifurcation loads are given by $f_c = 2 \pm \sqrt{2}$ and the bifurcation modes are

$$\eta = \frac{1}{\sqrt{4 \pm 2\sqrt{2}}} \begin{pmatrix} 1 \pm \sqrt{2} \\ 1 \end{pmatrix} \quad \text{(the double sign corresponds).}$$

Problem 2.12: For $k_1 = k(1+\varepsilon)$ and $k_2 = k$, the buckling condition in (A.3) becomes

$$\det J(0,0,f) = (kL^2)^2 \left[f^2 - (3+\varepsilon) f + 1 + \varepsilon \right] = 0,$$

from which the minimum bifurcation load is given by

$$\begin{aligned} f_c &= \frac{1}{2} \left[3 + \varepsilon - \sqrt{5} \left(1 + \frac{2\varepsilon}{5} + \frac{\varepsilon^2}{5} \right)^{1/2} \right] \\ &\approx \frac{1}{2} \left[3 + \varepsilon - \sqrt{5} \left(1 + \frac{\varepsilon}{5} \right) \right] \approx \frac{3 - \sqrt{5}}{2} + 0.276\varepsilon. \end{aligned}$$

For $k_1 = k$ and $k_2 = k(1+\varepsilon)$, we conduct a similar analysis to obtain the minimum buckling load

$$f_c \approx \frac{3 - \sqrt{5}}{2} + \varepsilon \left(1 - \frac{2\sqrt{5}}{5} \right) \approx \frac{3 - \sqrt{5}}{2} + 0.106\varepsilon.$$

Hence in the case of an increase of k_1, we have more increase in the minimum buckling load.

Problem 2.13: For $i = 1, 2, 3$, denote by θ_i the rotational angle of the ith member from the left. Then the geometrical condition $\theta_1 + \theta_2 + \theta_3 = 0$ should hold. The potential is evaluated to

$$U = \frac{k_\theta}{2} [(\theta_1 - \theta_2)^2 + (\theta_2 - \theta_3)^2] - fL(3 - \cos\theta_1 - \cos\theta_2 - \cos\theta_3).$$

The elimination of θ_2 from this equation using the geometrical condition $\theta_1 + \theta_2 + \theta_3 = 0$ gives

$$U(\theta_1, \theta_3, f) = \frac{k_\theta}{2} [(2\theta_1 + \theta_3)^2 + (\theta_1 + 2\theta_3)^2] - fL[3 - \cos\theta_1 - \cos(\theta_1 + \theta_3) - \cos\theta_3].$$

The governing equation

$$\begin{aligned} \frac{\partial U}{\partial \theta_1}(\theta_1, \theta_3, f) &= k_\theta(5\theta_1 + 4\theta_3) - fL[\sin\theta_1 + \sin(\theta_1 + \theta_3)] = 0, \\ \frac{\partial U}{\partial \theta_3}(\theta_1, \theta_3, f) &= k_\theta(4\theta_1 + 5\theta_3) - fL[\sin(\theta_1 + \theta_3) + \sin\theta_3] = 0 \end{aligned}$$

has the trivial solution path $\theta_1 = \theta_3 = 0$ and the Jacobian matrix on this path is given by

$$J(0,0,f) = \begin{pmatrix} 5k_\theta - 2fL & 4k_\theta - fL \\ 4k_\theta - fL & 5k_\theta - 2fL \end{pmatrix}.$$

The buckling condition

$$\det J(0,0,f) = 3(k_\theta - fL)(3k_\theta - fL) = 0$$

gives the buckling loads $f_A = k_\theta/L$ and $f_B = 3k_\theta/L$, and the (normalized) buckling modes $\boldsymbol{\eta}_A = \frac{1}{\sqrt{2}}(1,-1)^\top$ and $\boldsymbol{\eta}_B = \frac{1}{\sqrt{2}}(1,1)^\top$.

Chapter 3

Problem 3.1: The load versus displacement curve and the curves of eigenvalues obtained by the displacement control method are depicted in Fig. A.1.

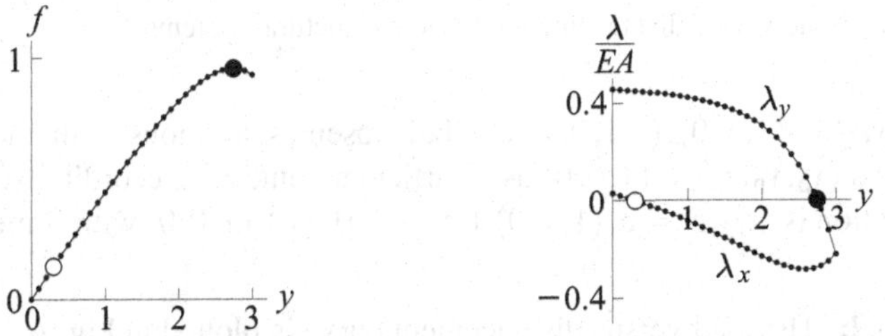

(a) Load versus displacement curve (b) Eigenvalues plotted against y

Figure A.1 Problem 3.1: The path tracing of the truss arch by the displacement control method ($\lambda = \lambda_x$, λ_y in (b))

Problem 3.2: (1) By the buckling analysis by changing the height h, we see that $f_c \geq 0.30$ holds for $h \geq 2.78$. (2) Similarly, we have $h \geq 2.40$. (3) For $h = 2.5$, the buckling loads are $f_c = 0.414$, 0.659, and 0.672; while those for $h = 3$ are $f_c = 0.248$, 0.755, and 0.784. Thus the case of $h = 2.5$ has a larger minimum buckling load of $f_c = 0.414$.

Chapter 4

Problem 4.1: (1) A row transformation is applied to this equation to arrive at

$$\begin{pmatrix} 1 & 0 & 4/5 \\ 0 & 1 & 2/5 \\ 0 & 0 & 0 \end{pmatrix} \begin{pmatrix} x \\ y \\ z \end{pmatrix} = \begin{pmatrix} 9/5 \\ 7/5 \\ 0 \end{pmatrix}.$$

Hence $(x,y,z) = (-4/5,-2/5,1)$ can be chosen as a solution to the homogeneous equation and $(x_p,y_p,z_p) = (9/5,7/5,0)$ as a particular solution. Accordingly, the solution to this equation is $(x,y,z) = c(-4/5,-2/5,1) + (9/5,7/5,0)$ with a parameter c. (2) A row transformation is applied to this equation to arrive at

$$\begin{pmatrix} 1 & -1 & 1 \\ 0 & 0 & 0 \\ 0 & 0 & 0 \end{pmatrix} \begin{pmatrix} x \\ y \\ z \end{pmatrix} = \begin{pmatrix} 1 \\ 0 \\ 0 \end{pmatrix}.$$

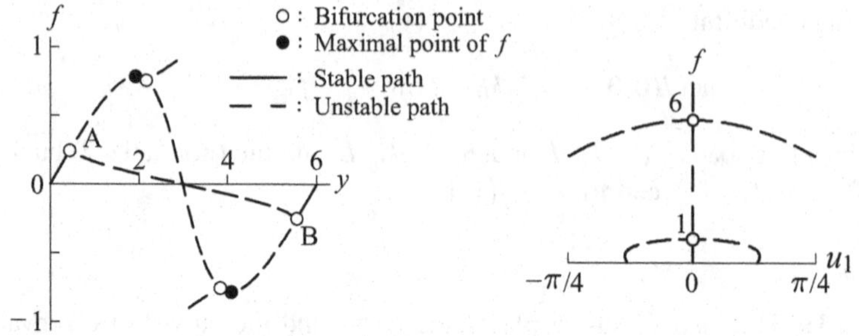

(a) Problem 4.2: Two-bar truss arch (b) Problem 4.3: Bar–spring system

Figure A.2 Load versus displacement curves of structural systems

Hence $(x,y,z) = (1,1,0), (-1,0,1)$ can be chosen as solutions to the homogeneous equation and $(x_\mathrm{p}, y_\mathrm{p}, z_\mathrm{p}) = (1,0,0)$ as a particular solution. Accordingly, the solution to this equation is $(x,y,z) = c_1(1,1,0) + c_2(-1,0,1) + (1,0,0)$ with some parameters c_1 and c_2.

Problem 4.2: The load versus displacement curve is plotted in Fig. A.2(a).

Problem 4.3: The potential function is

$$U(u_1,u_2,f) = kL^2\left[\frac{3}{2}\sin^2 u_1 + (\sin u_1 + \sin u_2)^2 - (2 - \cos u_1 - \cos u_2)f\right]$$

and the equilibrium equation is

$$kL^2\begin{pmatrix}\frac{5}{2}\sin 2u_1 + 2\cos u_1 \sin u_2 - f\sin u_1 \\ 2\sin u_1 \cos u_2 + \sin 2u_2 - f\sin u_2\end{pmatrix} = \begin{pmatrix}0\\0\end{pmatrix}.$$

On the trivial solution $(u_1,u_2) = (0,0)$, the tangent stiffness matrix becomes

$$J(0,0,f) = kL^2\begin{pmatrix}5-f & 2\\ 2 & 2-f\end{pmatrix}.$$

From the criticality condition $\det J(0,0,f) = 0$, we obtain the buckling loads $f_\mathrm{A} = 1$ and $f_\mathrm{B} = 6$. The associated buckling modes are $\boldsymbol{\eta}_\mathrm{A} = \frac{1}{\sqrt{5}}\begin{pmatrix}-1\\2\end{pmatrix}$ and $\boldsymbol{\eta}_\mathrm{B} = \frac{1}{\sqrt{5}}\begin{pmatrix}2\\1\end{pmatrix}$. By setting these buckling modes as the initial values in the path tracing iteration, we can obtain the equilibrium path in Fig. A.2(b).

Chapter 5

Problem 5.1: For the inflection point of f, we have $A_{20} = 0$ and $A_{30} \neq 0$ (cf., Fig. 5.2), and the bifurcation equation in (5.34) and the Jacobian, respectively, become

$$\hat{F}(w,\tilde{f}) \approx A_{30}w^3 + A_{01}\tilde{f} = 0, \quad \hat{J}(w,\tilde{f}) \approx 3A_{30}w^2. \tag{A.4}$$

This shows that the potential for $\tilde{f} = 0$ becomes

$$\hat{U}(w, 0) \approx \frac{1}{4} A_{30} w^4. \tag{A.5}$$

From (A.4) and (A.5), the inflection point and its neighboring equilibrium points are both stable for $A_{30} > 0$ and unstable for $A_{30} < 0$. Thus the stability does not change at an inflection point of f.

Problem 5.2: (1) From the equilibrium equation, we can obtain the fundamental path (trivial solution) $u = 0$ and the bifurcating path $f = \cos 2u + \frac{3}{4}\sin 2u$. As the intersection of these two paths, we obtain the bifurcation point $(u_c, f_c) = (0, 1)$. Since the bifurcating path is not symmetric with respect to $u \longmapsto -u$, this is an asymmetric bifurcation point. (2) In the neighborhood of this bifurcation point, define incremental variables $(w, \tilde{f})$ by

$$u = u_c + w = 0 + w = w, \qquad f = f_c + \tilde{f} = 1 + \tilde{f}.$$

The substitution of these expressions into the equilibrium equation and the use of approximations, such as $\sin w \approx w - w^3/6$, lead to asymptotic forms of the bifurcation equation and the Jacobian, respectively, as

$$\hat{F}(w, \tilde{f}) \approx 3w^2 - 2w\tilde{f} = 0, \qquad \hat{J}(w, \tilde{f}) = \frac{\partial \hat{F}}{\partial w} \approx 2(3w - \tilde{f}).$$

From the asymptotic form of the bifurcation equation, we obtain the trivial solution $w = 0$ and the bifurcating path $\tilde{f} \approx 3w/2$. On the trivial solution $w = 0$, we have $\hat{J}(0, \tilde{f}) \approx -2\tilde{f}$. Accordingly, the trivial solution is stable for $\tilde{f} < 0$ and is unstable for $\tilde{f} > 0$. On the bifurcating path $\tilde{f} \approx 3w/2$, we have $\hat{J}(w, 3w/2) \approx 3w$. Accordingly, the bifurcating path is unstable for $w < 0$ and stable for $w > 0$. These results are plotted in Fig. A.3(a).

Problem 5.3: The incremental variables $(w, \tilde{f})$ in the neighborhood of the bifurcation point $(u_c, f_c) = (0, 1)$ are defined by $w = u$ and $\tilde{f} = f - 1$, and are substituted into the equilibrium equation (5.65) to arrive at the asymptotic forms of the bifurcation equation and the tangent stiffness, respectively, as

$$\hat{F}(w, \tilde{f}) \approx kL^2 \left(\frac{3w^3}{2} - w\tilde{f} \right) = 0, \qquad \hat{J}(w, \tilde{f}) = \frac{\partial \hat{F}}{\partial w} \approx kL^2 \left(\frac{9w^2}{2} - \tilde{f} \right).$$

The bifurcation equation has the trivial solution $w = 0$ and the bifurcating path $\tilde{f} \approx 3w^2/2$. On the trivial solution $w = 0$, the tangent stiffness becomes $\hat{J}(0, \tilde{f}) \approx -kL^2\tilde{f}$. Accordingly, the trivial solution is stable for $\tilde{f} < 0$ and unstable for $\tilde{f} > 0$. On the bifurcating path $\tilde{f} \approx 3w^2/2$, the tangent stiffness becomes $\hat{J}(w, 3w^2/2) \approx 3kL^2w^2 > 0$ $(w \neq 0)$. Accordingly, the bifurcating path is stable. These results are plotted in Fig. A.3(b).

Problem 5.4: The total potential is given by

$$U(u,f) = kL^2\left[\frac{1}{2}\sin^2 u + \frac{\alpha}{4}\sin^4 u - f(1-\cos u)\right].$$

The increment of $U(u,1)$ around the bifurcation point $(u_c, f_c) = (0,1)$ with $f = f_c = 1$ is given by

$$U(u,1) - U(0,1) \approx \frac{kL^2}{24}(6\alpha - 3)u^4.$$

Note that

$$\frac{1}{2}\sin^2 u - f_c(1-\cos u) \approx \frac{1}{2}\left(u - \frac{u^3}{6}\right)^2 - \left(\frac{u^2}{2} - \frac{u^4}{24}\right) \approx -\frac{3}{24}u^4, \quad \frac{\alpha}{4}\sin^4 u \approx \frac{\alpha}{4}u^4.$$

Hence the bifurcation point is stable for $\alpha > 1/2$, for which the potential $U(u,1)$ takes a local minimum at $u = 0$.

Problem 5.5: The solution of the system is

$$\begin{cases} x = 0, \quad f = 3y, \quad y < 1, & \text{fundamental path} \\ y = 1 - (1/4 - x^2)^{1/2}, \quad f = 2 - (1/4 - x^2)^{1/2}, & \text{bifurcating path} \end{cases}$$

and the Jacobian matrix on the fundamental path is

$$J(0,y,f) = \begin{pmatrix} 2 - \dfrac{1}{1-y} & 0 \\ 0 & 3 \end{pmatrix}.$$

We have a bifurcation point at $(x_c, y_c, f_c) = (0, 1/2, 3/2)$, and the Jacobian matrix at this point is

$$J(0,1/2,3/2) = \begin{pmatrix} 0 & 0 \\ 0 & 3 \end{pmatrix}.$$

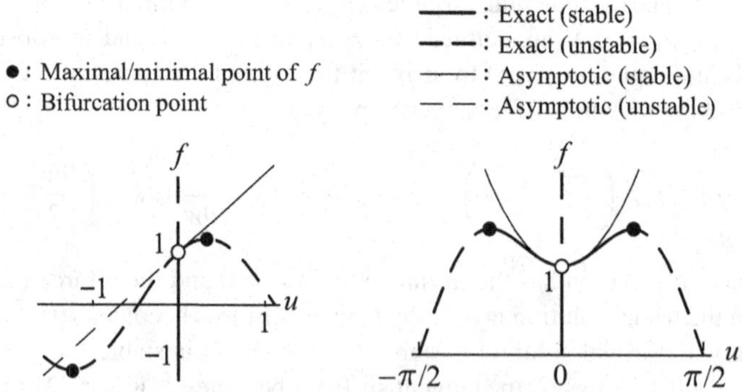

(a) Problem 5.2 (b) Problem 5.3

Figure A.3 Comparison of a load versus displacement curve and its asymptotic form

Define the incremental variables by $\tilde{x} = x$, $\tilde{y} = y - 1/2$, and $\tilde{f} = f - 3/2$ and expand the equilibrium equation in the neighborhood of the bifurcation point to obtain

$$F_1 \approx -4\tilde{x}\tilde{y} + 4\tilde{x}^3 = 0, \qquad F_2 \approx 3\tilde{y} - \tilde{f} - 2\tilde{x}^2 = 0.$$

The elimination of $\tilde{y}$ from these two equations leads to the asymptotic form of the bifurcation equation

$$\hat{F} \approx \frac{4}{3}(\tilde{x}\tilde{f} - \tilde{x}^3).$$

Problem 5.6: We expand $\tilde{f}$ into a power series of w as $\tilde{f} = \sum_{i\geq 1} a_i w^i$ and substitute this into (5.47) to obtain

$$\begin{aligned}\hat{F}(w,\tilde{f}) &\approx A_{30}w^3 + A_{11}w\sum_{i\geq 1} a_i w^i + A_{02}\left(\sum_{i\geq 1} a_i w^i\right)^2 \\ &\approx (A_{11}a_1 + A_{02}a_1{}^2)w^2 + (A_{30} + A_{11}a_2 + 2A_{02}a_1a_2)w^3.\end{aligned}$$

The condition for the vanishing of the coefficient $A_{11}a_1 + A_{02}a_1{}^2$ of the term of w^2 gives $a_1 = 0$ or $a_1 = -\dfrac{A_{11}}{A_{02}}$. When $a_1 = 0$, the condition for the vanishing of the coefficient of the term of w^3 gives $a_2 = -\dfrac{A_{30}}{A_{11}}$, and hence $\tilde{f} = -\dfrac{A_{30}}{A_{11}}w^2 + \mathrm{O}(w^3)$. When $a_1 = -\dfrac{A_{11}}{A_{02}} \neq 0$, we have $\tilde{f} = -\dfrac{A_{11}}{A_{02}}w + \mathrm{O}(w^2)$. This proves (5.56).

Problem 5.7: Using $\sin u \approx u - u^3/6$ and $\cos u \approx 1 - u^2/2 + u^4/24$, we can expand $U(u,f)$ with respect to u at the bifurcation point to obtain

$$\begin{aligned}U(u,1) &\approx kL^2\left\{\frac{1}{2}\left[\left(u - \frac{u^3}{6}\right)^2 + \left(u - \frac{u^3}{6}\right)^4\right] - \frac{u^2}{2} + \frac{u^4}{24}\right\} \\ &\approx kL^2\left(\frac{u^2}{2} - \frac{u^4}{6} + \frac{u^4}{2} - \frac{u^2}{2} + \frac{u^4}{24}\right) = \frac{3}{8}kL^2u^4 \geq 0.\end{aligned}$$

This function has a local minimum at $u = 0$ for the bifurcation point.

Problem 5.8: For the structural system in Fig. 5.11, we have

$$U(u,f) = \frac{1}{2}\alpha kL^2(2u)^2 + \frac{1}{2}\alpha kL^2u^2 + \frac{1}{2}2k(L\sin u)^2 - f\cdot 2L(1-\cos u), \tag{A.6}$$

$$F(u,f) = kL^2(5\alpha u + \sin 2u) - 2Lf\sin u. \tag{A.7}$$

(1) By setting $\alpha = 1$ in (A.7), we have

$$F(u,f) = kL^2(5u + \sin 2u) - 2Lf\sin u.$$

Accordingly, $F(u,f) = 0$ has the trivial solution $u = 0$ and the bifurcated solution $f = kL(5u + \sin 2u)/(2\sin u)$. By setting $u \to 0$ on this bifurcated solution, we can obtain the bifurcation load $f_c = 7kL/2$. (2) By (A.6) with $\alpha = 1$, the increment of the potential around the bifurcation point $(u_c, f_c) = (0, 7kL/2)$ is given by

$$U(u, f_c) - U(u_c, f_c) = \sum_{j=1} \frac{\partial^j U}{\partial u^j}(0, 7kL/2)\frac{u^j}{j!} \approx -kL^2 \frac{u^4}{4!} \leq 0.$$

Since $U(u, f_c)$ takes a local maximum at this bifurcation point, this point is unstable. The tangent stiffness is given by $J(u, f) = kL^2(5 + 2\cos 2u) - 2Lf\cos u$, and, on the bifurcated path just after bifurcation ($u \approx 0$), it becomes

$$\begin{aligned} J\left(u, kL\left(\frac{5u}{2\sin u} + \cos u\right)\right) &= kL^2\left[5\left(1 - \frac{u}{\tan u}\right) - 2\sin^2 u\right] \\ &\approx kL^2\left(\frac{5}{3}u^2 - 2u^2\right) = -\frac{kL^2}{3}u^2 \leq 0, \end{aligned}$$

where $u/\tan u \approx 1 - u^2/3$ is used. Accordingly, the bifurcated path is unstable just after bifurcation. (3) From (A.7), we have

$$f_c = \lim_{u \to 0} kL \frac{5\alpha u + \sin 2u}{2\sin u} = \frac{kL}{2}(5\alpha + 2).$$

Accordingly, $f_c \geq 10kL$ is satisfied by $\alpha \geq 18/5$.

Problem 5.9: We rely on Section 2.7 for details of the analysis of this arch. The equilibrium equation is given by (2.29) as

$$\begin{pmatrix} F_x \\ F_y \end{pmatrix} = \begin{pmatrix} \sum_{m=1}^{2} EA\left(\frac{1}{L} - \frac{1}{\hat{L}^{(m)}}\right)(x - x_m) \\ \sum_{m=1}^{2} EA\left(\frac{1}{L} - \frac{1}{\hat{L}^{(m)}}\right)(y - y_m) - EAf \end{pmatrix} = \begin{pmatrix} 0 \\ 0 \end{pmatrix},$$

where (x, y) denotes the displaced location of the top node 3, $(x_1, y_1) = (-1, 3)$ and $(x_2, y_2) = (1, 3)$ denote the initial locations of nodes 1 and 2, respectively; $L = \sqrt{10}$ and $\hat{L}^{(1)}$ and $\hat{L}^{(2)}$ are the member lengths after deformation. Recall that the fundamental path is obtained explicitly as

$$x = 0, \quad f = 2\left(\frac{1}{\sqrt{10}} - \frac{1}{\sqrt{1 + (y-3)^2}}\right)(y - 3)$$

by (2.32) with $h = 3$. We focus on the first critical point F with $(x_c, y_c, f_c) = (0, 0.4473, 0.2478)$ on the fundamental path shown in Fig. 2.5(d) and derive the bifurcation equation at this point. Define the incremental variables by $(\tilde{x}, \tilde{y}, \tilde{f}) = (x, y, f) - (x_c, y_c, f_c)$, and set $\boldsymbol{\eta}_1 = (-1, 0)^\top$ and $\boldsymbol{\eta}_2 = (0, 1)^\top$, where $\boldsymbol{\eta}_1$ is chosen in view of the consistency with sign of the coefficient of the bifurcation equation. Then the incremental equilibrium equation in (5.19) becomes

$$\begin{pmatrix} \tilde{F}_1(\tilde{x}, \tilde{y}, \tilde{f}) \\ \tilde{F}_2(\tilde{x}, \tilde{y}, \tilde{f}) \end{pmatrix} \equiv \begin{pmatrix} -F_x(x_c + \tilde{x}, y_c + \tilde{y}, f_c + \tilde{f}) \\ F_y(x_c + \tilde{x}, y_c + \tilde{y}, f_c + \tilde{f}) \end{pmatrix} = \begin{pmatrix} 0 \\ 0 \end{pmatrix}.$$

The first equation has an expanded form of

$$\tilde{F}_1(\tilde{x},\tilde{y},\tilde{f}) \approx -EA\left[\frac{2}{\hat{L}^5}(y_c-3)(\hat{L}^2-3)\tilde{x}\tilde{y}+\frac{1}{\hat{L}^7}(y_c-3)^2(\hat{L}^2-5)\tilde{x}^3\right],$$

where $\hat{L}=\sqrt{1+(y_c-3)^2}$ is the member length of both members at the bifurcation point. The second equation has an expanded form of

$$\tilde{F}_2(\tilde{x},\tilde{y},\tilde{f}) \approx EA\left[\frac{2}{L}\left(1-\frac{L}{\hat{L}^3}\right)\tilde{y}+\frac{1}{\hat{L}^5}(y_c-3)(\hat{L}^2-3)\tilde{x}^2-\tilde{f}\right].$$

The elimination of the passive coordinate $\tilde{y}$ from $\tilde{F}_1(\tilde{x},\tilde{y},\tilde{f})=0$ and $\tilde{F}_2(\tilde{x},\tilde{y},\tilde{f})=0$ leads to an asymptotic form of the bifurcation equation $c_1\tilde{x}^3+c_2\tilde{x}\tilde{f}=0$, where c_1 and c_2 are some constants.

Chapter 6

Problem 6.1: The multiplication table for D_4 is given below:

	e	r	r^2	r^3	s	sr	sr^2	sr^3
e	e	r	r^2	r^3	s	sr	sr^2	sr^3
r	r	r^2	r^3	e	sr^3	s	sr	sr^2
r^2	r^2	r^3	e	r	sr^2	sr^3	s	sr
r^3	r^3	e	r	r^2	sr	sr^2	sr^3	s
s	s	sr	sr^2	sr^3	e	r	r^2	r^3
sr	sr	sr^2	sr^3	s	r^3	e	r	r^2
sr^2	sr^2	sr^3	s	sr	r^2	r^3	e	r
sr^3	sr^3	s	sr	sr^2	r	r^2	r^3	e

Problem 6.2: The rule of hierarchical bifurcation of a D_4-symmetric reciprocal system is given in Fig. A.4

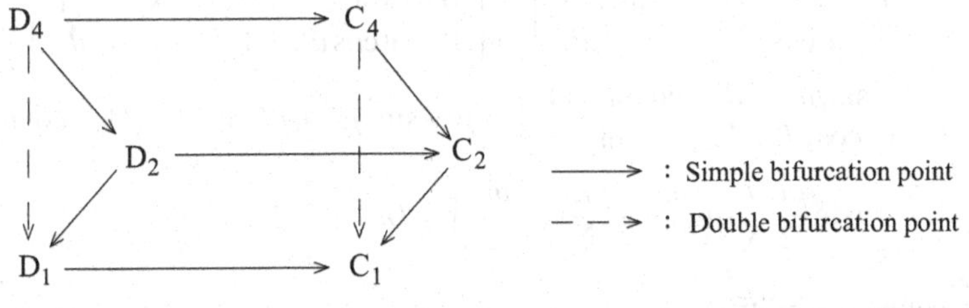

Figure A.4 Problem 6.2: Rule of hierarchical bifurcation of a D_4-symmetric reciprocal system

Chapter 7

Problem 7.1: (1) A particular solution is given by $y=x^2$. To obtain solutions to the homogeneous equation, we set $y=\exp(\lambda x)$ and substitute it into the homogeneous equation. Then we obtain $\lambda^2+3\lambda+2=(\lambda+2)(\lambda+1)=0$, which gives $\lambda=-1,\,-2$. Hence the general solution is given by $y=c_1\exp(-x)+c_2\exp(-2x)+x^2$ with some constants c_1 and c_2. (2) Obviously, the solution to this equation is given by

$$y(x)=c_1\sin(2\sqrt{f}x)+c_2\cos(2\sqrt{f}x).$$

Then it follows from the boundary conditions that

$$c_2=0,\quad 2\sqrt{f}c_1\cos(2\sqrt{f})=0.$$

Hence a nontrivial solution to this differential equation exists only when f satisfies $2\sqrt{f}=\pi(i-1/2)$ $(i=1,2,\ldots)$. Then

$$y(x)=c_1\sin\pi(i-1/2)x,\quad i=1,2,\ldots.$$

Problem 7.2: (1) For the boundary conditions $w(0)=w(L)=0$ and $w'(0)=w'(L)=0$ for both ends fixed, we have (7.22) with

$$A=\begin{pmatrix} 1 & 0 & 0 & 1\\ 0 & 1 & \mu & 0\\ 1 & L & \sin\mu L & \cos\mu L\\ 0 & 1 & \mu\cos\mu L & -\mu\sin\mu L \end{pmatrix}.$$

Then the buckling condition in (7.23) is evaluated to

$$\begin{aligned}
\det A &= \begin{vmatrix} 1 & 0 & 0 & 1\\ 0 & 1 & \mu & 0\\ 1 & L & \sin\mu L & \cos\mu L\\ 0 & 1 & \mu\cos\mu L & -\mu\sin\mu L \end{vmatrix} = \begin{vmatrix} 1 & 0 & 0 & 1\\ 0 & 1 & \mu & 0\\ 0 & L & \sin\mu L & \cos\mu L-1\\ 0 & 1 & \mu\cos\mu L & -\mu\sin\mu L \end{vmatrix}\\
&= \begin{vmatrix} 1 & \mu & 0\\ L & \sin\mu L & \cos\mu L-1\\ 1 & \mu\cos\mu L & -\mu\sin\mu L \end{vmatrix} = \begin{vmatrix} 1 & \mu & 0\\ 0 & \sin\mu L-\mu L & \cos\mu L-1\\ 0 & \mu(\cos\mu L-1) & -\mu\sin\mu L \end{vmatrix}\\
&= \mu\begin{vmatrix} \sin\mu L-\mu L & \cos\mu L-1\\ \cos\mu L-1 & -\sin\mu L \end{vmatrix} = \mu[-\sin^2\mu L+\mu L\sin\mu L-(1-\cos\mu L)^2]\\
&= -4\mu\sin\frac{\mu L}{2}\left(\sin\frac{\mu L}{2}-\frac{\mu L}{2}\cos\frac{\mu L}{2}\right)=0.
\end{aligned}$$

We have two possibilities

$$\sin\frac{\mu L}{2}=0\quad\text{or}\quad\sin\frac{\mu L}{2}-\frac{\mu L}{2}\cos\frac{\mu L}{2}=0.$$

The minimum value of μL (>0) for the former is 2π, while that for the latter is obtained numerically as 2.86π. The minimum buckling load for the smaller $\mu L=2\pi$ gives $P=EI\mu^2=4\pi^2EI/L^2$.

(2) For the boundary conditions $w(0) = w(L) = 0$, $w'(0) = 0$, and $w''(L) = 0$ for one end fixed and the other end pinned, similarly to (1), the buckling condition (7.23) reads

$$\begin{vmatrix} 1 & 0 & 0 & 1 \\ 0 & 1 & \mu & 0 \\ 1 & L & \sin\mu L & \cos\mu L \\ 0 & 0 & \sin\mu L & \cos\mu L \end{vmatrix} = \begin{vmatrix} 1 & 0 & 0 & 1 \\ 0 & 1 & \mu & 0 \\ 0 & L & 0 & -1 \\ 0 & 0 & \sin\mu L & \cos\mu L \end{vmatrix} = \sin\mu L - \mu L \cos\mu L = 0. \tag{A.8}$$

The minimum value of μL is obtained numerically as $4.493 \approx 1.43\pi$ and, in turn, the minimum buckling load is $P = EI\mu^2 \approx 4.493^2 EI/L^2 \approx \pi^2 EI/(0.70L)^2$.

Problem 7.3: (1) The boundary conditions are $w(0) = 0$, $w'(0) = 0$, and $EIw''(L) = 0$, as well as the elastic support condition $EIw'''(L) + Pw'(L) = kw(L)$. With the use of (7.34), the buckling condition (7.23) becomes

$$\begin{vmatrix} 1 & 0 & 0 & 1 \\ 0 & 1 & \mu & 0 \\ 1 & L-P/k & \sin\mu L & \cos\mu L \\ 0 & 0 & \sin\mu L & \cos\mu L \end{vmatrix} = \begin{vmatrix} 1 & 0 & 0 & 1 \\ 0 & 1 & \mu & 0 \\ 0 & L-P/k & 0 & -1 \\ 0 & 0 & \sin\mu L & \cos\mu L \end{vmatrix}$$
$$= \sin\mu L - \mu\left(L - \frac{P}{k}\right)\cos\mu L = 0.$$

(2) For $k \to 0$, the buckling condition becomes $\cos\mu L = 0$, coinciding with the buckling condition of the cantilever. In contrast, for $k \to \infty$, the buckling condition becomes $\sin\mu L - \mu L \cos\mu L = 0$, coinciding with the buckling condition in (A.8) when one end is fixed and the other end is pinned.

Problem 7.4: (1) The buckling condition (7.23) is obtained from (7.53) as

$$\begin{vmatrix} 0 & 1 & 0 & 1 \\ 0 & -\omega_-^2 & 0 & -\omega_+^2 \\ \sin\omega_- L & \cos\omega_- L & \sin\omega_+ L & \cos\omega_+ L \\ -\omega_-^2 \sin\omega_- L & -\omega_-^2 \cos\omega_- L & -\omega_+^2 \sin\omega_+ L & -\omega_+^2 \cos\omega_+ L \end{vmatrix}$$
$$= - \begin{vmatrix} 1 & 1 & 0 & 0 \\ -\omega_+^2 & -\omega_-^2 & 0 & 0 \\ \cos\omega_+ L & \cos\omega_- L & \sin\omega_+ L & \sin\omega_- L \\ -\omega_+^2 \cos\omega_+ L & -\omega_-^2 \cos\omega_- L & -\omega_+^2 \sin\omega_+ L & -\omega_-^2 \sin\omega_- L \end{vmatrix}$$
$$= - \begin{vmatrix} 1 & 1 \\ -\omega_+^2 & -\omega_-^2 \end{vmatrix} \times \begin{vmatrix} \sin\omega_+ L & \sin\omega_- L \\ -\omega_+^2 \sin\omega_+ L & -\omega_-^2 \sin\omega_- L \end{vmatrix}$$
$$= -(\omega_-^2 - \omega_+^2)^2 \sin\omega_- L \sin\omega_+ L = 0,$$

which gives the condition $\sin\omega_- L \sin\omega_+ L = 0$ in (7.55).

(2) From $\omega_\pm = n\pi/L$ in (7.56) and the definition of $\omega_\pm$ in (7.50), we have

$$\frac{n\pi}{L} = \sqrt{\frac{P}{2EI} \pm \sqrt{\left(\frac{P}{2EI}\right)^2 - \frac{k}{EI}}} \implies \left(\frac{n\pi}{L}\right)^2 = \frac{P}{2EI} \pm \sqrt{\left(\frac{P}{2EI}\right)^2 - \frac{k}{EI}}$$

$$\implies \left[\left(\frac{n\pi}{L}\right)^2 - \frac{P}{2EI}\right]^2 = \left(\frac{P}{2EI}\right)^2 - \frac{k}{EI} \implies \frac{P}{EI}\left(\frac{n\pi}{L}\right)^2 = \left(\frac{n\pi}{L}\right)^4 + \frac{k}{EI}.$$

Then we have $P = EI\left(\frac{n\pi}{L}\right)^2 + k\left(\frac{n\pi}{L}\right)^{-2}$. This gives (7.57).

Problem 7.5: (1) The substitution of the general solution (7.59) into the boundary condition (7.25) of simple support gives the conditions

$$w(0) = C_1 + C_4 = 0, \tag{A.9}$$

$$w''(0) = \frac{q}{P} - \mu^2 C_4 = 0, \tag{A.10}$$

$$w(L) = \frac{qL^2}{2P} + C_1 + LC_2 + C_3 \sin\mu L + C_4 \cos\mu L = 0, \tag{A.11}$$

$$w''(L) = \frac{q}{P} - \mu^2 (C_3 \sin\mu L + C_4 \cos\mu L) = 0. \tag{A.12}$$

The relation $C_4 = \dfrac{q}{\mu^2 P}$ is obtained from (A.10). Use of this relation in (A.9) and in (A.12) leads, respectively, to $C_1 = -\dfrac{q}{\mu^2 P}$ and $C_3 = \dfrac{q}{\mu^2 P}\dfrac{1-\cos\mu L}{\sin\mu L} = \dfrac{q}{\mu^2 P}\tan\dfrac{\mu L}{2}$. Substituting (A.12) into (A.11) and using $C_1 = -\dfrac{q}{\mu^2 P}$, we obtain

$$C_2 = -\frac{1}{L}\left(\frac{qL^2}{2P} - \frac{q}{\mu^2 P} + \frac{q}{\mu^2 P}\right) = -\frac{qL}{2P}.$$

The use of $C_1, \ldots, C_4$ obtained in this manner in (7.59) shows (7.60).

(2) By (7.60), the deflection at the midpoint $x = L/2$ is

$$\begin{aligned} w\left(\frac{L}{2}\right) &= -\frac{qL^2}{8P} + \frac{q}{\mu^2 P}\left(\tan\frac{\mu L}{2}\sin\frac{\mu L}{2} + \cos\frac{\mu L}{2} - 1\right) \\ &= \frac{q}{\mu^2 P}\left(\frac{1}{\cos\frac{\mu L}{2}} - 1 - \frac{\mu^2 L^2}{8}\right) \\ &= \frac{qL^4}{\pi^4 EI}\left(\frac{P}{P_{\rm E}}\right)^{-2}\left(\frac{1}{\cos\left(\frac{\pi}{2}\sqrt{\frac{P}{P_{\rm E}}}\right)} - 1 - \frac{\pi^2}{8}\frac{P}{P_{\rm E}}\right), \end{aligned}$$

thereby showing (7.61). Note that by $\mu = \sqrt{\frac{P}{EI}}$ and $P_{\rm E} = \frac{\pi^2 EI}{L^2}$, we have

$$\mu L = \sqrt{\frac{PL^2}{EI}} = \sqrt{\frac{P}{P_{\rm E}}\pi^2} = \pi\sqrt{\frac{P}{P_{\rm E}}}, \qquad \mu^2 P = \frac{P^2}{EI} = \frac{P_{\rm E}^2}{EI}\left(\frac{P}{P_{\rm E}}\right)^2 = \frac{\pi^4 EI}{L^4}\left(\frac{P}{P_{\rm E}}\right)^2.$$

Problem 7.6: The general solution in (7.21) for the homogeneous equation (7.19) gives the first four terms of (7.66). The last term is a particular solution to (7.65) as follows. For the ith term

$$w_i(x) = \frac{\varepsilon_i}{1 - \frac{1}{i^2}\frac{P}{P_E}} \sin \frac{i\pi}{L} x,$$

we obtain

$$\begin{aligned} & w_i'''' + \mu^2 w_i'' \\ &= \frac{\varepsilon_i}{1 - \frac{1}{i^2}\frac{P}{P_E}} \left[\left(\frac{i\pi}{L}\right)^4 - \left(\frac{i\pi}{L}\right)^2 \mu^2 \right] \sin \frac{i\pi}{L} x \\ &= \varepsilon_i \left[1 - \frac{1}{i^2} \left(\frac{L}{\pi}\right)^2 \frac{P}{EI} \right]^{-1} \left(\frac{i\pi}{L}\right)^4 \left[1 - \frac{1}{i^2} \left(\frac{L}{\pi}\right)^2 \frac{P}{EI} \right] \sin \frac{i\pi}{L} x = \varepsilon_i \left(\frac{i\pi}{L}\right)^4 \sin \frac{i\pi}{L} x, \end{aligned}$$

using $\frac{P}{P_E} = \left(\frac{L}{\pi}\right)^2 \frac{P}{EI}$. Then $\sum_{i=1}^{n} w_i(x)$ is equal to the right-hand side of (7.65) and hence the last term of (7.66) is a particular solution to (7.65).

Problem 7.7: We have

$$I_y = \frac{8 \times 10^3}{12} - 2 \times \frac{3 \times 8^3}{12} = \frac{1232}{3} \ (\text{cm}^4), \ I_z = 2 \times \frac{1 \times 8^3}{12} + \frac{8 \times 2^3}{12} = \frac{272}{3} \ (\text{cm}^4),$$

from which $I_y > I_z$ follows. Hence the z-axis is the weak axis. Note that the moment of inertia of a rectangular cross-section around the center is equal to $bh^3/12$ (b and h are the two sides of the rectangle and b is parallel to the axis for the moment of inertia).

Problem 7.8: (1) With $A = 9\pi$, $I = 369\pi/4$, and $L = 200$, (7.43) gives the slenderness ratio

$$\lambda = \frac{L}{r} = L\sqrt{\frac{A}{I}} = 200\sqrt{9\pi \frac{4}{369\pi}} \approx \frac{200}{3.2}.$$

From (7.71) the standard slenderness ratio is

$$\bar{\lambda} = \frac{k\lambda}{\pi}\sqrt{\frac{\sigma_Y}{E}} \approx \frac{k \times 200/3.2}{\pi}\sqrt{\frac{3.0 \times 10^3}{2.0 \times 10^6}} \approx 0.77k.$$

Hence the condition $\bar{\lambda} > 1$ for elastic buckling is satisfied by the boundary condition with one end fixed and the other end free, which corresponds to $k = 2$ and $\bar{\lambda} \approx 1.54 > 1$. Plastic buckling prevails for the other two boundary conditions with $k = 0.5$ and $k \approx 0.70$, for which $\bar{\lambda} < 1$ holds.

(2) The substitution of (7.73) for the stress versus strain relation and (7.74) for the tangent modulus into (7.72) leads to

$$\sigma_c = \sigma_Y \left[\frac{\varepsilon}{\varepsilon_Y} - \frac{1}{4}\left(\frac{\varepsilon}{\varepsilon_Y}\right)^2 \right] = \frac{E\pi^2}{k^2\lambda^2}\left(1 - \frac{1}{2}\frac{\varepsilon}{\varepsilon_Y}\right), \quad 0 \le \frac{\varepsilon}{\varepsilon_Y} \le 2. \tag{A.13}$$

Setting $k = 1$ (simple support) and the value of the slenderness ratio $\lambda \approx 200/3.2$ in this equation, we numerically obtain the strain $\varepsilon/\varepsilon_{\mathrm{Y}} = 1.07$ at the onset of buckling. The use of this value in (A.13) gives the buckling stress $\sigma_{\mathrm{c}}/\sigma_{\mathrm{Y}} = 0.784$.

Problem 7.9: Following Example 7.1, we denote the circular cross-section by subscript a and the cylindrical cross-section by subscript b.

By (7.44) with $\lambda^2 = (L/r)^2 = L^2A/I$ in (7.43), we obtain

$$\sigma_{\mathrm{Y}} = \sigma_{\mathrm{c}} = \frac{\pi^2 EI}{k^2L^2A}.$$

The two kinds of cross-sections have the same area A, the same coefficient k, and the same Young's modulus E but different moments of inertia I_{a} and I_{b} and different lengths L_{a} and L_{b}. Then we have

$$\sigma_{\mathrm{Y}} = \frac{\pi^2 E}{Ak^2}\frac{I_{\mathrm{a}}}{L_{\mathrm{a}}^2} = \frac{\pi^2 E}{Ak^2}\frac{I_{\mathrm{b}}}{L_{\mathrm{b}}^2}.$$

This equation, together with $I_{\mathrm{b}}/I_{\mathrm{a}} \approx 4.56$ in (7.41), gives

$$\frac{L_{\mathrm{a}}}{L_{\mathrm{b}}} = \sqrt{\frac{I_{\mathrm{a}}}{I_{\mathrm{b}}}} = \frac{1}{\sqrt{4.56}} \approx 0.468.$$

Chapter 8

Problem 8.1: Similarly to Section 8.3.3, the equilibrium equation of this cross truss structure is obtained as

$$\left(\sum_{m=1}^{4} k^{(m)}\right) \boldsymbol{u}_5 - \boldsymbol{f}_5 = \boldsymbol{0}; \qquad \boldsymbol{f}_m = -k^{(m)}\boldsymbol{u}_5, \quad m = 1,2,3,4. \tag{A.14}$$

With the use of the external force $\boldsymbol{f}_5 = (P/\sqrt{2}, P/\sqrt{2})^\top$ and

$$\sum_{m=1}^{4} k^{(m)} = \frac{EA}{L}\left[\begin{pmatrix}0 & 0\\ 0 & 1\end{pmatrix} + \begin{pmatrix}1 & 0\\ 0 & 0\end{pmatrix} + \begin{pmatrix}0 & 0\\ 0 & 1\end{pmatrix} + \begin{pmatrix}1 & 0\\ 0 & 0\end{pmatrix}\right] = \frac{2EA}{L}\begin{pmatrix}1 & 0\\ 0 & 1\end{pmatrix}$$

in (A.14), we obtain the displacement vector

$$\boldsymbol{u}_5 = \left(\sum_{m=1}^{4} k^{(m)}\right)^{-1} \boldsymbol{f}_5 = \frac{\sqrt{2}PL}{4EA}\begin{pmatrix}1\\ 1\end{pmatrix}.$$

Then from (A.14), the fixed end forces are evaluated to

$$\boldsymbol{f}_1 = \boldsymbol{f}_3 = -\frac{EA}{L}\begin{pmatrix}0 & 0\\ 0 & 1\end{pmatrix}\frac{\sqrt{2}PL}{4EA}\begin{pmatrix}1\\ 1\end{pmatrix} = \frac{\sqrt{2}P}{4}\begin{pmatrix}0\\ -1\end{pmatrix},$$

$$\boldsymbol{f}_2 = \boldsymbol{f}_4 = -\frac{EA}{L}\begin{pmatrix}1 & 0\\ 0 & 0\end{pmatrix}\frac{\sqrt{2}PL}{4EA}\begin{pmatrix}1\\ 1\end{pmatrix} = \frac{\sqrt{2}P}{4}\begin{pmatrix}-1\\ 0\end{pmatrix}.$$

Problem 8.2: (1) Using

$$k_{55} = k_{\mathrm{E}}^{(1)} + k_{\mathrm{E}}^{(2)} + k_{\mathrm{E}}^{(5)} + k_{\mathrm{E}}^{(7)}, \quad k_{66} = k_{\mathrm{E}}^{(3)} + k_{\mathrm{E}}^{(4)} + k_{\mathrm{E}}^{(6)} + k_{\mathrm{E}}^{(7)},$$

we have

$$K_{\mathrm{E}}^{\circ} = \begin{pmatrix} k_{\mathrm{E}}^{(1)} & O & O & O & -k_{\mathrm{E}}^{(1)} & O \\ O & k_{\mathrm{E}}^{(2)} + k_{\mathrm{E}}^{(3)} & O & O & -k_{\mathrm{E}}^{(2)} & -k_{\mathrm{E}}^{(3)} \\ O & O & k_{\mathrm{E}}^{(4)} & O & O & -k^{(4)} \\ O & O & O & k_{\mathrm{E}}^{(5)} + k_{\mathrm{E}}^{(6)} & -k_{\mathrm{E}}^{(5)} & -k_{\mathrm{E}}^{(6)} \\ -k_{\mathrm{E}}^{(1)} & -k_{\mathrm{E}}^{(2)} & O & -k_{\mathrm{E}}^{(5)} & k_{55} & -k_{\mathrm{E}}^{(7)} \\ O & -k_{\mathrm{E}}^{(3)} & -k_{\mathrm{E}}^{(4)} & -k_{\mathrm{E}}^{(6)} & -k_{\mathrm{E}}^{(7)} & k_{66} \end{pmatrix}.$$

(2) Using the boundary conditions

$$\boldsymbol{u}_1 = \boldsymbol{u}_2 = \boldsymbol{u}_3 = \boldsymbol{u}_4 = \boldsymbol{0}, \qquad \boldsymbol{f}_5 = \boldsymbol{f}_6 = EAf\boldsymbol{e}$$

with $\boldsymbol{e} = (0,0,1)^\top$, we have

$$\begin{pmatrix} k_{55} & -k_{\mathrm{E}}^{(7)} \\ -k_{\mathrm{E}}^{(7)} & k_{66} \end{pmatrix} \begin{pmatrix} \boldsymbol{u}_5 \\ \boldsymbol{u}_6 \end{pmatrix} = EAf \begin{pmatrix} \boldsymbol{e} \\ \boldsymbol{e} \end{pmatrix}.$$

Problem 8.3: We omit the superscript $(\cdot)^{(m)}$. From (8.6) and (8.8), we have

$$\frac{\partial \hat{L}}{\partial u_i} = -\hat{l}_x, \qquad \frac{\partial \hat{L}}{\partial u_j} = \hat{l}_x, \qquad \frac{\partial \hat{L}}{\partial v_i} = -\hat{l}_y, \qquad \cdots .$$

Also, from (8.4), (8.8), and (8.9), we have

$$\boldsymbol{n} = EA\left(\frac{1}{L} - \frac{1}{\hat{L}}\right) \begin{pmatrix} u_j - u_i + x_j - x_i \\ v_j - v_i + y_j - y_i \\ w_j - w_i + z_j - z_i \end{pmatrix}.$$

Setting $\boldsymbol{n} = (n_x, n_y, n_z)^\top$, we obtain

$$\begin{aligned} \frac{\partial n_x}{\partial u_i} &= EA(u_j - u_i + x_j - x_i)\frac{\partial}{\partial u_i}\left(\frac{1}{L} - \frac{1}{\hat{L}}\right) - EA\left(\frac{1}{L} - \frac{1}{\hat{L}}\right) \\ &= EA\hat{L}\hat{l}_x\frac{1}{\hat{L}^2}(-\hat{l}_x) - EA\frac{\hat{L}-L}{L}\frac{1}{\hat{L}} = -\frac{EA}{\hat{L}}\hat{l}_x^2 - \frac{N}{\hat{L}}, \\ \frac{\partial n_x}{\partial v_i} &= EA(u_j - u_i + x_j - x_i)\frac{\partial}{\partial v_i}\left(\frac{1}{L} - \frac{1}{\hat{L}}\right) = EA\hat{L}\hat{l}_x\frac{1}{\hat{L}^2}(-\hat{l}_y) = -\frac{EA}{\hat{L}}\hat{l}_x\hat{l}_y, \\ &\ \ \vdots \end{aligned}$$

This shows

$$-\frac{\partial \boldsymbol{n}}{\partial \boldsymbol{u}_i} = \frac{EA}{\hat{L}} \begin{pmatrix} \hat{l}_x^2 & \hat{l}_x\hat{l}_y & \hat{l}_x\hat{l}_z \\ \hat{l}_x\hat{l}_y & \hat{l}_y^2 & \hat{l}_y\hat{l}_z \\ \hat{l}_x\hat{l}_z & \hat{l}_y\hat{l}_z & \hat{l}_z^2 \end{pmatrix} + \frac{N}{\hat{L}} I_3 = \frac{EA}{\hat{L}} h + \frac{N}{\hat{L}} I_3. \tag{A.15}$$

The use of this relation in $k = -\dfrac{\partial \boldsymbol{n}}{\partial \boldsymbol{u}_i}$ in (8.14) shows (8.15).

Furthermore, (8.16) is proved as follows. From (8.4), we have

$$\frac{N}{\hat{L}} = EA\left(\frac{1}{L} - \frac{1}{\hat{L}}\right) \quad \Rightarrow \quad \frac{EA}{\hat{L}} = \frac{EA}{L} - \frac{N}{\hat{L}}.$$

Then the last expression in (A.15) is rewritten as

$$\frac{EA}{\hat{L}} h + \frac{N}{\hat{L}} I_3 = \frac{EA}{L} h + \frac{N}{\hat{L}} (I_3 - h).$$

This proves (8.16).

Problem 8.4: We omit the superscript (m) in $(\cdot)^{(m)}$. We have the following relation

$$\begin{aligned}
\frac{\hat{L}}{L} &= \frac{1}{L}[(u_j - u_i + x_j - x_i)^2 + (v_j - v_i + y_j - y_i)^2 + (w_j - w_i + z_j - z_i)^2]^{1/2} \\
&= \left[1 + \frac{2}{L^2}\{(u_j - u_i)(x_j - x_i) + (v_j - v_i)(y_j - y_i) + (w_j - w_i)(z_j - z_i)\}\right. \\
&\quad \left. + \frac{1}{L^2}\{(u_j - u_i)^2 + (v_j - v_i)^2 + (w_j - w_i)^2\}\right]^{1/2} \\
&\approx 1 + \frac{1}{L^2}\{(u_j - u_i)(x_j - x_i) + (v_j - v_i)(y_j - y_i) + (w_j - w_i)(z_j - z_i)\}.
\end{aligned}$$

Then we have

$$\begin{aligned}
N &= EA\frac{\hat{L} - L}{L} \\
&\approx \frac{EA}{L^2}\{(u_j - u_i)(x_j - x_i) + (v_j - v_i)(y_j - y_i) + (w_j - w_i)(z_j - z_i)\} \\
&= \frac{EA}{L}\left\{\frac{x_j - x_i}{L}(u_j - u_i) + \frac{y_j - y_i}{L}(v_j - v_i) + \frac{z_j - z_i}{L}(w_j - w_i)\right\} \\
&= \frac{EA}{L}\{l_x(u_j - u_i) + l_y(v_j - v_i) + l_z(w_j - w_i)\} \\
&= \frac{EA}{L}(-l_x, -l_y, -l_z, l_x, l_y, l_z)(u_i, v_i, w_i, u_j, v_j, w_j)^\top.
\end{aligned}$$

This proves the relation (8.33).

Problem 8.5: This truss arch can be treated as a special case of the analysis in Section 8.4.2. The details of the linear buckling analysis are obtained by setting $h = 1$

and $c = s = 1/\sqrt{2}$ in the results in Section 8.4.2. Note that the two buckling loads f_x and f_y in (8.58) coincide in this case.

Problem 8.6: (1) The axial forces of this truss structure are

$$N^{(1)} = -P, \quad N^{(2)} = N^{(4)} = -\sqrt{2}P, \quad N^{(3)} = N^{(6)} = P, \quad N^{(5)} = 2P.$$

(2) Members 1, 2, and 4 are subjected to compression and have possibility to undergo buckling. Member 1 with the length of L is subjected to the compression load of $-P$. Accordingly, this member undergoes buckling at the load

$$P = P_{\rm c}^{(1)} = \frac{\pi^2 EI}{L^2}.$$

Members 2 and 4 with the length of $\sqrt{2}L$ are both subjected to the same compression load of $-\sqrt{2}P$. The load $P = P_{\rm c}^{(2)} = P_{\rm c}^{(4)}$ for the buckling of these members is determined from

$$\sqrt{2}P_{\rm c}^{(2)} = \sqrt{2}P_{\rm c}^{(4)} = \frac{\pi^2 EI}{(\sqrt{2}L)^2},$$

which gives

$$P_{\rm c}^{(2)} = P_{\rm c}^{(4)} = \frac{\sqrt{2}}{4}\frac{\pi^2 EI}{L^2}.$$

The condition that no member buckles is given by

$$P < \min\{P_{\rm c}^{(1)}, P_{\rm c}^{(2)}, P_{\rm c}^{(4)}\} = \frac{\sqrt{2}}{4}\frac{\pi^2 EI}{L^2}.$$

Chapter 9

Problem 9.1: The relation (9.6) is rewritten as

$$\begin{pmatrix} w_i \\ \varphi_i L \\ w_j \\ \varphi_j L \end{pmatrix} = \begin{pmatrix} 1 & 0 & 0 & 0 \\ 0 & 1 & 0 & 0 \\ 1 & 1 & 1 & 1 \\ 0 & 1 & 2 & 3 \end{pmatrix} \begin{pmatrix} c_0 \\ c_1 L \\ c_2 L^2 \\ c_3 L^3 \end{pmatrix},$$

from which follows

$$\begin{pmatrix} c_0 \\ c_1 L \\ c_2 L^2 \\ c_3 L^3 \end{pmatrix} = \begin{pmatrix} 1 & 0 & 0 & 0 \\ 0 & 1 & 0 & 0 \\ 1 & 1 & 1 & 1 \\ 0 & 1 & 2 & 3 \end{pmatrix}^{-1} \begin{pmatrix} w_i \\ \varphi_i L \\ w_j \\ \varphi_j L \end{pmatrix} = \begin{pmatrix} 1 & 0 & 0 & 0 \\ 0 & 1 & 0 & 0 \\ -3 & -2 & 3 & -1 \\ 2 & 1 & -2 & 1 \end{pmatrix} \begin{pmatrix} w_i \\ \varphi_i L \\ w_j \\ \varphi_j L \end{pmatrix}.$$

Then

$$
\begin{aligned}
w(s) &= c_0 + c_1 s + c_2 s^2 + c_3 s^3 = (1, s, s^2, s^3)\boldsymbol{c} \\
&= (1, s, s^2, s^3)\begin{pmatrix} 1 & 0 & 0 & 0 \\ 0 & 1/L & 0 & 0 \\ -3/L^2 & -2/L^2 & 3/L^2 & -1/L^2 \\ 2/L^3 & 1/L^3 & -2/L^3 & 1/L^3 \end{pmatrix}\begin{pmatrix} w_i \\ \varphi_i L \\ w_j \\ \varphi_j L \end{pmatrix} \\
&= (1, s, s^2, s^3)\begin{pmatrix} 1 & 0 & 0 & 0 \\ 0 & 1 & 0 & 0 \\ -3/L^2 & -2/L & 3/L^2 & -1/L \\ 2/L^3 & 1/L^2 & -2/L^3 & 1/L^2 \end{pmatrix}\begin{pmatrix} w_i \\ \varphi_i \\ w_j \\ \varphi_j \end{pmatrix} \\
&= \left(1 - 3\frac{s^2}{L^2} + 2\frac{s^3}{L^3},\ s - 2\frac{s^2}{L} + \frac{s^3}{L^2},\ 3\frac{s^2}{L^2} - 2\frac{s^3}{L^3},\ -\frac{s^2}{L} + \frac{s^3}{L^2}\right)^{\top} \boldsymbol{u}^{(m)}.
\end{aligned}
$$

Problem 9.2: From (9.8), we have

$$
\frac{\mathrm{d}^2\boldsymbol{H}}{\mathrm{d}s^2} = \frac{2}{L}\left(\frac{3}{L}\left(-1 + 2\frac{s}{L}\right),\ -2 + 3\frac{s}{L},\ \frac{3}{L}\left(1 - 2\frac{s}{L}\right),\ -1 + 3\frac{s}{L}\right)^{\top}.
$$

Then from (9.9), the $(1,1)$, $(1,2)$, $(2,2)$ components of the stiffness matrix $K_{\mathrm{E}}^{(m)}$ become

$$
\begin{aligned}
(1,1) &: EI\int_0^L \frac{36}{L^4}\left(-1 + 2\frac{s}{L}\right)^2 \mathrm{d}s = \frac{36EI}{L^3}\int_0^1 (-1 + 2t)^2 \mathrm{d}t = \frac{12EI}{L^3}, \\
(1,2) &: EI\int_0^L \frac{12}{L^3}\left(-1 + 2\frac{s}{L}\right)\left(-2 + 3\frac{s}{L}\right) \mathrm{d}s \\
&= \frac{12EI}{L^2}\int_0^1 (-1 + 2t)(-2 + 3t)\mathrm{d}t = \frac{6EI}{L^2}, \\
(2,2) &: EI\int_0^L \frac{4}{L^2}\left(-2 + 3\frac{s}{L}\right)^2 \mathrm{d}s = \frac{4EI}{L}\int_0^1 (-2 + 3t)^2 \mathrm{d}t = \frac{4EI}{L}.
\end{aligned}
$$

Problem 9.3: From (9.8), we have

$$
\frac{\mathrm{d}\boldsymbol{H}}{\mathrm{d}s} = \left(\frac{6}{L}\left(-\frac{s}{L} + \frac{s^2}{L^2}\right),\ 1 - 4\frac{s}{L} + 3\frac{s^2}{L^2},\ \frac{6}{L}\left(\frac{s}{L} - \frac{s^2}{L^2}\right),\ -2\frac{s}{L} + 3\frac{s^2}{L^2}\right)^{\top}.
$$

Then from (9.10), the $(1,1)$, $(1,2)$, $(2,2)$ components of the geometric stiffness matrix $K_{\rm G}^{(m)}$ become

$$(1,1):N\int_0^L \frac{36}{L^2}\left(-\frac{s}{L}+\frac{s^2}{L^2}\right)^2 {\rm d}s=\frac{36N}{L}\int_0^1(-t+t^2)^2{\rm d}t=\frac{N}{30}\times\frac{36}{L},$$

$$\begin{aligned}(1,2):N&\int_0^L \frac{6}{L}\left(-\frac{s}{L}+\frac{s^2}{L^2}\right)\left(1-4\frac{s}{L}+3\frac{s^2}{L^2}\right){\rm d}s\\ &=6N\int_0^1(-t+t^2)(1-4t+3t^2){\rm d}t=\frac{N}{30}\times 3,\end{aligned}$$

$$(2,2):N\int_0^L\left(1-4\frac{s}{L}+3\frac{s^2}{L^2}\right)^2{\rm d}s=NL\int_0^1(1-4t+3t^2)^2{\rm d}t=\frac{N}{30}\times 4L.$$

Problem 9.4: The stiffness matrix and the geometric stiffness matrix are given, respectively, by (9.41) and (9.42) as

$$\hat{K}_{\rm E}^{\circ}=\frac{2EI}{L^2}\left(\begin{array}{rr|rr|rr} 6 & 3 & -6 & 3 & 0 & 0\\ 3 & 2 & -3 & 1 & 0 & 0\\ \hline -6 & -3 & 12 & 0 & -6 & 3\\ 3 & 1 & 0 & 4 & -3 & 1\\ \hline 0 & 0 & -6 & -3 & 6 & -3\\ 0 & 0 & 3 & 1 & -3 & 2 \end{array}\right),\quad \hat{K}_{\rm G}^{\circ}=\frac{-P}{30}\left(\begin{array}{rr|rr|rr} 36 & 3 & -36 & 3 & 0 & 0\\ 3 & 4 & -3 & 1 & 0 & 0\\ \hline -36 & -3 & 72 & 0 & -36 & 3\\ 3 & 1 & 0 & 8 & -3 & -1\\ \hline 0 & 0 & -36 & -3 & 36 & -3\\ 0 & 0 & 3 & -1 & -3 & 4 \end{array}\right).$$

In view of the boundary condition $w_1=\varphi_1=w_2=0$, we focus on the 4th, 5th, and 6th components associated with $(\varphi_2,w_3,\varphi_3)$ to obtain the stiffness and geometric stiffness matrices for the present case. Then the buckling condition reads

$$\begin{aligned}&\det\left[\frac{2EI}{L^2}\begin{pmatrix}4&-3&1\\-3&6&-3\\1&-3&2\end{pmatrix}-\frac{P}{30}\begin{pmatrix}8&-3&-1\\-3&36&-3\\-1&-3&4\end{pmatrix}\right]\\ &=\left(\frac{2EI}{L^2}\right)^3\times\det\left[\begin{pmatrix}4&-3&1\\-3&6&-3\\1&-3&2\end{pmatrix}-\alpha\begin{pmatrix}8&-3&-1\\-3&36&-3\\-1&-3&4\end{pmatrix}\right]=0\end{aligned}$$

with $\alpha=PL^2/(60EI)$. This equation is solved numerically as $\alpha=0.026$, 0.28, 0.83, which yield the buckling loads $P_1=1.57EI/L^2$, $P_2=16.8EI/L^2$, and $P_3=49.6EI/L^2$.

Problem 9.5: The frame in Fig. 9.13(a) does not undergo bifurcation and the frames in (b) and (c) undergo bifurcation. Buckling modes are sketched in Fig. A.5.

Problem 9.6: Recall the definition of nodal variables in Fig. 9.7(b). Since there is no axial force in the members, there is no geometrical stiffness. The geometrical boundary conditions are given by $u_1=w_1=0$, $\varphi_1=0$, $u_3=w_3=0$, and $\varphi_3=0$. There are further constraints of $u_2=w_2=0$ due to the assumption of no axial deformation. Accordingly, the rotation φ_2 is the only independent variable. The stiffness equation is one-dimensional and is given by $(8EI/L^2)\varphi_2=M/L$, since the external force is given as $M_2=M$ and the (4,4) component of $\hat{K}_{\rm E}$ in (9.74) is equal to $8EI/L^2$. Hence, we have $\varphi_2=ML/(8EI)$.

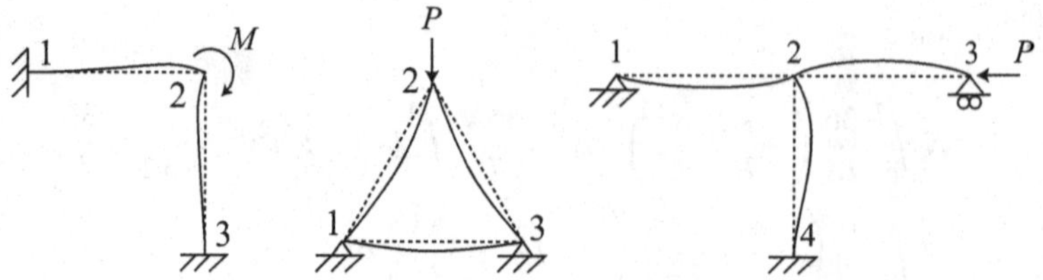

Figure A.5 Problem 9.5: Deformation or bifurcation mode of the frame structures in Fig. 9.13

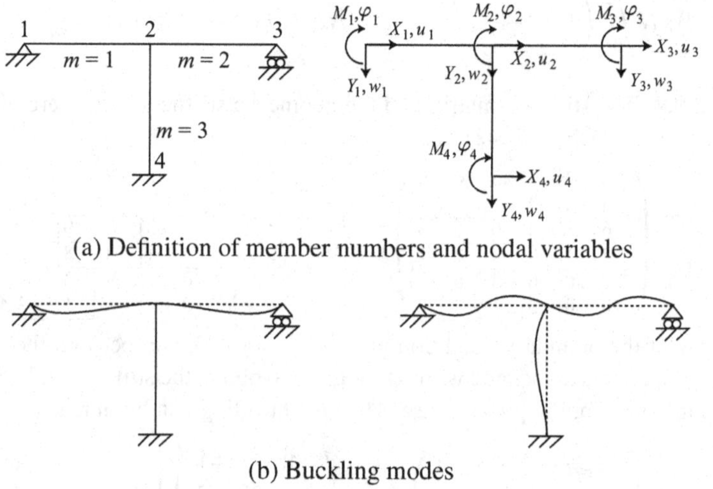

Figure A.6 Problem 9.7: The definition of member numbers and nodal variables and the second and the third buckling modes of the frame structure in Fig. 9.13(c)

Problem 9.7: We define the member numbers and nodal variables as shown in Fig. A.6(a). Recall the (normalized) member stiffness matrices in (9.27):

$$\hat{K}_{\mathrm{E}}^{(m)} = \frac{2EI^{(m)}}{L^2} \left(\begin{array}{cc|cc} 6 & 3 & -6 & 3 \\ 3 & 2 & -3 & 1 \\ \hline -6 & -3 & 6 & -3 \\ 3 & 1 & -3 & 2 \end{array} \right), \qquad \hat{K}_{\mathrm{G}}^{(m)} = \frac{N^{(m)}}{30} \left(\begin{array}{cc|cc} 36 & 3 & -36 & 3 \\ 3 & 4 & -3 & -1 \\ \hline -36 & -3 & 36 & -3 \\ 3 & -1 & -3 & 4 \end{array} \right).$$

Note that $EI^{(1)} = EI^{(2)} = EI^{(3)} = EI$, $N^{(1)} = N^{(2)} = -P$, and $N^{(3)} = 0$ for the present frame structure, where the vertical member is indexed by $m = 3$. The displacement boundary conditions are $u_1 = w_1 = 0$, $w_3 = 0$, $u_4 = w_4 = 0$, and $\varphi_4 = 0$. Since axial deformation is ignored, we have $u_2 = w_2 = u_3 = 0$. Therefore, the free displacements are the rotation angles $(\varphi_1, \varphi_2, \varphi_3)$.

Member end displacements in (9.3) for member stiffness matrices are as follows:

$$\begin{array}{ll} \text{Member 1:} & w_1, \varphi_1, w_2, \varphi_2, \\ \text{Member 2:} & w_2, \varphi_2, w_3, \varphi_3, \\ \text{Member 3:} & -u_2, \varphi_2, -u_4, \varphi_4. \end{array}$$

Thus we need to consider 9 variables $(w_1, \varphi_1, u_2, w_2, \varphi_2, w_3, \varphi_3, u_4, \varphi_4)$ in assembling structural matrices.

The member matrices for the members are assembled to arrive at the following 9×9 matrices

$$\hat{K}_{\mathrm{E}}^{\circ} = \frac{2EI}{L^2}\begin{pmatrix} 6 & 3 & 0 & -6 & 3 & 0 & 0 & 0 & 0 \\ 3 & 2 & 0 & -3 & 1 & 0 & 0 & 0 & 0 \\ 0 & 0 & 0 & 0 & 0 & 0 & 0 & 0 & 0 \\ -6 & -3 & 0 & 6 & -3 & 0 & 0 & 0 & 0 \\ 3 & 1 & 0 & -3 & 2 & 0 & 0 & 0 & 0 \\ 0 & 0 & 0 & 0 & 0 & 0 & 0 & 0 & 0 \\ 0 & 0 & 0 & 0 & 0 & 0 & 0 & 0 & 0 \\ 0 & 0 & 0 & 0 & 0 & 0 & 0 & 0 & 0 \\ 0 & 0 & 0 & 0 & 0 & 0 & 0 & 0 & 0 \end{pmatrix} + \frac{2EI}{L^2}\begin{pmatrix} 0 & 0 & 0 & 0 & 0 & 0 & 0 & 0 & 0 \\ 0 & 0 & 0 & 0 & 0 & 0 & 0 & 0 & 0 \\ 0 & 0 & 0 & 0 & 0 & 0 & 0 & 0 & 0 \\ 0 & 0 & 0 & 6 & 3 & -6 & 3 & 0 & 0 \\ 0 & 0 & 0 & 3 & 2 & -3 & 1 & 0 & 0 \\ 0 & 0 & 0 & -6 & -3 & 6 & -3 & 0 & 0 \\ 0 & 0 & 0 & 3 & 1 & -3 & 2 & 0 & 0 \\ 0 & 0 & 0 & 0 & 0 & 0 & 0 & 0 & 0 \\ 0 & 0 & 0 & 0 & 0 & 0 & 0 & 0 & 0 \end{pmatrix}$$

$$+ \frac{2EI}{L^2}\begin{pmatrix} 0 & 0 & 0 & 0 & 0 & 0 & 0 & 0 & 0 \\ 0 & 0 & 0 & 0 & 0 & 0 & 0 & 0 & 0 \\ 0 & 0 & 6 & 0 & -3 & 0 & 0 & -6 & -3 \\ 0 & 0 & 0 & 0 & 0 & 0 & 0 & 0 & 0 \\ 0 & 0 & -3 & 0 & 2 & 0 & 0 & 3 & 1 \\ 0 & 0 & 0 & 0 & 0 & 0 & 0 & 0 & 0 \\ 0 & 0 & 0 & 0 & 0 & 0 & 0 & 0 & 0 \\ 0 & 0 & -6 & 0 & 3 & 0 & 0 & 6 & 3 \\ 0 & 0 & -3 & 0 & 1 & 0 & 0 & 3 & 2 \end{pmatrix} = \frac{2EI}{L^2}\begin{pmatrix} 6 & 3 & 0 & -6 & 3 & 0 & 0 & 0 & 0 \\ 3 & 2 & 0 & -3 & 1 & 0 & 0 & 0 & 0 \\ 0 & 0 & 6 & 0 & -3 & 0 & 0 & -6 & -3 \\ -6 & -3 & 0 & 12 & 0 & -6 & 3 & 0 & 0 \\ 3 & 1 & -3 & 0 & 6 & -3 & 1 & 3 & 1 \\ 0 & 0 & 0 & -6 & -3 & 6 & -3 & 0 & 0 \\ 0 & 0 & 0 & 3 & 1 & -3 & 2 & 0 & 0 \\ 0 & 0 & -6 & 0 & 3 & 0 & 0 & 6 & 3 \\ 0 & 0 & -3 & 0 & 1 & 0 & 0 & 3 & 2 \end{pmatrix},$$

$$\hat{K}_{\mathrm{G}}^{\circ} = -\frac{P}{30}\begin{pmatrix} 36 & 3 & 0 & -36 & 3 & 0 & 0 & 0 & 0 \\ 3 & 4 & 0 & -3 & -1 & 0 & 0 & 0 & 0 \\ 0 & 0 & 0 & 0 & 0 & 0 & 0 & 0 & 0 \\ -36 & -3 & 0 & 36 & -3 & 0 & 0 & 0 & 0 \\ 3 & -1 & 0 & -3 & 4 & 0 & 0 & 0 & 0 \\ 0 & 0 & 0 & 0 & 0 & 0 & 0 & 0 & 0 \\ 0 & 0 & 0 & 0 & 0 & 0 & 0 & 0 & 0 \\ 0 & 0 & 0 & 0 & 0 & 0 & 0 & 0 & 0 \\ 0 & 0 & 0 & 0 & 0 & 0 & 0 & 0 & 0 \end{pmatrix} - \frac{P}{30}\begin{pmatrix} 0 & 0 & 0 & 0 & 0 & 0 & 0 & 0 & 0 \\ 0 & 0 & 0 & 0 & 0 & 0 & 0 & 0 & 0 \\ 0 & 0 & 0 & 0 & 0 & 0 & 0 & 0 & 0 \\ 0 & 0 & 0 & 36 & 3 & -36 & 3 & 0 & 0 \\ 0 & 0 & 0 & 3 & 4 & -3 & -1 & 0 & 0 \\ 0 & 0 & 0 & -36 & -3 & 36 & -3 & 0 & 0 \\ 0 & 0 & 0 & 3 & -1 & -3 & 4 & 0 & 0 \\ 0 & 0 & 0 & 0 & 0 & 0 & 0 & 0 & 0 \\ 0 & 0 & 0 & 0 & 0 & 0 & 0 & 0 & 0 \end{pmatrix}$$

$$= -\frac{P}{30}\begin{pmatrix} 36 & 3 & 0 & -36 & 3 & 0 & 0 & 0 & 0 \\ 3 & 4 & 0 & -3 & -1 & 0 & 0 & 0 & 0 \\ 0 & 0 & 0 & 0 & 0 & 0 & 0 & 0 & 0 \\ -36 & -3 & 0 & 72 & 0 & -36 & 3 & 0 & 0 \\ 3 & -1 & 0 & 0 & 8 & -3 & -1 & 0 & 0 \\ 0 & 0 & 0 & -36 & -3 & 36 & -3 & 0 & 0 \\ 0 & 0 & 0 & 3 & -1 & -3 & 4 & 0 & 0 \\ 0 & 0 & 0 & 0 & 0 & 0 & 0 & 0 & 0 \\ 0 & 0 & 0 & 0 & 0 & 0 & 0 & 0 & 0 \end{pmatrix}.$$

Recall that free displacements are the rotation angles $(\varphi_1, \varphi_2, \varphi_3)$. Then by extracting the 2nd, 5th, and 7th row and column components of the above matrices, we can construct the buckling condition of this frame as

$$\det\left[\frac{2EI}{L^2}\begin{pmatrix} 2 & 1 & 0 \\ 1 & 6 & 1 \\ 0 & 1 & 2 \end{pmatrix} - \frac{P}{30}\begin{pmatrix} 4 & -1 & 0 \\ -1 & 8 & -1 \\ 0 & -1 & 4 \end{pmatrix}\right] = 0.$$

By setting $\alpha = PL^2/(60EI)$, we obtain $\alpha = 1/2$, $(11 \pm \sqrt{46})/15$, from which the buckling loads are obtained as

$$P_1 = 16.9\frac{EI}{L^2}, \qquad P_2 = 30.0\frac{EI}{L^2}, \qquad P_3 = 71.1\frac{EI}{L^2}.$$

The corresponding buckling modes in terms of $\hat{\eta} = (\mathrm{d}\varphi_1, \mathrm{d}\varphi_2, \mathrm{d}\varphi_3)^\top$ are

$$\hat{\eta}_1 = \begin{pmatrix} 1.46 \\ -1.00 \\ 1.46 \end{pmatrix}, \qquad \hat{\eta}_2 = \begin{pmatrix} 1 \\ 0 \\ -1 \end{pmatrix}, \qquad \hat{\eta}_3 = \begin{pmatrix} 0.80 \\ 1.00 \\ 0.80 \end{pmatrix}.$$

These modes are depicted in Figs. A.5 and A.6(b).

Problem 9.8: The analysis is similar to that for the two-beam arch in Section 9.6.2. By assembling the member stiffness equations of members 1 and 2 and using the boundary conditions of nodes 1 and 3 being fixed, we can obtain the structural stiffness equation $(K_{\mathrm{E}} + K_{\mathrm{G}})\boldsymbol{u} = \boldsymbol{p}$ in $\boldsymbol{u} = \boldsymbol{u}_2 = (u_2, w_2, \varphi_2)^\top$ with $\boldsymbol{p} = \boldsymbol{p}_2 = (0, P, 0)^\top$. The matrices K_{E} and K_{G} are obtained as

$$K_{\mathrm{E}} = \sum_{m=1}^{2} T^{(m)\top}\bar{k}_{\mathrm{E}22}^{(m)}T^{(m)}, \qquad K_{\mathrm{G}} = \sum_{m=1}^{2} T^{(m)\top}\bar{k}_{\mathrm{G}22}^{(m)}T^{(m)}, \tag{A.16}$$

similarly to (9.103). Here we have $\theta^{(1)} = -\pi/3$ and $\theta^{(2)} = \pi/3$, and, accordingly, the transformation matrices are

$$T^{(1)} = \begin{pmatrix} \sqrt{3}/2 & -1/2 & 0 \\ 1/2 & \sqrt{3}/2 & 0 \\ 0 & 0 & 1 \end{pmatrix}, \qquad T^{(2)} = \begin{pmatrix} \sqrt{3}/2 & 1/2 & 0 \\ -1/2 & \sqrt{3}/2 & 0 \\ 0 & 0 & 1 \end{pmatrix}.$$

Using $EA/L = 12EI/L^3$, the submatrices $\bar{k}_{\mathrm{E}22}^{(m)}$ are obtained from (9.50) and (9.51) with $\beta = 6$ as

$$\bar{k}_{\mathrm{E}22}^{(1)} = \frac{2EI}{L^2}\begin{pmatrix} 6/L & 0 & 0 \\ 0 & 6/L & -3 \\ 0 & -3 & 2L \end{pmatrix}, \qquad \bar{k}_{\mathrm{E}22}^{(2)} = \frac{2EI}{L^2}\begin{pmatrix} 6/L & 0 & 0 \\ 0 & 6/L & 3 \\ 0 & 3 & 2L \end{pmatrix}.$$

Then, from (A.16), we have

$$\begin{aligned} K_{\mathrm{E}} &= \sum_{m=1}^{2} T^{(m)\top}\bar{k}_{\mathrm{E}22}^{(m)}T^{(m)} \\ &= \begin{pmatrix} \sqrt{3}/2 & -1/2 & 0 \\ 1/2 & \sqrt{3}/2 & 0 \\ 0 & 0 & 1 \end{pmatrix}^\top \frac{2EI}{L^2}\begin{pmatrix} 6/L & 0 & 0 \\ 0 & 6/L & -3 \\ 0 & -3 & 2L \end{pmatrix}\begin{pmatrix} \sqrt{3}/2 & -1/2 & 0 \\ 1/2 & \sqrt{3}/2 & 0 \\ 0 & 0 & 1 \end{pmatrix} \\ &\quad + \begin{pmatrix} \sqrt{3}/2 & 1/2 & 0 \\ -1/2 & \sqrt{3}/2 & 0 \\ 0 & 0 & 1 \end{pmatrix}^\top \frac{2EI}{L^2}\begin{pmatrix} 6/L & 0 & 0 \\ 0 & 6/L & 3 \\ 0 & 3 & 2L \end{pmatrix}\begin{pmatrix} \sqrt{3}/2 & 1/2 & 0 \\ -1/2 & \sqrt{3}/2 & 0 \\ 0 & 0 & 1 \end{pmatrix} \\ &= \frac{2EI}{L^2}\begin{pmatrix} 12/L & 0 & -3 \\ 0 & 12/L & 0 \\ -3 & 0 & 4L \end{pmatrix}. \end{aligned}$$

Similarly, we can obtain the geometric stiffness matrix as

$$K_G = \frac{N}{30}\begin{pmatrix} 18/L & 0 & -3 \\ 0 & 54/L & 0 \\ -3 & 0 & 8L \end{pmatrix}.$$

Here we used $N^{(1)} = N^{(2)} = N$ due to the bilateral symmetry.

By the small displacement analysis, we obtain the unknown displacement vector $\boldsymbol{u} = (u_2, w_2, \varphi_2)^\top$ as

$$\boldsymbol{u} = (K_E)^{-1}\boldsymbol{p} = \left(0, \frac{PL^3}{24EI}, 0\right)^\top$$

and the member end force vector $\bar{\boldsymbol{p}}_2^{(1)} = (\bar{N}_2^{(1)}, \bar{V}_2^{(1)}, \bar{M}_2^{(1)})^\top$ of member 1 as

$$\bar{\boldsymbol{p}}_2^{(1)} = \bar{k}_{E22}^{(1)} T^{(1)} \boldsymbol{u} = \frac{P}{8}\left(-2,\ 2\sqrt{3},\ -L\right)^\top.$$

Then the member axial force is obtained as $N^{(1)} = N^{(2)} = N = -P/4$.

The buckling condition is

$$\det(K_E + K_G) = \left(\frac{2EI}{L^2}\right)^3 \frac{18}{L}(2-9\alpha)(45\alpha^2 - 50\alpha + 13) = 0$$

with $\alpha = PL^2/(240EI)$. Accordingly, buckling occurs for $\alpha = 0.222,\ 0.415,\ 0.696$ and the buckling loads are given by

$$P_1 = 53.3\frac{EI}{L^2}, \qquad P_2 = 99.6\frac{EI}{L^2}, \qquad P_3 = 167.1\frac{EI}{L^2}.$$

The corresponding buckling modes in terms of $\boldsymbol{\eta} = (\mathrm{d}u_2, \mathrm{d}w_2, \mathrm{d}\varphi_2)^\top$ are

$$\boldsymbol{\eta}_1 = \begin{pmatrix} 0 \\ 1 \\ 0 \end{pmatrix}, \qquad \boldsymbol{\eta}_2 = \begin{pmatrix} 1 \\ 0 \\ -2.58/L \end{pmatrix}, \qquad \boldsymbol{\eta}_3 = \begin{pmatrix} 1 \\ 0 \\ 0.581/L \end{pmatrix}.$$

The buckling mode $\boldsymbol{\eta}_1$ for the minimum buckling load P_1 corresponds to snap-through buckling.

Problem 9.9: By assembling the member stiffness equations and using the boundary conditions of nodes 1 and 4 being fixed, we obtain the structural stiffness equation $(K_E + K_G)\boldsymbol{u} = \boldsymbol{p}$, where $\boldsymbol{u} = (u_2, w_2, \varphi_2, u_3, w_3, \varphi_3)^\top$ and $\boldsymbol{p} = (0, P, 0, 0, P, 0)^\top$. In the transformation matrix in (9.89), we use $\theta^{(1)} = -\pi/6$, $\theta^{(2)} = 0$, and $\theta^{(3)} = \pi/6$ for the tilt angles of the members. The stiffness matrix is given by

$$K_{\rm E} = \begin{pmatrix} T^{(1)\top}\bar{k}^{(1)}_{\rm E22}T^{(1)} + T^{(2)\top}\bar{k}^{(2)}_{\rm E22}T^{(2)} & T^{(2)\top}\bar{k}^{(2)}_{\rm E23}T^{(2)} \\ T^{(2)\top}\bar{k}^{(2)}_{\rm E32}T^{(2)} & T^{(2)\top}\bar{k}^{(2)}_{\rm E33}T^{(2)} + T^{(3)\top}\bar{k}^{(3)}_{\rm E33}T^{(3)} \end{pmatrix}$$

$$= \frac{12EI}{L^2}\left(\begin{array}{ccc|ccc} 2/L & 0 & -1/4 & -1/L & 0 & 0 \\ 0 & 2/L & \frac{2-\sqrt{3}}{4} & 0 & -1/L & 1/2 \\ -1/4 & \frac{2-\sqrt{3}}{4} & 2L/3 & 0 & -1/2 & L/6 \\ \hline -1/L & 0 & 0 & 2/L & 0 & -1/4 \\ 0 & -1/L & -1/2 & 0 & 2/L & -\frac{2-\sqrt{3}}{4} \\ 0 & 1/2 & L/6 & -1/4 & -\frac{2-\sqrt{3}}{4} & 2L/3 \end{array}\right), \tag{A.17}$$

where $\bar{k}^{(m)}_{{\rm E}ij}$ in (9.94) and (9.95) are used. Since $N^{(1)} = N^{(3)}$ holds from the bilateral symmetry of the frame structure, the geometric stiffness matrix is given by

$$K_{\rm G} = \begin{pmatrix} T^{(1)\top}\bar{k}^{(1)}_{\rm G22}T^{(1)} + T^{(2)\top}\bar{k}^{(2)}_{\rm G22}T^{(2)} & T^{(2)\top}\bar{k}^{(2)}_{\rm G23}T^{(2)} \\ T^{(2)\top}\bar{k}^{(2)}_{\rm G32}T^{(2)} & T^{(2)\top}\bar{k}^{(2)}_{\rm G33}T^{(2)} + T^{(3)\top}\bar{k}^{(3)}_{\rm G33}T^{(3)} \end{pmatrix}$$

$$= \frac{N^{(1)}}{30}\left(\begin{array}{ccc|ccc} 9/L & 9\sqrt{3}/L & -3/2 & & & \\ 9\sqrt{3}/L & 27/L & -3\sqrt{3}/2 & & O & \\ -3/2 & -3\sqrt{3}/2 & 4L & & & \\ \hline & & & 9/L & -9\sqrt{3}/L & -3/2 \\ & O & & -9\sqrt{3}/L & 27/L & 3\sqrt{3}/2 \\ & & & -3/2 & 3\sqrt{3}/2 & 4L \end{array}\right)$$

$$+ \frac{N^{(2)}}{30}\left(\begin{array}{ccc|ccc} 0 & 0 & 0 & 0 & 0 & 0 \\ 0 & 36/L & 3 & 0 & -36/L & 3 \\ 0 & 3 & 4L & 0 & -3 & -L \\ \hline 0 & 0 & 0 & 0 & 0 & 0 \\ 0 & -36/L & -3 & 0 & 36/L & -3 \\ 0 & 3 & -L & 0 & -3 & 4L \end{array}\right), \tag{A.18}$$

where (9.55) and (9.56) are used for $\bar{k}^{(m)}_{{\rm G}ij}$.

By the small displacement analysis, the unknown nodal displacement vector $\boldsymbol{u} = \begin{pmatrix} \boldsymbol{u}_2 \\ \boldsymbol{u}_3 \end{pmatrix}$ with $\boldsymbol{u}_2 = (u_2, w_2, \varphi_2)^\top$ and $\boldsymbol{u}_3 = (u_3, w_3, \varphi_3)^\top$ is evaluated to

$$\boldsymbol{u} = (K_{\rm E})^{-1}\boldsymbol{p} = \frac{PL^3}{168EI}\left(\sqrt{3},\ 23,\ \frac{12\sqrt{3}}{L},\ -\sqrt{3},\ 23,\ -\frac{12\sqrt{3}}{L}\right)^\top.$$

The member end forces of member 1 at node 2 are

$$\bar{\boldsymbol{p}}^{(1)}_2 = \bar{k}^{(1)}_{\rm E22}T^{(1)}\boldsymbol{u}_2 = \frac{P}{7}(-5,\ 3\sqrt{3},\ -3\sqrt{3}L)^\top,$$

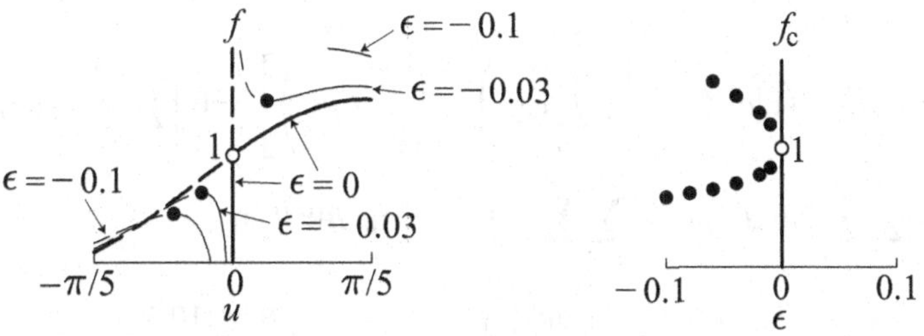

Figure A.7 Problem 10.2: Load versus displacement curves and imperfection sensitivity

giving a member axial force of $N^{(1)} = -5P/7$ in its first component. The member end forces of member 2 at node 3 are

$$\bar{\boldsymbol{p}}_3^{(2)} = \bar{k}_{\mathrm{E32}}^{(2)} T^{(2)} \boldsymbol{u}_2 + \bar{k}_{\mathrm{E33}}^{(2)} T^{(2)} \boldsymbol{u}_3 = \frac{\sqrt{3}P}{7}(-1,\ 0,\ -L)^{\top},$$

giving a member axial force of $N^{(2)} = -\sqrt{3}P/7$ in its first component. The explicit form of the geometric stiffness matrix K_{G} is obtained by using these axial forces $N^{(1)}$ and $N^{(2)}$ in (A.18), while the explicit form of K_{E} is given in (A.17). The use of the explicit forms of K_{E} and K_{G} in $\det(K_{\mathrm{E}} + K_{\mathrm{G}}) = 0$ leads to the buckling condition.

Chapter 10

Problem 10.1: The potential and the governing equation are given, respectively, by

$$U(u,f,\varepsilon) = kL^2 \left[\frac{1}{2}\sin^2 u - f(1-\cos u) - \varepsilon \sin u \right],$$
$$F(u,f,\varepsilon) = kL^2 (\sin u \cos u - f \sin u - \varepsilon \cos u).$$

Problem 10.2: The potential and the governing equation are given, respectively, by

$$U(u,f,\varepsilon) = kL^2 \left[\frac{1}{2}(\sin u - \sin\varepsilon)^2 + \frac{1}{2}(\sin u - \sin\varepsilon)^3 - f(\cos\varepsilon - \cos u) \right],$$
$$F(u,f,\varepsilon) = kL^2 \left[(\sin u - \sin\varepsilon)\cos u + \frac{3}{2}(\sin u - \sin\varepsilon)^2 \cos u - f \sin u \right].$$

The equilibrium paths determined by $F = 0$ for $\varepsilon = 0$, -0.03, and -0.1 are plotted at the left of Fig. A.7, and the imperfection sensitivity of this system is plotted at the right.

Problem 10.3: The average of a is given by

$$\mathrm{E}[a] = \mathrm{E}\left[\sum_{i=1}^{p} c_i d_i \right] = \sum_{i=1}^{p} c_i \mathrm{E}[d_i] = 0,$$

and the variance is

$$\tilde{\sigma}^2 = \mathrm{E}[a^2] - (\mathrm{E}[a])^2 = \mathrm{E}\left[\left(\sum_{i=1}^{p} c_i d_i\right) \times \left(\sum_{j=1}^{p} c_j d_j\right)\right] = \mathrm{E}\left[\sum_{i=1}^{p}\sum_{j=1}^{p} c_i d_i c_j d_j\right]$$

$$= \sum_{i=1}^{p}\sum_{j=1}^{p} c_i \mathrm{E}[d_i d_j] c_j = \sum_{i=1}^{p}\sum_{j=1}^{p} c_i W_{ij} c_j = \boldsymbol{\eta}_{\mathrm{c}}^{\top} B W B^{\top} \boldsymbol{\eta}_{\mathrm{c}},$$

where $W = (W_{ij} \mid i,j = 1,\ldots,p)$ and $(c_1,\ldots,c_p) = \boldsymbol{\eta}_{\mathrm{c}}^{\top} B$ in (10.31).

Problem 10.4: Since $A_{100} = A_{010} = 0$ and $A_{200} \neq 0$ hold for an asymmetric bifurcation point (Section 5.4.2 and (10.10)), the bifurcation equation (10.7) and the criticality condition, respectively, become

$$\hat{F}(w,\tilde{f},\varepsilon) \approx A_{200}w^2 + A_{110}w\tilde{f} + A_{020}\tilde{f}^2 + A_{001}\varepsilon = 0, \tag{A.19}$$

$$\hat{J}(w,\tilde{f},\varepsilon) \approx 2A_{200}w + A_{110}\tilde{f} = 0. \tag{A.20}$$

Figure A.8 shows the solution curves of (A.19).

When the condition (5.45) of non-degeneracy and $A_{200}A_{001}\varepsilon > 0$ hold, an imperfect system has a maximal/minimal point $(w_{\mathrm{c}}, \tilde{f}_{\mathrm{c}})$. The location of this point is obtained from (A.19) and (A.20) as

$$\tilde{f}_{\mathrm{c}} \approx \pm\left(\frac{4\,|A_{200}A_{001}|}{A_{110}{}^2 - 4A_{200}A_{020}}\right)^{1/2} |\varepsilon|^{1/2}, \tag{A.21}$$

$$w_{\mathrm{c}} \approx \mp\frac{A_{110}}{A_{200}}\left(\frac{|A_{200}A_{001}|}{A_{110}{}^2 - 4A_{200}A_{020}}\right)^{1/2} |\varepsilon|^{1/2},$$

where the double sign corresponds. The case of $\tilde{f}_{\mathrm{c}} < 0$ in (A.21), which is associated with sign "$-$" of the double sign $\pm$, is of engineering interest.

Problem 10.5: We employ the two-thirds power law: $\tilde{f}_{\mathrm{c}} = f_{\mathrm{c}} - f_{\mathrm{c}}^0 \approx a\varepsilon^{2/3}$. (1) From the given conditions, we have

$$2.45 - f_{\mathrm{c}}^0 = a(1/10)^{2/3}, \quad 2.60 - f_{\mathrm{c}}^0 = a(1/1000)^{2/3},$$

which yield $a = -0.73$ and $f_{\mathrm{c}}^0 = 2.607$. (2) From $2.50 \leq 2.607 - 0.73\varepsilon^{2/3}$, the tolerance of ε is 0.056.

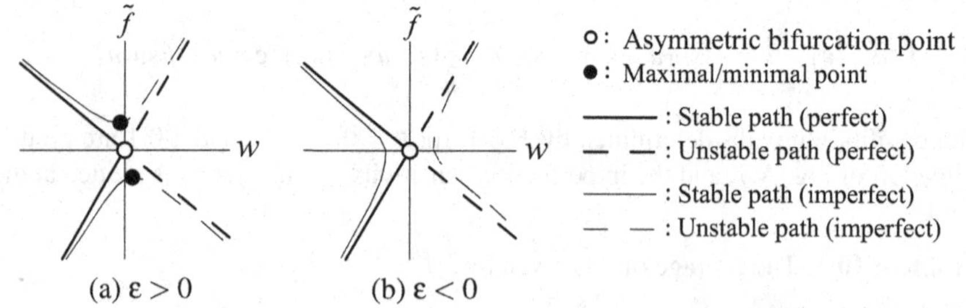

Figure A.8 Problem 10.4: Solution curves in the neighborhood of an asymmetric bifurcation point ($A_{200} > 0$, $A_{020} < 0$, $A_{001} > 0$)

Index

Printed in the United States
by Baker & Taylor Publisher Services